D0438622

Vascular Flora of the Southeastern United States

Vascular Flora of the Southeastern United States

Vascular Flora of the Southeastern United States

Volume I

ASTERACEAE

by

Arthur Cronquist

The New York Botanical Garden

THE UNIVERSITY OF NORTH CAROLINA PRESS
CHAPEL HILL

Manufactured in the United States of America
ISBN 0-8078-1362-1
Library of Congress Catalog Card Number 79-769

Library of Congress Cataloging in Publication Data

Main entry under title:

Vascular flora of the southeastern United States.

Includes index.
CONTENTS: v. 1. Cronquist, A. Asteraceae.
1. Botany—Southern States. I. Cronquist, Arthur.
QK125.V37 582'.0975 79-769
ISBN 0-8078-1362-1

CONTENTS

PREFACE

The Asteraceae of the Southeastern United States, by Arthur Cronquist of The New York Botanical Garden, with the tribe *Vernonieae* written by Samuel B. Jones of the University of Georgia, is the first of five planned volumes on the "Vascular Flora of the Southeastern United States."

The general objectives for the study of the vascular plants in the southeastern United States are as follows:

1. To survey floristically the forested region of the southeastern United States west to the prairie and north to the southernmost terminal moraines. This region includes Delaware, Maryland, Virginia, North Carolina, South Carolina, Georgia, Florida, Alabama, Mississippi, Louisiana, Arkansas, Tennessee, Kentucky, and West Virginia. Notation of presence in the adjacent states of Texas, Oklahoma, Missouri, Illinois, Indiana, Ohio, Pennsylvania, and New Jersey will be made. (No effort has been made to include total distribution beyond the adjacent states.)
2. To produce a Manual that will includes keys, descriptions, habitats, distributional data, and pertinent synonymy to every vascular species growing without cultivation in the southeastern United States.

The following inclusions on format, abbreviations, geography, ecology, and synonymy are presented for effective and efficient use of the treatments of the vascular plant taxa found in the southeastern United States. These inclusions represent excerpts from the *Contributors' Guide for the Vascular Flora of the Southeastern United States* prepared for the author contributors by Radford et al., 1967, in a private publication.

FORMAT

1. Family number and name (a, b, f*). Authority. Common name (c).
2. Family description (d).
3. Key to genera (e).
4. Generic number and name (a, f, i). Authority. Common name (c).
5. Generic description (d, f).
6. Key to species (e).
7. Species number and name (i). Authority. Common name (c).
8. Species description (d).
9. Infraspecific taxon number and name (j). Authority. Common name (c).
10. Infraspecific taxon description (g, h, j).

(a) If a family or genus has a single species within our area, the family and/or generic descriptions are omitted. The species description is modified to include the family and/or generic characteristics for our material.

(b) Family names are according to the Cronquist system of classification.

(c) The common name, if available, should be the most commonly used one (rarely more) for the taxon within our area.

(d) All taxon descriptions should follow the detailed arrangement in Appendix 1*, examples in Appendix 2*, abbreviations in Appendix 3, geography-ecology in Appendix 4, and synonymy in Appendix 5.

(e) Keys to all taxa are to be strictly dichotomous, indented, and each couplet successively numbered. Each line or heading of a couplet should start with the same noun or morphological character. All characters used should have the contrasting condition in the second couplet entry. Nouns usually come before adjectives. No more than three characters should be used within a couplet to distinguish any two taxa or groupings. The character easiest to determine should be used first in the couplet. Vegetative, flower, and fruit characters should be used whenever possible. Quantitative measurements or ratios should be used, not qualitative statements. Positive statements should be used whenever possible, negative only when the positive contrasting characters occupy too much space. If combined flowering and fruiting or seasonal and sexual keys become too cumbersome, write keys for flowering material and fruiting material, for winter condition and summer condition, for male and female plants, and so forth. Do not use scientific names or geographical localities as key characters.

ABBREVIATIONS

(Appendix 3)
Nomenclature

sp., spp.	species (sing.), species (pl.)
ssp., sspp.	subspecies (sing.), subspecies (pl.)
var., vars.	variety (sing.), varieties (pl.)
f.	form (sing.), forms (pl.)

Authority abbreviations—follow those in the *Manual of the Vascular Flora of the Carolinas*. In case of doubt, write out in full.

Common name for species which duplicates that of genus, use capital first letter for generic common name, e.g., OAK for *Quercus*, WHITE O. for *Q. alba.*

Description

Without period		*With period*	
mm	millimeter(s)	ca.	circa, about
cm	centimeter(s)	cult.	cultivated
dm	decimeter(s)	diam.	diameter
m	meter(s)	freq.	frequent

Flowering times—No abbreviations; do not use months. Spell out spring, summer, fall, all year; late or early can be used as modifiers where appropriate.

*These appendixes and f and g of the Format are not included in these excerpts because the treatments are self-explanatory.

Distribution

Compass directions (small letters, no periods)—n, ne, e, se, s, sw, w, nw

SE (caps, no period)—general distribution throughout all states of the southeastern United States

States (standard abbreviations without periods or internal spacing)—Ala, Ark, Del, Fla, Ga, Ky, La, Md, Miss, NC, SC, Tenn, Va, WVa, Tex, Okla, Mo, Ill, Ind, Ohio, Pa, NJ

Physiographic provinces (caps, no period)

SA Southern Appalachian Highlands
BR Blue Ridge
RV Ridge and Valley
CP Coastal Plain (both Atlantic and Gulf and Mississippi Embayment)
AC Atlantic Coastal Plain
GC Gulf Coastal Plain
ME Mississippi Embayment
PP Piedmont Plateau
IP Interior Low Plateau
CU Cumberland Plateau
AP Alleghany Plateau
CA Cumberland-Alleghany Plateau
OH Ozark-Ouachita Highlands
OZ Ozarks
OU Ouachitas

Synonyms from Manuals

S Small, 1933
F Fernald, 1950
G Gleason and Cronquist, 1963
R Radford et al., 1968

Incl. If name used includes synonym, e.g., "incl. *A. discolor* Pursh—F, S" under *Aesculus pavia.*

Journal abbreviations—Follow those in Schwarten, L. & H. W. Rickett. 1958. Abbreviations of titles of Serials cited by Botanists. Bull. Torrey Club 85:277–300.

GEOGRAPHY-ECOLOGY

(Appendix 4)
Geography

Distribution is to be by major physiographic provinces, unless restricted (see below), within the southeastern states, and by state within the adjacent states. The symbols for the major provinces are as follows: *mts.* for mountains, which includes all subprovinces under SA and OH; *pied.* for the piedmont and interior low plateaus; and *cp.* for the coastal plain, which includes the subprovinces AC, GC, and ME (see abbreviations).

Province data will follow semicolon after habitat, e.g., Rich woods; mts. If a species is found in two provinces, the provinces will be connected by "and," e.g., Rich woods; mts. and pied. If a species is found in all major provinces in the southeastern states, "all prov." is used, e.g., Rich woods; all prov. If a species is typically found in one or two provinces and occasionally in one or two others, province data should be preceded by "chiefly," e.g., Rich woods; chiefly mts. and pied. This means that the species is occasionally found in the cp., but primarily in the mts. and pied. If "chiefly mts." had been indicated, then the species also might be found occasionally in the pied. and/or cp.

Distribution by states within the southeast will be a separate entry with the states alphabetized, e.g., Rich woods; mts. Ala, Ark, NC, SC. If distribution is in each state of the southeastern United States, the symbol "SE" is used. If distribution is throughout the southeastern except one, two, or three states, then is followed by "except" and the one, two, or three states listed alphabetically.

Fields; all prov. SE except Fla and WVa.
Fields; all prov. SE except Del, Fla, WVa.

Distribution in adjacent states will be a separate entry to be shown by abbreviations, in brackets [ ], from southwest to northeast, e.g., Tex, Okla, Mo, Ill, Ind, Ohio, Pa, NJ. If the species is present in all eight adjacent states, use the symbol [ALL].

Rich woods; mts. Ky, Va, WVa. [Ind, Ohio, Pa, NJ]
Fields; all prov. SE [ALL]

If distribution is restricted, then the author should give precise data using the appropriate symbols from Appendix 3.

Juncus trifidus—Rocky ledges; BR of NC.
Heuchera rudgei—Rocky woods; IP. Ky and Tenn. [Ohio]
Avicennia nitida—Mangrove swamps; Keys. Fla.

If a species is found in two or three major provinces within the southeastern United States but not in the two or three major provinces for each state listed, then the province(s) in which it is found within a state is given after the state symbol.

Planta herba—Rich woods; mts. and pied. Ala mts., Ga, NC, SC.

This means that the species is found in the mts. and pied. provinces of the three-state area, but in Ala only in the mts.

Planta herba—Fields; all prov. Ala, Ark cp., Del, Fla, Ga, Ky.

This means that the species is found in all three major prov. of the southeastern states listed but in Ark only in the cp.

Distribution information is to be based on specimens examined by the author in several major herbaria and from publications by competent authorities. Do not base distribution on general ranges given in manuals.

These geographic instructions do not cover all distribution possibilities, so each author is enjoined to be consistent in describing range exceptions.

Ecology-Habitats

Designations of *frequency* or *density* should not be used unless the author is entirely confident of statements such as rare, infrequent, common, abundant.

Endemism should be stated where certain, e.g., *Shortia galacifolia*, endemic to BR of Ga, SC, NC.

Specific habitats should be used for highly localized species when definitely known—i.e., Sand dunes (*Uniola paniculata*), Shale barrens (*Oenothera argillicola*), Limestone or dolomite outcrops (*Asplenium resiliens*), Hammocks (*Clusia flava*), etc.

HABITAT DESIGNATIONS

Aquatic

river	lake
stream	pond or pool
springhead	marsh
ditches	bog

Use modifiers where certain:

brackish	brackish marsh
marine shores	peat bog
acid (pH5 and below)	shrub bog
rapid	cypress swamp

Fields (use modifiers when needed)

wet	clay
dry	loam
low	rocky
upland	old
sandy	cult.

Balds
heath balds
grassy balds

Disturbed habitats

waste places	gardens
roadsides	pastures
railroads	meadows
fence rows, hedge rows	

Savannahs (use modifiers when needed)
wet
dry
pine

Outcrops (use modifiers where certain)
granite
sandstone
limestone

Woods (use modifiers to describe condition and also characteristic species where needed)

evergreen	pine, spruce-fir, longleaf pine
deciduous	oak woods, mixed hardwoods, oak-pine
low (alluvial or swamp)	*Taxodium, Nyssa*
rich	oak woods, mixed mesophytic
dry	oak-hickory, oak-pine
dry sandy	turkey oak
dry clayey	blackjack-post oak
open dense	pine
	cove hardwoods

SYNONYMY

(Appendix 5)

Synonyms are to be included only when the names employed differ from those used in recent manuals of the eastern United States. These recent manuals and their abbreviations are in chronological order by author:

S J. K. Small. 1933. Manual of the Southeastern Flora. Published by the author, New York.

F M. L. Fernald. 1950. Gray's Manual of Botany. 8th edition. American Book Company, New York.

G H. A. Gleason and A. Cronquist. 1963. Manual of the Vascular Plants of the Northeastern United States and Adjacent Canada. D. Van Nostrand Company, Inc., Princeton, New Jersey.

R A. E. Radford, H. E. Ahles and C. R. Bell. 1968. Manual of the Vascular Flora of the Carolinas. University of North Carolina Press, Chapel Hill, North Carolina.

Previous segregates accepted in the above recent manuals but not by the contributor should be listed as follows:

[Under *Cassia nictitans* L.]
"Incl. *Chamaecrista multipinnata* (Pollard) Greene—S; *Cassia n.* var. *hebecarpa* Fernald, *C. n.* var. *leiocarpa* Fernald—F, G."

Taxa previously included as synonyms in the above recent manuals but regarded as distinct by the contributor should be listed in synonymy as follows:

[Under *Desmodium nuttallii* (Schindler) Schubert]
". . . *Meibomia viridiflora* (Nuttall) Kuntze—S, in part; *D. viridiflorum* (L.) DC.—G, in part."
[Under *Cassia obtusifolia* L.]
". . . *Emelista tora* (L.) Britton & Rose—S; *C. tora* L.—F, G."

Taxa previously passing under the name of other taxa should be indicated as follows:

[Under *Amorpha georgiana* Wilbur]
"*A. cyanostachya* Curtis—sensu S, not Curtis."
[Under *Aeschynomene indica* L.]
"*A. virginica* (L.) BSP—sensu S, in large part."
[Under *Sabatia stellaris* Pursh]
"*S. campanulata* (L.) Torrey—sensu S, in part."

Obvious synonyms from the above recent manuals should be listed:
[Under *Diamorpha smallii* Britton]
"incl. *D. cymosa* (Nuttall) Britton ex Small—S, *Sedum smallii* (Britton) Ahles—R."
[Under *Psoralea lupinellus* Michaux]
"*Rhytidomene lupinellus* (Michaux) Rydberg—S."
[Under *Lotus helleri* Britton]
"*Acmispon helleri* (Britton) Small—S."

Note: Synonymy is not categorized as mechanically as the above might indicate. The above is meant merely as a tentative guide, and the contributor is encouraged to convey the facts accurately and concisely. Contributors are strongly urged to publish their nomenclatural research in full elsewhere. Scholarship demands more than the legal minimum of the basionym.

The Editors

ACKNOWLEDGMENTS

Many people helped me in one way or another during the preparation of this manuscript. Dr. R. K. Godfrey and Albert E. Radford read most of the manuscript seriatim over a period of several years. Individual bits of information, assistance, and advice were provided by T. M. Barkley, Janice C. Beatley, C. Ritchie Bell, Mark W. Bierner, R. B. Channell, Wilbur H. Duncan, F. R. Fosberg, Charles B. Heiser, Jr., Almut G. Jones, Samuel B. Jones, Robert Kral, Sidney McDaniel, Gary H. Morton, Dan H. Nicolson, Robert E. Perdue, Jr., Victoria I. Sullivan, R. Dale Thomas, and Daniel B. Ward. The help of other botanists is acknowledged in the treatments of individual genera. The curators of the herbaria at the following institutions graciously permitted me to use their collections: University of Florida, Florida State University, Gray Herbarium of Harvard University, University of Georgia, Northeastern Louisiana University, Southwestern Louisiana University, University of Mississippi, Mississippi State University, University of North Carolina, Oxford University, Smithsonian Institution, University of Tennessee, and Vanderbilt University. Lucy F. Kluska typed interim drafts and the final manuscript. Laurie S. Radford kindly checked the entire manuscript for completeness and accuracy of presentation. My wife, Mabel A. Cronquist, provided the domestic milieu conducive to my work, and she accompanied me much of the time in the field. To all of these people, and to others whom I should not have forgotten, I am duly thankful. None of them is to be held responsible for any of the taxonomic interpretations here presented; these are my own, by birth or adoption.

Partial support from the National Science Foundation is also gratefully acknowledged. Over a period of several years I was permitted to use funds that would otherwise have gone to support work on another floristic project.

A. Cronquist

Vascular Flora of the Southeastern United States

THE ASTERACEAE Dumort. 1822 THE ASTER FAMILY

Annual, biennial, or perennial herbs, or less commonly shrubs, only some extralimital spp. becoming trees; plants commonly producing sesquiterpene lactones, or polyacetylenes, or both, sometimes with other chemical defenses in addition or instead. Leaves simple to sometimes compound or dissected, exstipulate, variously opposite, alternate, or even whorled, often opposite below and alternate above. Flowers sessile in a compact head on a common receptacle, sometimes individually subtended by a small bract, and nearly always (including all our spp.) collectively surrounded by an involucre of few to many bracts; secondary inflorescence (the arrangement of heads) sometimes strictly determinate (the terminal head blooming first, followed by the terminal heads on the main branches, etc.), or often mixed (the terminal head blooming first, but the sequence thereafter more complex), only very rarely (as in a few spp. of *Eupatorium*) strictly indeterminate; sequence of flowering *within* a head strictly indeterminate, the marginal flowers opening first; individual flowers epigynous, perfect or unisexual (or some of them neutral), sympetalous, regular or irregular, commonly 5-merous; calyx represented by a pappus of scales, awns, short setae, long, capillary or plumose bristles, or a hyaline to chartaceous or coriaceous crown or ring, or by some combination of these, or seldom wholly wanting; stamens as many as the corolla lobes and alternate with them; filaments attached well down in the corolla tube; anthers elongate, connate into a tube, rarely merely connivent; ovary inferior, of 2 carpels, unilocular, with a single basal, erect, anatropous, unitegmic, tenuinucellar ovule; style usually 2-cleft (often undivided in functionally staminate flowers), the stigmatic surfaces variously restricted, so that each branch often has a sterile (nonstigmatic) appendage. Fruit an achene, crowned by the persistent (less often deciduous) pappus. (Compositae)

The Asteraceae are one of the two largest families of flowering plants, with certainly more than 15,000 species. The only other family of comparable size is the Orchidaceae. The Asteraceae are cosmopolitan in distribution, but partial to open or semiopen habitats rather than deep woods. In most parts of the temperate zone, including our region, they are by far the largest family. Many genera and spp. are cultivated for ornament. The family is one of the easiest to recognize, but many of the genera are poorly defined or confluent.

The flower heads vary from small to large, and are often brilliantly colored. The number of flowers in a head is seldom less than 5, and ranges upward into the hundreds or even more than a thousand, as in the common cultivated sunflower. A few species have only a single flower in each head. *Echinops* and some other genera have one-flowered, individually involucrate heads aggregated into a secondary head with a secondary involucre. Compound heads with more than one flower in each individual head also occur in some genera, such as *Elephantopus*.

The *involucral bracts* are usually herbaceous or subherbaceous in texture, varying to scarious, hyaline, coriaceous, or cartilaginous. They may be few and in a single row, or numerous and imbricate, or modified into spines, or even (as in *Xanthium*) concrescent into a spiny bur.

The *receptacle* may be *chaffy*, with a bract behind each flower (as in many Heliantheae), or may be covered with long, stout bristles (as in most Cynareae), or may be *naked*, without chaff or bristles. When naked it may sometimes be minutely pitted, with slender, chaffy partitions separating the pits, and is then said to be *alveolate*. It may even be softly hairy, as in some spp. of *Artemisia*.

The flowers are of several general types. In one type they are perfect (or functionally staminate) and the corolla is tubular or trumpet-shaped or goblet-shaped, with typically 5 short terminal lobes or teeth. This type of flower is called a *disk flower*. A head composed wholly of disk flowers is said to be *discoid*.

In another type the flower is pistillate or neutral (without a style), and the corolla is tubular only at the very base, above which it is flat and usually bent backward so as to spread away from the center of the head. The flattened part of a corolla of this type is called a *ray* or *ligule*, and the flower bearing it is called a *ray flower* or *ligulate flower*. Often the ligule exhibits traces of two or three corolla lobes as small terminal teeth. Except for the pistillate heads of a few dioecious groups, the head is never composed solely of flowers of this type. Instead, these pistillate or neutral ray flowers are found at the margin of the head, the center being occupied by disk flowers. Such a head, with both ray flowers and disk flowers, is said to be *radiate*.

In some spp. the ray or ligule of the marginal, pistillate flowers does not develop, so that the corolla is tubular. In addition to not bearing stamens, a corolla of this type differs from the corolla of an ordinary disk flower in the absence of the regular terminal teeth, and often also in being more slender. A head in which the pistillate flowers lack rays is said to be *disciform*, although the term discoid is sometimes loosely extended to cover this type.

Another type of flower superficially resembles the ray flower of a radiate head, but differs in being perfect and in usually having 5 terminal teeth on the ligule. The heads of the tribe Lactuceae consist wholly of flowers of this type and are called *ligulate heads*. Ligulate perfect flowers are rare in other tribes, and almost never make up the whole head. Among our genera, *Stokesia*, in the Vernonieae, has the marginal flowers perfect and ligulate.

Still another type of flower, found only in the Mutisieae, has a bilabiate corolla, with the outer lip generally the larger. These bilabiate flowers are generally perfect, and differ from ordinary disk flowers only in the shape of the corolla.

In some spp. of *Centaurea* the marginal flowers are neutral and have an enlarged, irregular, raylike corolla.

The anthers are coherent by their lateral margins, or rarely merely connivent (as in members of the subtribe Ambrosiinae of the Heliantheae). The base of the anthers varies from obtuse or subtruncate to broadly rounded, sagittate, or distinctly caudate (tailed). The anthers dehisce introrsely, and the pollen is pushed out through the anther tube by growth of the style. The style branches commonly diverge above the anther tube, have various distinctive forms and textures, and tend to be stigmatic only on limited parts of their surface. The characteristic style branches of the various tribes are to be sought only in the fertile disk flowers. The styles of ray flowers are mostly very similar in all groups, and those of the sterile disk flowers are often reduced and undivided. The sterile disk flowers, when present, are said to be *functionally staminate*. Strictly staminate flowers, with no pistillate parts, do not normally occur in the Asteraceae, because the style is necessary as a piston or plunger to eject the pollen.

Conspectus of the Tribes

1 Flowers all ligulate and perfect; juice usually milky .... Tribe 10. Lactuceae.

1 Flowers, or some of them with tubular, eligulate corolla; ligulate flowers, when present, marginal in the head, surrounding the central disk flowers, and never perfect except in *Stokesia*, of the Vernonieae; juice usually watery.

2 Heads mostly radiate, or not infrequently disciform, or in scattered genera, species, and individuals discoid; corollas predominantly yellow, although various anthocyanic shades and white are not uncommon.

3 Lower (or all) leaves in most genera opposite, although a number of genera have wholly alternate leaves; pappus chaffy, or of a few firm awns, or none, only very rarely capillary; receptacle chaffy or less often naked; involucral bracts tending to be herbaceous and several-seriate, but varying in groups of genera to uniseriate, or subchartaceous, or even concrescent into a bur; style branches often more or less hispidulous and with the stigmatic lines poorly differentiated, varying as in the Astereae or Senecioneae; anthers not tailed; rays tending to be relatively broad, often much larger than is usual for the other tribes .... Tribe 1. Heliantheae.

3 Leaves all alternate (except notably *Arnica*, in the Senecioneae); receptacle naked in most genera, but sometimes chaffy; style branches mostly flattened and with well-defined ventromarginal stigmatic lines; involucre, pappus, anthers, and rays various.

4 Anthers obtuse to merely sagittate, but not tailed; styles diverse, but not as in the Inuleae; heads radiate or less often disciform or discoid; pappus and receptacle various.

5 Involucral bracts rather dry and scarcely herbaceous, becoming hyaline-scarious toward the margins and tip; pappus none, or minute and chaffy or coroniform (of definite small scales in *Hymenopappus*); style branches mostly truncate and penicillate; receptacle chaffy or naked; leaves generally more or less dissected, varying to sometimes entire and then mostly small; plants mostly with a characteristic odor ...... Tribe 2. Anthemideae.

5 Involucral bracts herbaceous to chartaceous, rarely evidently hyaline-scarious toward the margins and tip; pappus capillary, chaffy, or of awns, or occasionally none; style branches of diverse sorts; receptacle naked; leaves usually entire or merely toothed; plants mostly inodorous, or at least not with the odor of the Anthemideae.

6 Involucral bracts mostly equal and subuniseriate, a calyculus of very much reduced outer bracts often also present; style branches very often truncate and penicillate, without an appendage, but in a few genera with a definite terminal appendage; pappus capillary ...... Tribe 3. Senecioneae.

6 Involucral bracts mostly in several series, imbricate or subequal (or the outer larger), calyculus wanting; style branches usually with a terminal appendage; pappus capillary or chaffy or of awns, or seldom none .... Tribe 4. Astereae.

4 Anthers more or less strongly tailed at the base; style branches generally glabrous or merely papillate, subtruncate to broadly rounded at the tip, the stigmatic lines often confluent around the end; heads very often disciform or discoid, the rays when present yellow like the disk; pappus mostly capillary (sometimes plumose), rarely chaffy or none; receptacle chaffy or much more often naked .... Tribe 5. Inuleae.

2 Heads mostly discoid, or with some of the corollas bilabiate (marginal flowers ligulate and perfect in *Stokesia*, or subradiate and neutral in a few spp. of *Centaurea*); corollas mostly anthocyanic or white, yellow only in a few spp. of Cynareae.

7 Anthers obtuse to sagittate at the base; plants not spiny; receptacle naked or rarely chaffy, not densely bristly.

8 Style branches cylindric to often clavate, obtuse, minutely papillate, not hairy, with inconspicuous ventromarginal stigmatic lines near the base; leaves mostly opposite, less often whorled, only occasionally alternate .... Tribe 6. Eupatorieae.

8 Style branches slender and gradually attenuate, minutely hispidulous outside, smooth and stigmatic inside; leaves alternate .... Tribe 7. Vernonieae.

7 Anthers more or less distinctly tailed at the base; plants variously spiny or unarmed; receptacle naked or often densely bristly, not chaffy.

9 None of the corollas bilabiate, although some of the marginal ones may be enlarged and subradiate (the flower then neutral); plants very often with

the leaves or involucre more or less spiny, but sometimes wholly unarmed; receptacle mostly densely bristly, seldom naked; style with a thickened, often hairy ring below the branches, changing abruptly in texture at that point and papillate thence to the tip, the branches commonly more or less connate at least below .... Tribe 8. Cynareae.

9 Some of the corollas bilabiate (in our genus the heads radiate, with marginal, pistillate, ray flowers, intermediate pistillate flowers with short, tubular corolla, and central, perfect or functionally staminate flowers with bilabiate corolla); plants unarmed; receptacle naked in our genus; style branches in our genus short, rounded, externally papillate-hairy, with inconspicuous ventromarginal stigmatic lines, sometimes tardily or scarcely separating .... Tribe 9. Mutisieae.

Key to the Genera of Heliantheae

1 Receptacle chaffy at least near the margin (chaff bristlelike in *Eclipta*, which has white flowerheads).

2 Receptacle chaffy throughout, or sometimes naked in the center; other features various.

3 Pistillate flowers with evident corolla, usually provided with a ligule, or the flowers all perfect; other features various.

4 Disk flowers fertile, with divided style.

5 Pappus a cup or crown, or a few teeth or awns, or of both scales (or minute bristles) and awns, or none, but not of 5–many similar scales or plumose awns.

6 Achenes of the disk radially flattened or not evidently flattened; receptacular bracts mostly concave-convex and clasping (bristlelike in *Eclipta*); involucral bracts various, but not biseriate and dimorphic.

7 Rays (when present) deciduous, not persistent on the achene (subtribe Verbesininae).

8 Leaves all alternate; receptacle strongly conic or columnar, commonly elongating in fruit.

9 Rays not subtended by receptacular bracts.

10 Receptacular bracts not spinescent, although sometimes shortly awn-pointed; disk corollas narrowed to a slender base .... 1. *Rudbeckia*.

10 Receptacular bracts spinescent, surpassing the disk corollas; disk corollas slightly thickened at the base, rather than narrowed .... 2. *Echinacea*.

9 Rays subtended by receptacular bracts.

11 Achenes flattened, with 2 sharp and often 2 very blunt angles; leaves pinnatifid or bipinnatifid; perennial .... 3. *Ratibida*.

11 Achenes subterete; leaves simple, cordate-clasping, entire or merely toothed; annual .... 4. *Dracopis*.

8 Leaves variously opposite (or even ternate) or alternate, but if all alternate then the receptacle flat or convex to merely low-conic.

12 Achenes thick or quadrangular to subterete, not strongly flattened.

13 Heads white or gray.

14 Receptacular bracts narrow, awnlike or bristlelike; rays minute but evident .... 5. *Eclipta*.

14 Receptacular bracts of ordinary type, concave and more or less clasping the disk flowers; rays wanting .... 6. *Melanthera*.

13 Heads yellow (sometimes pale) or orange.

15 Rays pistillate and fertile.

16 Involucre of 4 (5) broad, herbaceous bracts; receptacle conic .... 7. *Tetragonotheca*.

16 Involucre of more than 5 bracts; receptacle flat or convex.

17 Achenes broad-topped, not narrowed to the summit; shrubs .... 8. *Borrichia*.

17 Achenes more or less narrowed to the summit, or shouldered so that the morphologically marginal pappus appears to be set well in from the lateral margins; ours herbs .... 9. *Wedelia*.

15 Rays sterile, usually neutral.
18 Pappus in our spp. none; receptacle in our spp. distinctly conic .... 10. *Viguiera.*
18 Pappus generally of a pair of marginal awns, these often somewhat paleaceous-dilated toward the base, sometimes also with some intermediate shorter scales or minute bristles; receptacle flat to low-conic.
19 Peduncles notably fistulous; ours a coarse, soft shrub .... 11. *Tithonia.*
19 Peduncles not fistulous; herbs.
20 Pappus of 2 caducous awn scales, rarely with a few additional short scales; plants not producing horizontal tubers (except in *H. tuberosus*), although often with tuberous-thickened, vertical roots .... 12. *Helianthus.*
20 Pappus of 1 or 2 persistent (often short) awn scales or evident teeth and an indefinite, large number of minute, more or less coalescent bristles; plants producing horizontal tubers, or with the rhizomes tuberous-thickened at intervals .... 13. *Phoebanthus.*
12 Achenes strongly flattened, often also wing-margined.
21 Receptacle flat to shortly conic; style branches with an acute appendage; leaves opposite or alternate .... 14. *Verbesina.*
21 Receptacle elongate, conic or subcylindric; style branches truncate, exappendiculate; leaves opposite .... 15. *Spilanthes.*
7 Rays persistent on the achenes and becoming papery (subtribe Zinniinae).
22 Achenes quadrangular, scarcely compressed; involucral bracts herbaceous at least distally, in 1–3 subequal series, or the outer ones enlarged and leafy; our spp. perennial .... 16. *Heliopsis.*
22 Achenes radially compressed; involucral bracts imbricate in several series, largely chartaceous or coriaceous, stramineous, with a darker band near the rounded and erose-ciliate tip; our spp. annual .... 17. *Zinnia.*
6 Achenes at least of the disk flowers more or less flattened parallel to the involucral bracts, or occasionally tetragonal; receptacular bracts flat or only slightly concave-convex; involucral bracts often biseriate and dimorphic (subtribe Coreopsidinae).
23 Involucral bracts all about alike, or the inner passing into those of the receptacle; heads small and few-flowered, with inconspicuous rays only 1–3 mm long.
24 Achenes all about alike, wingless or somewhat thick-winged distally, not lacerate .... 18. *Calyptocarpus.*
24 Achenes markedly dimorphic, those of the rays with conspicuous, lacerate wing margins, the others wingless .... 19. *Synedrella.*
23 Involucral bracts obviously biseriate and dimorphic; heads various, but not wholly as above.
25 Bracts of the involucre all separate or nearly so.
26 Achenes winged; pappus awns, when present, antrorsely barbed or barbless .... 20. *Coreopsis.*
26 Achenes wingless; pappus often (not always) of retrorsely barbed awns.
27 Achenes beakless, flattened to sometimes quadrangular.
28 Rays pink or white, or yellow with red-brown base; pappus obsolete or nearly so .... spp. of *Coreopsis.*
28 Rays yellow, or wanting, or if white then the pappus of retrorsely barbed awns; pappus commonly of retrorsely barbed awns, less often of antrorsely barbed awns or sometimes obsolete .... 21. *Bidens.*
27 Achenes beaked, not much flattened .... 22. *Cosmos.*
25 Inner bracts of the involucre connate for at least 1/4 of their length .... 23. *Thelesperma.*
5 Pappus, at least of the disk flowers, of 5–many similar scales or plumose awns (subtribe Galinsoginae).
29 Leaves opposite.
30 Plants annual; heads small, the involucre up to 4 mm high and

the disk up to 6 mm wide; pappus scales somewhat fimbriate, but not plumose; involucral bract subtending each ray achene tending to be basally connate with 2 adjacent receptacular bracts, the 3 falling as a unit with the subtended achene .... 24. *Galinsoga.*

30 Plants perennial (ours); heads larger, ours with the involucre 7–10 mm high and the disk 7–15 mm wide; pappus in ours of 20 long, stout, conspicuously plumose bristles; involucral bracts free from the receptacular bracts .... 25. *Tridax.*

29 Leaves alternate (or all basal).

31 Heads radiate, at least the rays yellow; receptacle honeycombed; pappus scales 7–12 .... 38. *Balduina.*

31 Heads discoid, anthocyanic to white; receptacle not honeycombed; pappus scales 5 (6) .... 26. *Marshallia.*

4 Disk flowers sterile, with undivided style (subtribe Melampodiinae).

32 Achenes strongly flattened parallel to the bracts of the involucre; pappus present or absent.

33 Ray achenes in 2–3 series, even at flowering time, each achene completely free from the nearby bracts and falling separately .... 27. *Silphium.*

33 Ray achenes in a single series, each achene more or less enclosed by and often attached to the subtending involucral bract and 2–3 adjacent receptacular bracts, all commonly falling as a unit.

34 Heads yellow, conspicuously radiate; receptacle flat.

35 Leaves opposite, plants fibrous-rooted .... 28. *Chrysogonum.*

35 Leaves alternate (or largely basal); plants taprooted .... 29. *Berlandiera.*

34 Heads white, very shortly and inconspicuously radiate; receptacle conic or convex .... 30. *Parthenium.*

32 Achenes thick, not much flattened; pappus none.

36 Achenes merely subtended by the unarmed bracts; perennial ...... 31. *Polymnia.*

36 Achenes embraced and enclosed by the prickly inner bracts of the involucre; annual .... 32. *Acanthospermum.*

3 Pistillate flowers nearly or quite without corolla; staminate flowers with undivided style; perfect flowers none; heads small, discoid, often unisexual; anthers scarcely united; filaments monadelphous; pappus none (subtribe Ambrosiinae).

37 Staminate and pistillate flowers in the same heads; involucre of a few rounded, not at all tuberculate or spiny bracts .... 33. *Iva.*

37 Staminate and pistillate flowers in separate heads, the staminate generally uppermost; involucre of the pistillate heads nutlike or burlike.

38 Involucral bracts of the staminate heads united; involucre of the pistillate heads with 1 or several series of tubercles or straight spines ..... 34. *Ambrosia.*

38 Involucral bracts of the staminate heads separate; involucre of the pistillate heads a bur with numerous hooked prickles .... 35. *Xanthium.*

2 Receptacle chaffy only near the margin; involucral bracts uniseriate, equal, strongly carinate, each enfolding a ray-achene; heavy-scented annual (subtribe Madiinae) .... 36. *Madia.*

1 Receptacle essentially naked (bristly in one sp. of *Gaillardia*, this with partly or wholly red-purple rays).

39 Involucre and leaves not dotted with oil glands, although the leaves may be finely glandular-punctate.

40 Pappus of separate scales or bristles; involucral bracts several or many; leaves opposite or alternate (subtribe Heleniinae).

41 Involucral bracts herbaceous, neither petallike nor scarious toward the tip; heads mostly radiate in our spp. with numerous (more than 50) flowers.

42 Leaves opposite; pappus of slender, quickly deciduous capillary bristles .... 37. *Jamesianthus.*

42 Leaves alternate, or all basal; pappus of awnless or awn-tipped scales.

43 Style branches with a more or less elongate appendage .... 39. *Gaillardia.*

43 Style branches exappendiculate, the tip dilated, subtruncate, penicillate .... 40. *Helenium.*

41 Involucral bracts somewhat petaloid, colored, and usually more or less scarious, at least toward the tip; heads discoid in our spp., with relatively few (up to ca. 30) flowers (subtribe Baeriinae) .... 41. *Palafoxia.*

40 Pappus none; involucral bracts 2–8; leaves opposite (subtribe Flaveriinae) .... 42. *Flaveria.*

39 Involucre and generally also the leaves dotted with oil glands (subtribe Tagetinae).

44 Involucral bracts dimorphic, 2–3 seriate .... 43. *Dyssodia.*
44 Involucral bracts all about alike, uniseriate.
45 Involucral bracts distinct; style branches very short .... 44. *Pectis.*
45 Involucral bracts connate most of their length; style branches elongate .. 45. *Tagetes.*

Key to the Genera of Anthemideae

1 Pappus of short but definite and discrete scales; style branches shortly appendiculate; heads discoid .... 46. *Hymenopappus.*
1 Pappus a minute crown, or none; style branches truncate, exappendiculate; heads variously radiate, disciform, or discoid.
2 Receptacle chaffy, at least toward the middle.
3 Heads relatively large, terminating the branches; rays elongate; achenes terete or 4–5-angled, or sometimes more or less compressed but not callous-margined .... 47. *Anthemis.*
3 Heads small, numerous in a short, broad, paniculate-corymbiform inflorescence; rays short and broad; achenes strongly flattened, callous-margined .... 48. *Achillea.*
2 Receptacle naked (long-hairy in one sp. of *Artemisia*).
4 Heads disciform, marginal flowers numerous, pistillate, with elongate, indurate-persistent style and without corolla; disk flowers few, functionally staminate; achenes winged .... 53. *Soliva.*
4 Heads variously radiate to discoid or disciform but the flowers all with definite corolla, and the style not indurate-persistent; disk flowers fertile except in one sp. of *Artemisia*; achenes terete or angular or sometimes flattened, but not winged.
5 Inflorescence corymbiform, or the heads solitary and terminating the branches; heads variously large or small, radiate or rayless, with or without pappus.
6 Receptacle flat or somewhat convex.
7 Heads radiate .... 49. *Chrysanthemum.*
7 Heads disciform, the pistillate flowers with short, tubular corolla .... 50. *Tanacetum.*
6 Receptacle high, hemispheric or conic at least at maturity .... 51. *Matricaria.*
5 Inflorescence spiciform, racemiform, or paniculiform; heads small, rayless, epappose .... 52. *Artemisia.*

Key to the Genera of Senecioneae

1 Disk flowers perfect and fertile; habit diverse.
2 Leaves (except the uppermost) opposite (in our spp. mostly basal) .... 54. *Arnica.*
2 Leaves all alternate.
3 Heads radiate or discoid, the pistillate flowers, if present, with a definite ligule.
4 Heads yellow to orange or sometimes red or anthocyanic, radiate or discoid; disk corollas merely toothed.
5 Style branches truncate, penicillate; heads radiate or less often discoid, yellow to orange or rarely flame-red .... 55. *Senecio.*
5 Style branches with a definite, slender, hispidulous appendage; heads strictly discoid.
6 Style appendage short, less than 1 mm long; plants not purple-hairy; heads pink to purple or red, seldom orange .... 56. *Emilia.*
6 Style appendage elongate (ca. 2 mm); our spp. conspicuously purple-hairy, and with orange heads .... 57. *Gynura.*
4 Heads white to ochroleucous or chloroleucous, strictly discoid; disk corollas deeply lobed .... 58. *Cacalia.*
3 Heads disciform, with 2–several marginal rows of pistillate flowers that have a tubular-filiform, eligulate corolla; flowers whitish .... 59. *Erechtites.*
1 Disk flowers sterile, with slightly lobed or undivided style; leaves large and basal, the stem merely bracteate; plants blooming as or before the leaves develop.
7 Heads solitary, all alike, yellow .... 60. *Tussilago.*
7 Heads several or (in ours) numerous, in some plants chiefly or wholly with pistillate flowers, in others chiefly of sterile hermaphrodite (disk) flowers; flowers not yellow, in our spp. purple .... 61. *Petasites.*

Key to the Genera of Astereae

1 Rays present, yellow.
2 Pappus of the disk flowers composed of more or less numerous capillary bristles (sometimes short), with or without a short outer series of bristles or scales; pappus of the rays like that of the disk flowers, or sometimes reduced or wanting.
3 Pappus simple, not differentiated into an inner and outer series.
4 Herbs.
5 Plants taprooted, our spp. annual or casually biennial; rays more or less elongate, in our spp. commonly 4–14 mm long .... 62. *Haplopappus.*
5 Plants fibrous-rooted from a perennial caudex or rhizome; rays relatively short, not over ca. 5 mm long.
6 Plants with corymbiform, flat-topped inflorescence; rays more numerous than the disk flowers; leaves more or less evidently glandular-punctate; receptacle fimbrillate .... 69. *Euthamia.*
6 Plants not *at once* with corymbiform, flat-topped inflorescence *and* with the rays more numerous than the disk flowers, although these features occur separately in a few spp.; leaves, except in *S. odora*, not glandular-punctate; receptacle merely alveolate .... 66. *Solidago.*
4 Shrubs .... 68. *Chrysoma.*
3 Pappus distinctly double, the outer series much shorter than the inner.
7 Ray flowers with evident pappus like that of the disk flowers .... 63. *Chrysopsis.*
7 Ray flowers essentially epappose .... 64. *Heterotheca.*
2 Pappus of less than 10 scales or firm awns, or that of the rays reduced to a minute, toothed crown.
8 Heads relatively large, the rays mostly 15–40, 7–16 mm long, the disk mostly 1–2 cm wide; leaves mostly well over 3 mm wide .... 65. *Grindelia.*
8 Heads small, the rays mostly 6–10, 3–5 mm long, the disk up to ca. 5 mm wide; leaves up to ca. 3 mm wide .... 71. *Gutierrezia.*
1 Rays either some other color than yellow, or wanting.
9 Plants dioecious; shrubs; pappus of numerous capillary bristles .... 75. *Baccharis.*
9 Plants producing perfect and pistillate flowers in the same head; herbs; pappus various.
10 Pappus of more or less numerous capillary bristles (or that of the ray flowers sometimes reduced).
11 Heads strictly discoid, the flowers all tubular and perfect; flowers 2–16 in each head.
12 Flowers yellow, 2–8 in each head.
13 Heads in a terminal, flat-topped, corymbiform inflorescence .... 70. *Bigelowia.*
13 Heads in an open-paniculiform inflorescence, with several elongate, divaricate, recurved-secund branches (*S. brachyphylla*) .... 66. *Solidago.*
12 Flowers white or anthocyanic, 9–16 in each head .... 67. *Brintonia.*
11 Heads radiate to disciform, the marginal flowers pistillate, with or sometimes without a ligule, the central ones tubular and perfect; flowers often more than 16 in each head.
14 Pistillate flowers few to numerous, in one or several series, usually but not always with an evident ligule; disk flowers few to generally numerous.
15 Involucral bracts either subequal and the outer leafy, or more commonly evidently imbricate, with chartaceous base and evident green tip, sometimes chartaceous throughout; style appendages lanceolate or narrower, acute or acuminate, ordinarily more than 0.3 mm long; plants blooming in late summer and fall.
16 Disk flowers, as well as ray flowers, white (*S. ptarmicoides*) .... 66. *Solidago.*
16 Disk flowers yellow (sometimes pale) or anthocyanic; rays white to anthocyanic.
17 Rays small, only ca. 2–3 mm long, white, mostly 7–14; heads numerous in an elongate, thyrsoid inflorescence (*S. bicolor*) ..... 66. *Solidago.*
17 Rays larger, or more numerous, or anthocyanic; inflorescence various, but seldom elongate and thyrsoid .... 72. *Aster.*
15 Involucral bracts subequal to more or less imbricate, often green in part, but neither definitely leafy nor with chartaceous base and herbaceous green tip; style appendages lanceolate or broader, acute to obtuse, ours 0.3 mm long or less, or obsolete; plants blooming chiefly in spring and early summer .... 73. *Erigeron.*

14 Pistillate flowers numerous, filiform, with short, narrow, inconspicuous ligule scarcely if at all surpassing the disk; disk flowers in our spp. ca. 21 or generally fewer; annual weeds, blooming in late summer and fall; involucre and styles of *Erigeron* .... 74. *Conyza.*

10 Pappus not of numerous capillary bristles, instead consisting of minute bristles, often with 2–4 longer awns, or of a small, scaly, lacerate crown, or of short, unequal scales, or of a few short scales and a few longer awns, or wanting.

18 Pappus evident (at 10×), though small.

19 Plants fibrous-rooted perennials, often more than 5 dm tall; achenes strongly compressed, 2-nerved, evidently to obscurely wing-margined .... 76. *Boltonia.*

19 Plants taprooted annuals (seldom overwintering), 1–5 dm tall; achenes several-nerved, not winged.

20 Heads relatively narrow and few-flowered, ours with 5–15 rays, the disk only 2–3 mm wide; pappus of 5 short scales and usually also 5 longer awns .... 77. *Chaetopappa.*

20 Heads wider and more numerously flowered, ours with 15–45 rays, the disk 7–13 mm wide; pappus an irregular, scaly, lacerate crown, or of unequal scales less than 1 mm long .... 78. *Aphanostephus.*

18 Pappus essentially wanting.

21 Plant a leafy-stemmed annual; involucral bracts with green or greenish midstrip and chartaceous or scarious margins; receptacle convex, cushionlike .... 79. *Astranthium.*

21 Plant a scapose perennial; involucral bracts wholly herbaceous; receptacle definitely conic .... 80. *Bellis.*

Key to the Genera of Inuleae

1 Heads radiate .... 81. *Inula.*

1 Heads discoid or disciform.

2 Receptacle chaffy (sometimes naked in the center); at least the outer flowers without pappus.

3 Many of the flowers (the inner ones) with a well-developed capillary pappus; heads woolly at the base, but not completely invested, the involucre evident; receptacle naked in the center .... 82. *Filago.*

3 Flowers all epappose; heads completely invested in wool, the involucre hidden; receptacle chaffy throughout .... 83. *Evax.*

2 Receptacle naked; flowers all pappiferous.

4 Pappus of long-plumose capillary bristles .... 87. *Facelis.*

4 Pappus of capillary bristles, these sometimes barbellate or distally expanded and serrate, but not at all plumose.

5 Heads numerous in a very dense, terminal, spicate, acropetally flowering inflorescence, or in several such inflorescences, each on a short branch; stem conspicuously winged by the long-decurrent leaf bases .... 86. *Pterocaulon.*

5 Heads in various other sorts of inflorescences, but not as in *Pterocaulon*; stem mostly wingless, but evidently winged in a few spp.

6 Involucral bracts dry (or partly herbaceous) and sometimes firmly scarious, but neither hyaline nor shining, glabrous or often hairy or glandular over much or all of their surface.

7 Plants leafy-stemmed .... 84. *Pluchea.*

7 Plants scapose, the foliage leaves all in a basal rosette .... 85. *Sachsia.*

6 Involucral bracts largely or wholly scarious or hyaline and shining, usually glabrous above the often woolly base.

8 Heads all alike, the outer flowers pistillate, the inner ones perfect and fertile; heads sessile in variously arranged glomerules; our spp. annual or biennial .... 88. *Gnaphalium.*

8 Heads of two sorts, those on some plants wholly staminate in function, those on other plants largely or wholly pistillate; heads or most of them individually short-pedunculate (or solitary); perennial.

9 Pistillate heads with a few central functionally staminate flowers; plants rhizomatous but not stoloniferous, the leaves all cauline .. 89. *Anaphalis.*

9 Pistillate heads without central staminate flowers; our spp. stoloniferous and often mat-forming, the basal leaves and those at the ends of the stolons the largest, forming rosettes or mats .... 90. *Antennaria.*

Key to the Genera of Eupatorieae

1 Achenes 3–5-angled and -ribbed; pappus various.
  2 Pappus of more or less numerous (more than 10) capillary bristles.
    3 Involucre of more than 4 bracts, and in most species containing more than 4 flowers; herbs or shrubs, sometimes scrambling, but not twining ...... 91. *Eupatorium*.
    3 Involucre of 4 bracts, containing 4 flowers; ours twining, herbaceous vines .... 92. *Mikania*.
  2 Pappus of several (commonly 4–6) scales or awns, or a short crown, or none.
    4 Leaves opposite or whorled, chiefly or wholly cauline, not basally disposed.
      5 Leaves opposite, more than 5 mm wide; terrestrial plants, sometimes of wet places .... 93. *Ageratum*.
      5 Leaves whorled, up to ca. 2 mm wide; emergent aquatics .... 94. *Sclerolepis*.
    4 Leaves alternate, basally disposed, the cauline ones progressively reduced upward .... 95. *Hartwrightia*.
1 Achenes ca. 10-ribbed; pappus of more or less numerous bristles (these sometimes plumose).
    6 Involucral bracts arranged in 5 vertical ranks of 3 (4) each; shrub .... 96. *Garberia*.
    6 Involucral bracts not arranged in vertical ranks; ours herbs.
      7 Involucral bracts, or at least the inner ones, evidently few-striate longitudinally; leaves chiefly or wholly cauline, well distributed along the stem, not basally disposed.
        8 Pappus bristles merely barbellate .... 97. *Brickellia*.
        8 Pappus bristles evidently plumose .... 98. *Kuhnia*.
      7 Involucral bracts not striate; leaves tending to be basally disposed, the basal or lower cauline ones evidently larger than the progressively reduced (though often numerous) middle and upper ones.
        9 Inflorescence broadly corymbiform or sometimes thyrsoid-paniculate; plants with a stout, simple or branched caudex, crown, or short rhizome, the base neither cormlike nor suggesting a fleshy-thickened taproot nor suggesting a cluster of fingerlike, fleshy-thickened roots; involucre 3.5–12 mm high .... 99. *Carphephorus*.
        9 Inflorescence mostly spiciform or racemiform, or seldom with spiciform or racemiform branches, only rarely corymbiform; plants with a thickened, usually cormlike rootstock, or the rootstock suggesting a fleshy-thickened taproot or a cluster of fingerlike, fleshy-thickened roots; involucre often more than 12 mm high .... 100. *Liatris*.

Key to the Genera of Vernonieae

1 Heads discoid.
  2 Pappus double, the inner of numerous long, slender bristles, the outer of short bristles or short scales; heads not glomerate, our spp. with 9–many flowers .... 101. *Vernonia*.
  2 Pappus of 5–8 flattened scales, in our spp. prolonged at the summit into a terminal bristle; heads glomerate into secondary heads; primary heads with 1–5 (in our spp. 4) flowers .... 102. *Elephantopus*.
1 Heads radiate, many-flowered, not glomerate; pappus of 4–5 awnlike chaffy scales .... 103. *Stokesia*.

Key to the Genera of Cynareae

1 Achenes attached to the receptacle by the base; flowers all alike, the marginal ones not enlarged.
  2 Heads with more or less numerous flowers, not united into a secondary head.
    3 Receptacle densely bristly; plants not at once with the stem strongly spiny-winged and with the herbage tomentose throughout.
      4 Filaments separate; leaves not white-mottled.
        5 Involucral bracts hooked at the tip; leaves not bristly or spiny .... 104. *Arctium*.
        5 Involucral bracts not hooked; leaves bristly or spiny on the margin.
          6 Pappus bristles merely barbellate, not plumose; stem conspicuously spiny-winged .... 105. *Carduus*.
          6 Pappus bristles evidently plumose; stem in most species not conspicuously spiny-winged (*C. vulgare* the notable exception) .... 106. *Cirsium*.
      4 Filaments united below; leaves white-mottled .... 107. *Silybum*.

3 Receptacle fleshy, conspicuously honeycombed, not bristly or only very shortly so; plants with strongly spiny-winged stem and wholly tomentose herbage .... 108. *Onopordum*.
2 Heads 1-flowered, united into a secondary head .... 109. *Echinops*.
1 Achenes obliquely attached to the receptacle; marginal flowers often enlarged and neutral, or otherwise modified, but sometimes just like the inner ones.
7 Achenes without horny apical teeth, and with a pappus of several series of short (seldom elongate) scales or bristles, or the pappus wanting; leaves not prickly; flowers in most species anthocyanic (or white), seldom yellow .... 110. *Centaurea*.
7 Achenes with 10 horny teeth at the summit, and with a biseriate pappus of 10 long awns alternating with 10 shorter inner ones; leaves prickly-margined; flowers yellow .... 111. *Cnicus*.

Key to the Genera of Mutisieae

A single genus in our area .... 112. *Chaptalia*.

Key to the Genera of Lactuceae

1 Pappus of simple capillary bristles only.
2 Achenes terete or prismatic, only slightly or not at all flattened.
3 Flowers pink or purple to white or ochroleucous.
4 Plants rushlike, the leaves linear, up to ca. 3 mm wide, or reduced to mere scales; heads erect, terminating the branches .... 113. *Lygodesmia*.
4 Plants not rushlike, the leaves broader, mostly well over 1 cm wide; heads in a corybiform or paniculiform to thyrsoid or narrowly subracemiform inflorescence, often nodding .... 114. *Prenanthes*.
3 Flowers bright yellow to orange or orange-red.
5 Achenes with a long, slender beak from half as long to longer than the body.
6 Beak of the achene with a ring of soft, white, reflexed hairs at the summit, just beneath the pappus .... 126. *Pyrrhopappus*.
6 Beak of the achene without a ring of hairs at the summit.
7 Plants caulescent, with some more or less well-developed cauline leaves, branched above and with several or many heads.
8 Achene without scales; flowers in ours numerous; inner involucral bracts in ours 9–13 .... 120. *Crepis*.
8 Achene with a circle of small scales at the base of the beak; flowers few, in ours mostly 9–12; inner involucral bracts in ours ca. 8 ... 123. *Chondrilla*.
7 Plants strictly scapose, each scape with a solitary terminal head ..... 124. *Taraxacum*.
5 Achenes beakless, or with a short stout beak less than half as long as the body.
9 Plants perennial from an elongate or very short rhizome or from a short caudex or crown, strictly fibrous-rooted .... 119. *Hieracium*.
9 Plants either taprooted, or annual, or both.
10 Pappus bristles of unequal thickness, but not of 2 distinct sorts; leaves not prickly toothed; inflorescence open, corymbiform or paniculiform, the erect heads terminating the ultimate branches.
11 Heads few to numerous, larger, the involucre 5–12 mm high, often with more than 20 flowers, the achenes often more than 2.5 mm long .... 120. *Crepis*.
11 Heads numerous and small, the involucre 3.5–5 mm high, with ca. 10–20 flowers, the achenes 1.5–2.5 mm long .... 121. *Youngia*.
10 Pappus bristles of 2 sorts, a few of them a little coarser and more persistent than the very numerous other ones; leaves in ours finely prickly toothed; heads in ours shortly slender-pedunculate, scattered along the elongate principal branches of the inflorescence .... 117. *Launaea*.
2 Achenes more or less strongly flattened.
12 Achenes beaked or beakless, in either case somewhat enlarged at the summit where the pappus is attached; heads relatively few-flowered (ca. 8–56 in our spp.).
13 Erect, more or less leafy-stemmed plants; involucre in ours imbricate ... 115. *Lactuca*.
13 Delicate, creeping plants with erect scapes or peduncles up to ca. 1 dm long bearing 1 or 2 heads each; involucre calyculate .... 116. *Ixeris*.

12 Achenes beakless, without any enlarged, pappiferous disk at the summit; heads many-flowered (ca. 80–250 in our spp.) .... 118. *Sonchus*.

1 Pappus of plumose bristles, or bristles and scales, or only scales, or none, the scales sometimes minute and inconspicuous in spp. of *Krigia*.

14 Pappus of scales, or bristles and scales, or none, not plumose.

15 Corollas blue or occasionally pink or white; pappus of minute narrow scales only .... 125. *Cichorium*.

15 Corollas yellow or orange; pappus various.

16 Involucre calyculate; achenes 3–5 mm long .... 122. *Lapsana*.

16 Involucre not calyculate, the bracts all equal; achenes mostly less than 3 mm long .... 127. *Krigia*.

14 Pappus of plumose bristles, at least in part.

17 Plume branches of the pappus not interwebbed; involucre either imbricate, or calyculate, or biseriate with enlarged and foliaceous outer bracts; leaves not grasslike.

18 Leaves basally disposed, the cauline leaves few and progressively reduced, or wanting; hairs not forked at the summit.

19 Receptacle chaffy-bracted .... 128. *Hypochoeris*.

19 Receptacle naked .... 129. *Leontodon*.

18 Leaves mainly cauline, well distributed along the stem; many of the hairs forked at the summit, shaped like an anchor or a grappling hook ...... 130. *Picris*.

17 Plume branches of the pappus interwebbed; involucre uniseriate; leafy-stemmed plants with grasslike leaves .... 131. *Tragopogon*.

Artificial Key to the Genera of Asteraceae

1 Flowers all ligulate and perfect; juice milky .... Tribe Lactuceae; see previous key.

1 Flowers not all ligulate; ray (ligulate) flowers when present marginal, either pistillate or neutral; juice ordinarily watery.

2 Heads radiate.

3 Rays yellow or orange (sometimes marked with purple or brown at the base).

4 Receptacle chaffy or bristly, at least toward the margin .... Group I.

4 Receptacle naked .... Group II.

3 Rays white to pink, purple, red, or blue, but not yellow or orange .... Group III.

2 Heads discoid or disciform, without rays (some plants with very small and inconspicuous rays are keyed here as well as in the radiate group).

5 Plants thistles, with more or less spiny-margined leaves and usually also with spiny involucre, the stem sometimes spiny-winged .... Group IV.

5 Plants not thistlelike, the leaves not spiny-margined (though seldom with axillary spines), the involucre spiny only in a few genera, otherwise innocuous, the stem not spiny-winged.

6 Pappus partly or wholly of numerous (more than 10) capillary (sometimes plumose) bristles (wanting from the outermost flowers of *Filago*) ...... Group V.

6 Pappus of scales, or a few awns, or very short, chaffy bristles, or a mere crown, or none, never capillary or plumose .... Group VI.

Group I

(Heads with yellow or orange rays and chaffy or bristly receptacle)

1 Disk flowers sterile, with undivided style, their ovaries much smaller than those of the fertile ray flowers.

2 Inner bracts of the involucre (or outer bracts of the receptacle) conspicuously prickly; plants annual .... 32. *Acanthospermum*.

2 Involucral and receptacular bracts not prickly; plants perennial.

3 Ray achenes in 2–3 series, even at flowering time; rays relatively large, at least 1 cm long, often more than 2 cm long .... 27. *Silphium*.

3 Ray achenes uniseriate; rays mostly smaller, not more than 2 cm long, sometimes less than 1 cm.

4 Leaves alternate .... 29. *Berlandiera*.

4 Leaves opposite.

5 Coarse plants, mostly (6) 10–30 dm tall, with large, more or less lobed leaves commonly well over 1 dm long .... 31. *Polymnia*.

5 More slender, smaller plants, mostly 0.5–5 dm tall, with merely crenate leaves seldom as much as 1 dm long .... 28. *Chrysogonum*.

1 Disk flowers fertile, with divided style, their ovaries as large as, or larger than, those of the ray flowers, which may be either fertile or sterile.
6 Shrubs.
7 Heads large, the rays mostly 3–6 cm long; leaves large, mostly more than 5 cm wide .... 11. *Tithonia*.
7 Heads smaller, the rays less than 2 cm long; leaves smaller, rarely more than 5 cm wide.
8 Pappus a crown around the margin of the broad-topped achene; leaves succulent, not scabrous .... 8. *Borrichia*.
8 Pappus a crown at the strongly constricted center of the top of the achene, well removed from the margins; leaves scabrous, not succulent .... 9. *Wedelia*.
6 Herbs.
9 Receptacle with a single series of bracts between the ray and disk flowers, otherwise naked; involucral bracts uniseriate, each enfolding a ray achene .... 36. *Madia*.
9 Receptacle generally chaffy or bristly (or deeply honeycombed) throughout; involucral bracts, if uniseriate, not enfolding the ray achenes.
10 Leaves alternate.
11 Receptacle merely bristly; pappus of 6–10 awned scales .... 39. *Gaillardia*.
11 Receptacle definitely chaffy; pappus otherwise.
12 Receptacle merely convex or low-conic.
13 Receptacular bracts connate to form a toothed or cleft honeycomb in which the flowers are set; pappus a ring of 7–12 small scales .... 38. *Balduina*.
13 Receptacular bracts distinct, not forming a honeycomb; pappus otherwise.
14 Achenes flattened parallel to the involucral bracts, at right angles to the radii of the head; involucral bracts evidently biseriate and dimorphic .... 20. *Coreopsis*.
14 Achenes either not flattened, or flattened at right angles to the involucral bracts in alignment with the radii of the head; involucral bracts not biseriate and dimorphic.
15 Leaves narrow, seldom more than 5 mm wide; achenes thick, not strongly flattened .... 13. *Phoebanthus*.
15 Leaves wider, seldom less than 2 cm wide; achenes strongly flattened .... 14. *Verbesina*.
12 Receptacle strongly conic or columnar.
16 Ray flowers subtended by receptacular bracts.
17 Leaves simple, cordate-clasping .... 4. *Dracopis*.
17 Leaves pinnatifid .... 3. *Ratibida*.
16 Ray flowers not subtended by receptacular bracts.
18 Receptacular bracts spinescent, surpassing the disk flowers .... 2. *Echinacea*.
18 Receptacular bracts not spinescent, although sometimes shortly awn-pointed .... 1. *Rudbeckia*.
10 Leaves, or at least the lower ones, opposite or whorled (the middle and upper leaves in some spp. are consistently alternate).
19 Rays persistent on the achenes and becoming papery.
20 Achenes quadrangular, scarcely compressed; plants perennial .... 16. *Heliopsis*.
20 Achenes radially compressed; plants annual .... 17. *Zinnia*.
19 Rays deciduous from the achenes at maturity.
21 Pappus of 20 long, stout, conspicuously plumose bristles .... 25. *Tridax*.
21 Pappus distinctly otherwise, never plumose.
22 Receptacular bracts flat or nearly so (sometimes very narrow), not clasping the achenes; achenes mostly flattened parallel to the involucral bracts, at right angles to the radii of the heads, sometimes only slightly or scarcely so, but in any case not radially flattened.
23 Involucral bracts obviously biseriate and dimorphic, the outer ones more herbaceous than the inner; habit and heads various.
24 Bracts of the involucre all separate or nearly so.
25 Achenes beaked, not much flattened, mostly 1.5–3 cm long; leaves dissected .... 22. *Cosmos*.
25 Achenes beakless, seldom more than 1 cm long, more or less strongly flattened parallel to the involucral bracts, at right angles to the radii of the head; leaves simple to dissected.

26 Achenes mostly wing-margined, rarely wingless; pappus awns, when present, antrorsely barbed or barbless ..... 20. *Coreopsis*.
26 Achenes wingless; pappus commonly of retrorsely barbed awns, less often of antrorsely barbed awns, or sometimes obsolete ..... 21. *Bidens*.
24 Inner bracts of the involucre connate for about 1/4 to 1/2 their length ..... 23. *Thelesperma*.
23 Involucral bracts not obviously biseriate and dimorphic; annuals with small, few-flowered heads and short rays only 1–3 mm long.
27 Achenes all about alike, wingless or somewhat thick-winged distally, not lacerate ..... 18. *Calyptocarpus*.
27 Achenes markedly dimorphic, those of the rays with conspicuous, lacerate wing margins, the others wingless ..... 19. *Synedrella*.
22 Receptacular bracts more or less folded or concave-convex and clasping the achenes; achenes not flattened, or often more or less flattened radially.
28 Involucre of 4 (5) broad, herbaceous bracts ..... 17. *Tetragonotheca*.
28 Involucre of more than 5 bracts.
29 Achenes (at least of the disk flowers) strongly flattened, often also wing-margined.
30 Receptacle flat to shortly conic; style branches with an acute appendage ..... 14. *Verbesina*.
30 Receptacle elongate, conic or subcylindric; style branches truncate, exappendiculate ..... 15. *Spilanthes*.
29 Achenes thick or quadrangular to subterete, not strongly flattened, and not wing-margined.
31 Rays pistillate and fertile ..... 9. *Wedelia*.
31 Rays sterile, usually neutral.
32 Pappus none ..... 10. *Viguiera*.
32 Pappus present.
33 Pappus of 1 or 2 persistent (often short) awn scales or evident teeth and an indefinite large number of minute, more or less coalescent bristles; plants producing horizontal tubers, or with the rhizomes tuberous-thickened at intervals ..... 13. *Phoebanthus*.
33 Pappus of 2 caducous awn scales, rarely with a few short additional scales; plants (except *H. tuberosus*) not producing horizontal tubers, though often with tuberous-thickened vertical roots ..... 12. *Helianthus*.

Group II

(Heads with yellow or orange rays and naked receptacle)

1 Involucre and leaves beset with relatively large, scattered, embedded oil glands.
2 Leaves entire ..... 44. *Pectis*.
2 Leaves pinnately compound or pinnatifid or bipinnatifid.
3 Pappus of 10–20 scales, each cleft into several bristles ..... 43. *Dyssodia*.
3 Pappus of several very unequal scales, not divided into bristles ..... 45. *Tagetes*.
1 Involucre and leaves without embedded oil glands, or the leaves sometimes more finely glandular-punctate.
4 Pappus of scales, or of a few firm awns, or virtually none.
5 Leaves opposite; pappus none ..... 42. *Flaveria*.
5 Leaves alternate; pappus present, except in *Chrysanthemum*.
6 Pappus wanting or nearly so ..... 49. *Chrysanthemum*.
6 Pappus well developed, evident.
7 Pappus of 2–several firm, deciduous awns, not of scales ..... 65. *Grindelia*.
7 Pappus of evident scales, these sometimes awn-tipped.
8 Involucral bracts erect and appressed; heads small, the disk ca. 5 mm wide or less, the rays only 3–5 mm long ..... 71. *Gutierrezia*.
8 Involucral bracts spreading or deflexed; heads larger, the disk 6–30 mm wide, the rays 5–30 mm long.
9 Style branches with a subulate appendage ..... 39. *Gaillardia*.
9 Style branches truncate, exappendiculate ..... 40. *Helenium*.
4 Pappus (at least of the disk flowers) of capillary bristles (these few and quickly deciduous in *Jamesianthus*).

10 Leaves mainly or all opposite.
11 Leaves chiefly basal (but still opposite) .... 54. *Arnica.*
11 Leaves chiefly or all cauline .... 37. *Jamesianthus.*
10 Leaves all alternate (sometimes mostly basal).
12 Shrub .... 68. *Chrysoma.*
12 Herbs.
13 Involucral bracts uniseriate, equal, commonly with a few very much shorter outer ones at the base.
14 Disk flowers fertile; stem usually more or less leafy .... 55. *Senecio.*
14 Disk flowers sterile; stem merely bracteate .... 60. *Tussilago.*
13 Involucral bracts more or less imbricate in several series.
15 Heads very large, the involucre 2–2.5 cm high, the disk 3–5 cm wide, the numerous slender rays mostly 1.5–2.5 cm long .... 81. *Inula.*
15 Heads smaller, the involucre not more than 1.5 cm high, the disk less than 2 cm wide, the rays not more than 1.5 cm long.
16 Pappus double, the outer of short bristles, much shorter than the inner.
17 Ray flowers with evident pappus like that of the disk flowers .. 63. *Chrysopsis.*
17 Ray flowers essentially epappose .... 64. *Heterotheca.*
16 Pappus simple, not differentiated into an inner and outer series.
18 Plants taprooted, annual or casually biennial; rays more or less elongate, mostly 4–14 mm long .... 62. *Haplopappus.*
18 Plants fibrous-rooted from a perennial caudex or rhizome; rays short, up to ca. 5 mm long.
19 Plants with a corymbiform, flat-topped inflorescence; rays more numerous than the disk flowers; leaves more or less evidently glandular-punctate .... 69. *Euthamia.*
19 Plants not *at once* with corymbiform, flat-topped inflorescence *and* with the rays more numerous than the disk flowers, although these features occur separately in a few spp.; leaves, except in *S. odora*, not glandular-punctate .... 66. *Solidago.*

Group III

(Heads radiate, rays not yellow or orange)

1 Receptacle chaffy or bristly, at least toward the center.
2 Leaves, or many of them, opposite.
3 Pappus of several or many (more than 5) well-developed scales or long, plumose bristles.
4 Heads small, the involucre up to 4 mm high and the disk up to 6 mm wide; pappus scales often somewhat fimbriate, but not plumose .... 24. *Galinsoga.*
4 Heads larger, the involucre 7–10 mm high and the disk 7–15 mm wide; pappus of 20 long, stout, conspicuously plumose bristles .... 25. *Tridax.*
3 Pappus of a few awns or teeth, or a short crown, or none.
5 Disk flowers sterile, with undivided style .... 31. *Polymnia.*
5 Disk flowers fertile, with divided style.
6 Rays persistent on the achenes and becoming papery .... 17. *Zinnia.*
6 Rays deciduous from the achenes.
7 Rays numerous and short, only ca. 1 mm long, pistillate and fertile; disk white .... 5. *Eclipta.*
7 Rays few, commonly ca. 5 or 8, generally well over 1 mm long, sterile and usually neutral; disk yellow or anthocyanic.
8 Rhizomatous perennial (*C. rosea*) .... 20. *Coreopsis.*
8 Annual.
9 Achenes distinctly beaked .... 22. *Cosmos.*
9 Achenes sometimes narrowed above, but not beaked (*B. pilosa*) .... 21. *Bidens.*
2 Leaves alternate.
10 Heads only falsely radiate, the enlarged marginal corollas irregularly 5-lobed above the tubular base; receptacle bristly .... 110. *Centaurea.*
10 Heads radiate; receptacle chaffy.
11 Rays few, commonly 1–5, short, less than 1 cm long, usually white.
12 Disk flowers sterile, with undivided style .... 30. *Parthenium.*
12 Disk flowers fertile, with divided style.
13 Leaves pinnately dissected .... 48. *Achillea.*

13 Leaves merely toothed (*V. virginica*) .... 14. *Verbesina*.
11 Rays either more than 5, or more than 1 cm long, or both, often colored.
14 Pappus of 6–10 awned scales; receptacle merely bristly .... 39. *Gaillardia*.
14 Pappus otherwise, often a short crown, sometimes of 2 teeth or awns, or wanting; receptacle chaffy.
15 Leaves pinnatifid or pinnately dissected.
16 Rays brown-purple .... 3. *Ratibida*.
16 Rays white .... 47. *Anthemis*.
15 Leaves entire or merely toothed.
17 Achenes strongly flattened parallel to the involucral bracts; leaves junciform, only 1–2 mm thick (*C. nudata*) .... 20. *Coreopsis*.
17 Achenes not flattened, or flattened in alignment with the radii of the head; leaves flat, slightly to much wider.
18 Rays elongate, 2–8 cm long .... 2. *Echinacea*.
18 Rays shorter, mostly 1–1.5 cm long (*R. graminifolia*) .... 1. *Rudbeckia*.
1 Receptacle naked.
19 Pappus of several to usually more or less numerous capillary bristles, sometimes with some short outer bristles or small scales as well.
20 Central flowers of the head with distinctly bilabiate corolla, the outer lip 3-toothed, the inner with 1 or 2 teeth; scapose herbs with a basal rosette of leaves .... 112. *Chaptalia*.
20 Disk flowers all with regular, tubular, (4) 5-toothed corolla; habit various.
21 Taprooted annuals with very short rays, up to ca. 2 mm long.
22 Involucre mostly 3–4 mm high .... 74. *Conyza*.
22 Involucre 5–8 mm high (*A. subulatus*) .... 72. *Aster*.
21 Plants either fibrous-rooted perennials, or with the rays distinctly more than 2 mm long, or often both.
23 Involucral bracts subequal to more or less imbricate, often green in part, but neither definitely leafy nor with chartaceous base and herbaceous green tip; style appendages lanceolate or broader, acute to obtuse, 0.3 mm long or less, or obsolete; plants blooming chiefly in spring and early summer .... 73. *Erigeron*.
23 Involucral bracts either subequal and the outer leafy, or more commonly evidently imbricate, with chartaceous base and evident green tip, sometimes chartaceous throughout; style appendages lanceolate or narrower, ordinarily more than 0.3 mm long; plants blooming in late summer and fall.
24 Rays small, only 2–3 mm long, white, mostly 7–14; heads numerous in an elongate, thyrsoid inflorescence (*S. bicolor*) .... 66. *Solidago*.
24 Rays generally larger, or more numerous, or anthocyanic; inflorescence various, but seldom elongate and thyrsoid.
25 Disk flowers white (*S. ptarmicoides*) .... 66. *Solidago*.
25 Disk flowers yellow or anthocyanic .... 72. *Aster*.
19 Pappus of scales, or a few stout awns, or a short crown, or none, not of capillary bristles.
26 Heads only falsely radiate, the enlarged marginal corollas irregularly 5-lobed above the tubular base .... 103. *Stokesia*.
26 Heads radiate.
27 Heads small and few-flowered, with 5–15 rays not more than ca. 5 mm long, and with the disk only 2–3 mm wide .... 77. *Chaetopappa*.
27 Heads evidently larger or more numerously flowered, or both.
28 Pappus of 6–10 well-developed awn scales .... 39. *Gaillardia*.
28 Pappus otherwise.
29 Leaves all basal; pappus none .... 80. *Bellis*.
29 Leaves cauline as well as basal, or all cauline; pappus present or absent.
30 Receptacle elevated, hemispheric or conic.
31 Leaves entire or nearly so .... 76. *Boltonia*.
31 Leaves pinnatifid or pinnately dissected. .... 51. *Matricaria*.
30 Receptacle flat to merely convex or low-conic.
32 Plants evidently perennial; style branches truncate, penicillate .... 49. *Chrysanthemum*.
32 Plants annual; style branches with a more or less well-developed terminal appendage.

33 Pappus an irregular, scaly crown, or of short, unequal scales; achenes quadrate-columnar, several-nerved .... 78. *Aphanostephus*.
33 Pappus obsolete; achenes more or less flattened, 2-nerved .... 79. *Astranthium*.

Group IV

(Heads discoid; thistles, the leaves spiny-margined, the involucre usually spiny, and the stem sometimes also spiny-winged)

1 Heads individually 1-flowered, united into a secondary head .... 109. *Echinops*.
1 Heads with more or less numerous flowers, sometimes closely crowded, but not united into a secondary head.
2 Pappus of numerous plumose bristles .... 106. *Cirsium*.
2 Pappus otherwise, never plumose.
3 Heads yellow; pappus biseriate, the outer of 10 awns, these alternating with as many much shorter, minutely hairy and sparsely pectinate inner ones .... 111. *Cnicus*.
3 Heads pink or purple; pappus of numerous capillary or somewhat chaffy-flattened bristles.
4 Leaves white-mottled; stem not spiny-winged .... 107. *Silybum*.
4 Leaves green or sometimes covered with white tomentum, not mottled; stem spiny-winged.
5 Herbage densely white-tomentose; achenes sunken in the fleshy, honey-combed receptacle .... 108. *Onopordum*.
5 Herbage glabrous or more loosely hairy; achenes not sunken in the receptacle .... 105. *Carduus*.

Group V

(Heads discoid or disciform; plants not thistlelike; pappus partly or wholly of numerous capillary bristles)

1 Receptacle chaffy or densely bristly, at least toward the margins.
2 Heads disciform, with several series of pistillate flowers and a few central disk flowers .... 82. *Filago*.
2 Heads discoid (the flowers all perfect and fertile) or falsely subradiate (the outer flowers neutral and with enlarged, irregular corolla).
3 Involucral bracts terminating in an inwardly hooked prickle; leaves large, mostly cordate or cordate-based, the larger ones generally well over 10 cm wide .... 104. *Arctium*.
3 Involucral bracts innocuous or sometimes spine-tipped, but not ending in an inwardly hooked prickle; leaves smaller and not at all cordate, mostly less than 10 cm wide.
4 Achenes basally attached to the receptacle, 10-ribbed; involucral bracts not notably modified toward the tip .... 99. *Carphephorus*.
4 Achenes obliquely or laterally attached to the receptacle, seldom evidently nerved; involucral bracts with modified, erose or lacerate to pectinate or spinose tip .... 110. *Centaurea*.
1 Receptacle naked.
5 Flowers all perfect and fertile, the heads strictly discoid.
6 Flowers bright yellow or orange or even orange-red.
7 Involucre of equal, essentially uniseriate bracts, often with a few much smaller bracteoles at the base.
8 Style branches truncate, penicillate .... 55. *Senecio*.
8 Style branches with a definite, slender, minutely hispidulous appendage.
9 Style appendages short, less than 1 mm long; plants not purple-hairy .... 56. *Emilia*.
9 Style appendages elongate, ca. 2 mm long; plants conspicuously purple-hairy .... 57. *Gynura*.
7 Involucre of unequal bracts more or less strongly imbricate in several series.
10 Heads in a flat-topped, corymbiform inflorescence .... 70. *Bigelowia*.
10 Heads in an open-paniculiform inflorescence, with several elongate, divaricate, recurved-secund branches (*S. brachyphylla*) .... 66. *Solidago*.
6 Flowers white or ochroleucous to anthocyanic.

11 Pappus distinctly double, the inner of long bristles, the outer of very short bristles .... 101. *Vernonia.*
11 Pappus simple, of similar bristles.
12 Style branches asteroid or senecionoid, distinctly flattened, with evident ventromarginal stigmatic lines and either truncate-penicillate at the tip or with a well-defined terminal appendage.
13 Involucre of equal, essentially uniseriate bracts, often with a few much smaller bracteoles at the base.
14 Flowers white to ochroleucous or chloroleucous; plants perennial .... 58. *Cacalia.*
14 Flowers pink or lavender to purple or crimson or orange-red; plants annual .... 56. *Emilia.*
13 Involucre of several series of imbricate bracts .... 67. *Brintonia.*
12 Style branches eupatorioid, elongate and linear or linear-clavate, blunt, papillate, with obscure stigmatic lines toward the base.
15 Achenes 3–5-angled and -ribbed.
16 Involucre of 4 bracts, containing 4 flowers; twining herbaceous vines .... 92. *Mikania.*
16 Involucre of more than 4 bracts, in most species containing more than 4 flowers; herbs or shrubs, sometimes scrambling, but not twining .... 91. *Eupatorium.*
15 Achenes ca. 10-ribbed.
17 Shrubs; involucral bracts arranged in 5 vertical ranks of 3 (4) each .... 96. *Garberia.*
17 Herbs; involucral bracts not arranged in vertical ranks.
18 Involucral bracts, or at least the inner ones, evidently few-striate longitudinally; leaves chiefly or wholly cauline, well distributed along the stem, not basally disposed.
19 Pappus bristles merely barbellate .... 97. *Brickellia.*
19 Pappus bristles evidently plumose .... 98. *Kuhnia.*
18 Involucral bracts not striate; leaves tending to be basally disposed, the basal or lower cauline ones evidently larger than the progressively reduced (though often numerous) middle and upper ones.
20 Inflorescence broadly corymbiform or sometimes thyrsoid-paniculate; involucre only 3.5–6 mm high .... 99. *Carphephorus.*
20 Inflorescence mostly spiciform or racemiform, or seldom with spiciform or racemiform branches, only rarely corymbiform, and then the involucres very large; involucre in most species more than 6 mm high .... 100. *Liatris.*
5 Outer flowers, or all the flowers of some heads, pistillate.
21 Pappus of long-plumose capillary bristles .... 87. *Facelis.*
21 Pappus of capillary bristles, these sometimes barbellate or distally expanded and serrate, but not at all plumose.
22 Shrubs.
23 Heads unisexual, the plants dioecious .... 75. *Baccharis.*
23 Heads all alike, with marginal pistillate and central perfect flowers (*P. odorata*) .... 84. *Pluchea.*
22 Herbs.
24 Pistillate flowers with a short but definite, often circinately outrolled ligule (*A. subulatus*) .... 72. *Aster.*
24 Pistillate flowers without a ligule, the corolla merely tubular.
25 Leaves chiefly or wholly basal or in mats at ground level, the erect flowering stems leafless or with smaller leaves.
26 Leaves reniform or cordate, becoming 10 cm wide or more at maturity; involucre of equal, essentially uniseriate bracts, often with a few much smaller basal bracteoles .... 61. *Petasites.*
26 Leaves oblanceolate to obovate or elliptic, up to ca. 3.5 cm wide; involucral bracts well imbricate in several series.
27 Heads all alike, with marginal pistillate and central functionally staminate flowers .... 85. *Sachsia.*
27 Heads unisexual, the plants dioecious .... 90. *Antennaria.*
25 Leaves chiefly or all cauline, not forming a basal rosette or mat.
28 Stem conspicuously winged by the decurrent leaf bases.

29 Heads in one or several very dense, spiciform, acropetally flowering inflorescences; plants conspicuously tomentose .... 86. *Pterocaulon*.
29 Heads borne in broad, more or less flat-topped or rounded, corymbiform terminal cymes; plants not conspicuously tomentose (*P. suaveolens*) .... 84. *Pluchea*.
28 Stem not winged.
30 Involucral bracts varying from more or less herbaceous to dry and partly scarious, but neither hyaline nor shining, the surface glabrous to often hairy or glandular.
31 Involucre of equal, essentially uniseriate bracts, often with a few much smaller bracteoles at the base .... 59. *Erechtites*.
31 Involucre of 2–several series of bracts, these often well imbricate.
32 Leaves ample, chiefly lanceolate to ovate, at least the larger ones generally 1 cm wide or more .... 84. *Pluchea*.
32 Leaves narrow, chiefly linear or linear-oblanceolate, seldom more than 1 cm wide (to 2 cm in *C. bonariensis*) .... 74. *Conyza*.
30 Involucral bracts largely or wholly scarious or hyaline and shining, usually glabrous above the often woolly base.
33 Heads all alike, the outer flowers pistillate, the inner ones perfect and fertile; annual or biennial .... 88. *Gnaphalium*.
33 Heads of two sorts, those on some plants wholly staminate in function, those on other plants largely pistillate; rhizomatous perennial .... 89. *Anaphalis*.

Group VI

(Heads discoid or disciform; plants not thistlelike; pappus of a few awns or scales, or wanting)

1 Heads unisexual; pistillate involucre nutlike or burlike.
2 Involucre of the pistillate heads with one or more series of tubercles or straight spines; involucral bracts of the staminate heads united .... 34. *Ambrosia*.
2 Involucre of the pistillate heads a bur with numerous hooked prickles; involucral bracts of the staminate heads separate .... 35. *Xanthium*.
1 Heads bisexual; involucre various.
3 Receptacle chaffy or bristly.
4 Involucral bracts evidently biseriate and dimorphic, the outer ones more or less herbaceous, the inner ones membranous or indurate.
5 Inner involucral bracts indurate, individually enclosing the achenes, each forming a prickly bur; pappus none .... 32. *Acanthospermum*.
5 Inner involucral bracts innocuous, merely membranous; pappus of 2–4 retrorsely (seldom antrorsely) barbed awns, or rarely obsolete .... 21. *Bidens*.
4 Involucral bracts in 1–several series, but not biseriate and dimorphic.
6 Involucral bracts, or some of them, with an evidently modified tip, ending in a hooked prickle, or in a straight spine, or in an enlarged, scarious or hyaline, erose to lacerate or pectinate appendage, or at least with a pectinate or lacerate fringe toward the tip.
7 Involucral bracts ending in an inwardly hooked prickle; leaves large, the larger ones with more or less cordate or deltoid blade generally more than 10 cm wide .... 104. *Arctium*.
7 Involucral bracts of diverse sorts, not ending in a hooked prickle; leaves smaller and not cordate or deltoid, generally less than 10 cm wide .... 110. *Centaurea*.
6 Involucral bracts all innocuous and without a specially modified tip.
8 Heads strictly discoid, the flowers all tubular and perfect.
9 Flowers in each head few, mostly 7–10 .... 95. *Hartwrightia*.
9 Flowers in each head numerous, more than 20.
10 Pappus mostly of 5 short scales; achenes 5-angled, 10-ribbed .... 26. *Marshallia*.
10 Pappus otherwise, mostly of 2–several awns or awn scales; achenes quadrangular or compressed.
11 Achenes very strongly flattened and generally wing-margined .... 14. *Verbesina*.
11 Achenes often somewhat compressed, but not strongly flattened and not wing-margined.
12 Heads white .... 6. *Melanthera*.

12 Heads deep purple (*H. radula*) .... 12. *Helianthus*.
8 Heads disciform, the central flowers perfect or functionally staminate, the marginal ones pistillate and fertile.
13 Small, white-woolly annuals, less than 2 dm tall, with the heads completely invested in wool, the involucre hidden .... 83. *Evax*.
13 Plants not white-woolly, often more than 2 dm tall.
14 Disk flowers perfect, with divided style; heads white or whitish .... 5. *Eclipta*.
14 Disk flowers sterile, with undivided style; heads yellow or dull greenish or nondescript in color.
15 Leaves large, the larger ones mostly 8–30 cm wide.
16 Fibrous-rooted perennial .... 31. *Polymnia*.
16 Taprooted annual (*I. xanthifolia*) .... 33. *Iva*.
15 Leaves smaller, up to ca. 7 cm wide .... 33. *Iva*.
3 Receptacle naked (or with some soft hairs in *Artemisia absinthium*).
17 Leaves, or most of them, opposite or whorled.
18 Leaves whorled .... 94. *Sclerolepis*.
18 Leaves opposite.
19 Corollas yellow, not more than about 10 per head .... 42. *Flaveria*.
19 Corollas blue-purple or seldom white, more than 20 per head .... 93. *Ageratum*.
17 Leaves, or most of them, alternate, or the leaves all basal.
20 Leaves, or many of them, pinnatifid or pinnately dissected.
21 Pappus of evident scales.
22 Involucral bracts erect-appressed, at least the inner with wide, obtuse, scarious, somewhat petaloid, whitish or yellowish tip .... 46. *Hymenopappus*.
22 Involucral bracts more or less spreading, becoming reflexed in fruit, not petaloid .... 39. *Gaillardia*.
21 Pappus virtually obsolete.
23 Heads disciform, the marginal flowers numerous, pistillate, with elongate, indurate-persistent style and without corolla, the disk flowers few, functionally staminate .... 53. *Soliva*.
23 Heads variously discoid or disciform, the flowers all with definite corolla, and the style not indurate-persistent; disk flowers fertile and with divided style except in one sp. of *Artemisia*.
24 Inflorescence corymbiform, or the heads solitary and terminating the branches.
25 Receptacle flat or somewhat convex; rhizomatous perennial, 4–15 dm tall .... 50. *Tanacetum*.
25 Receptacle elevated, conic; taprooted annual, 0.5–4 dm tall .... 51. *Matricaria*.
24 Inflorescence spiciform, racemiform, or paniculiform .... 52. *Artemisia*.
20 Leaves all entire or merely toothed.
26 Pappus wanting.
27 Inflorescence corymbiform; herbage glandular, not tomentose .... 95. *Hartwrightia*.
27 Inflorescence compactly elongate-paniculiform; herbage more or less tomentose, not glandular .... 52. *Artemisia*.
26 Pappus more or less evident, though sometimes small.
28 Heads 4-flowered, aggregated into involucrate secondary heads .... 102. *Elephantopus*.
28 Heads with more than 4 flowers, not aggregated into secondary heads.
29 Involucral bracts, or many of them, conspicuously spinulose-ciliate .... 103. *Stokesia*.
29 Involucral bracts unarmed.
30 Flowers relatively few, less than 50 per head.
31 Flowers mostly 7–12 per head.
32 Larger (lower) leaves mostly 1–8 cm wide .... 95. *Hartwrightia*.
32 Leaves all less than 0.5 cm wide .... 41. *Palafoxia*.
31 Flowers mostly (13) 15–30 per head .... 41. *Palafoxia*.
30 Flowers numerous, more than 100 per head.
33 Style branches with a more or less elongate appendage .... 39. *Gaillardia*.
33 Style branches exappendiculate, the tip dilated, subtruncate, penicillate .... 40. *Helenium*.

1. RUDBECKIA L. CONEFLOWER

Herbs. Leaves alternate, entire to pinnatifid or subbipinnatifid. Heads radiate, the rays neutral, yellow or orange to sometimes purple, red, or maroon, mostly 5–21; involucral bracts subequal or irregularly unequal, green and more or less herbaceous, mostly spreading or reflexed, in 2–3 series; receptacle enlarged, conic or columnar, chaffy, its bracts partly enfolding the achenes, equaling or shorter than the disk corollas (a little longer in *R. triloba*), sometimes shortly awn-pointed (*R. triloba*) but not spinescent; disk corollas numerous, narrowed to a more or less distinct tube at the base; style branches flattened with short or (*R. hirta*) elongate, externally hairy appendage, without well-marked stigmatic lines. Achenes quadrangular, glabrous; pappus a short, often toothed or irregular crown, or (*R. hirta, R. mollis*) none. (x = 18, 19)

Spp. 1–10 belong to section *Rudbeckia*, with mostly sunflower-yellow to orange or red rays, mostly equably quadrangular, basally attached achenes completely filled by the seed, compact, not much elongating disk with receptacular bracts 1.5–2 times as long as the mature achene, and x = 19. Spp. 11–14 belong to the section *Macrocline*, with lemon-yellow or pale yellow rays, somewhat compressed-quadrangular, basilaterally attached achenes not filled by the seed, loose, often conspicuously elongating disk with receptacular bracts only slightly if at all surpassing the mature achenes, and x = 18. All spp. are sharply delimited.

1 Leaves narrow, grasslike, the blade mostly at least 10 times as long as wide and not more than 1 cm wide; herbage glabrous or somewhat strigose.
- 2 Head solitary; rays dark red to orange-red, 1–1.5 cm long; herbage slightly hairy .... 10. *R. graminifolia*.
- 2 Heads several; rays yellow, 1.5–3.5 cm long; herbage glabrous .... 11. *R. mohrii*.

1 Leaves wider, the larger ones with the blade mostly less than 10 times as long as wide and more than 1 cm wide (or a bit narrower in no. 2, with conspicuously spreading-hairy herbage).
- 3 Style appendages elongate, subulate; pappus none; herbage coarsely hirsute .... 1. *R. hirta*.
- 3 Style appendages short and blunt; pappus present (except in no. 6) but often minute; pubescence various.
  - 4 Receptacular bracts glabrous, or ciliolate on the margins, rarely with a few appressed hairs on the back, not at all canescent.
    - 5 Receptacular bracts obtuse or acute, not awn-pointed; leaves merely toothed or subentire.
      - 6 Leaves narrow, the basal ones up to 2 cm wide; herbage strongly hirsute .... 2. *R. missouriensis*.
      - 6 Leaves wider, some of them more than 2 cm wide; herbage seldom strongly hirsute.
        - 7 Plants very robust, commonly 2–3 m tall; middle and lower cauline leaves evidently auriculate-clasping .... 4. *R. auriculata*.
        - 7 Plants smaller, up to 1 (1.3) m tall; cauline leaves sessile or petiolate, but not auriculate-clasping .... 3. *R. fulgida*.
    - 5 Receptacular bracts conspicuously awn-pointed; some of the leaves generally trilobed or even pinnatifid .... 5. *R. triloba*.
  - 4 Receptacular bracts canescent near the tip with short, viscidulous hairs.
    - 8 Leaves entire or merely toothed (rare forms of *R. subtomentosa* with undivided leaves would be sought here).
      - 9 Herbage more or less pubescent (varying to subglabrous in no. 8); disk not much elongating in fruit.
        - 10 Principal cauline leaves sessile or broadly short-petiolate, often clasping, relatively small, not more than ca. 10 cm long overall and 2 (2.5) cm wide; stem densely and conspicuously woolly-pilose below .... 6. *R. mollis*.
        - 10 Principal cauline leaves slender-petiolate, larger, generally at least (10) 15 cm long (petiole included) and (2.5) 3 cm wide; stem glabrous

to rather coarsely spreading-hairy below, but not densely pilose-woolly.
11 Heads relatively small, the disk 1–1.5 cm wide, the rays mostly 1.5–2.5 cm long; leaves sparsely strigose or nearly glabrous .... 8. *R. heliopsidis*.
11 Heads larger, the disk mostly 1.5–2.5 cm wide; the rays mostly 3–5 cm long; leaves hirsute or scabrous-puberulent on both sides .... 9. *R. grandiflora*.
9 Herbage essentially glabrous; disk often conspicuously elongating in fruit.
12 Plants glaucous, robust, mostly 1–2.5 m tall, with large leaves (the larger ones with the blade 7–16 cm wide and 2/5–3/5 as wide as long) and large heads, the disk usually elongating to 4–8 cm at maturity .... 13. *R. maxima*.
12 Plants not glaucous, smaller, mostly 0.5–1.3 m tall, with smaller leaves (the larger ones with the blade up to 6.5 cm wide and up to half as wide as long) and smaller heads, the disk up to 4.5 cm long at maturity .... 12. *R. nitida*.
8 Leaves, or some of them, generally pinnatifid or trilobed.
13 Stem essentially glabrous; leaves usually subglabrous at least on the upper side, seldom more evidently hairy; disk pale .... 14. *R. laciniata*.
13 Stem densely short-hairy at least above the middle; leaves pubescent on both sides; disk dark .... 7. *R. subtomentosa*.

1. R. hirta L. BLACK-EYED SUSAN. Taprooted annual to more often biennial or fibrous-rooted perennial, 3–10 dm tall, more or less hispid or hirsute throughout. Heads more or less long-pedunculate, the hemispheric or ovoid disk dark purple or brown, 1.2–2 cm wide; involucral bracts copiously hirsute or hispid, sometimes much elongate; rays orange or orange-yellow, sometimes with darker base, 2–4 cm long; receptacular bracts acute, more or less hispid or hispidulous distally; style appendages elongate, subulate. Pappus none. (n = 19) Spring–fall. Widespread in e US. Three regional vars.:

Stems obviously leafy and generally branched at or above the middle, or in small plants unbranched and the peduncle then not more than 1/3 the height of the plant; never annual.
Leaves coarsely toothed to occasionally entire, relatively large and broad, the basal ones with the blade 2.5–7 cm wide and about twice as long, the cauline ones lance-ovate to pandurate; relatively undisturbed, open or forested habitats in mt. provinces from Pa to w SC and n Ga and Ala; *R. amplectens* T. V. Moore—S; *R. brittonii* Small—S; *R. hirta* var. *brittonii* (Small) Fern.—F; *R. monticola* Small—S. .... var. *hirta*
Leaves entire or sometimes slightly toothed, narrower, the basal ones with lanceolate or oblanceolate blade 1–2.5 (3) cm wide and (3) 4–5 times as long as wide, the cauline ones spatulate or oblanceolate to broadly linear; widespread and weedy; *R. longipes* T. V. Moore—S; *R. serotina* Nutt.—F; *R. serotina* var. *corymbifera* (Fern.) Fern. & Schub.—F; *R. serotina* var. *sericea* (T. V. Moore) Fern. & Schub.—F. .... var. *pulcherrima* Farw.
Stems tending to be leafy chiefly toward the base and more or less naked above, branched at or below the middle, or unbranched and the peduncle then at least half the height of the plant; plants sometimes annual; CP from s Ga to Fla and Tex; *R. bicolor* Nutt. (the annual phase)—F, S; *R. divergens* T. V. Moore—S; *R. floridana* T. V. Moore—S; *R. longipes* T. V. Moore—S. .... var. *angustifolia* (T. V. Moore) Perdue.

2. R. missouriensis Engelm. Densely spreading-hirsute throughout, including the involucral bracts; plants not stoloniferous; branches closely ascending. Leaves narrow, the basal ones broadly linear to lance-spatulate, up to 2 cm wide, the cauline ones linear-spatulate, entire. Otherwise much like *R. fulgida*. (n = 19) Late spring–early fall. Mostly in dry, open places; sw Ark (OU) to Mo, Ill, and Tex.

3. R. fulgida Aiton. Perennial, 3–10 (–13) dm tall, commonly stoloniferous, sparsely to moderately hairy, seldom densely hirsute. Lower leaves lanceolate to cordate, long-petiolate, the others similar or gradually reduced, short-petiolate or sessile. Heads commonly long-pedunculate, the hemispheric or ovoid disk dark purple or brown, 1–1.8 cm wide; rays yellow to partly or wholly orange; receptacular bracts obtuse or acute, smooth or more or less ciliolate-margined, rarely with a few appressed hairs on the back. Achenes 2–3 mm long, shorter than the disk corollas. Pappus an inconspicuous low crown. (n = 19, 38) Summer–fall. Widespread in e US. Three morphologically and geographically overlapping regional vars. that seem sharply distinct at some points of contact:

Rays mostly 2.5–4 cm long; leaves usually sharply toothed, varying in shape from nearly as in var. *umbrosa* to as in the wider-leaved forms of var. *fulgida*; streambanks, swamps, and other wet, open or shady places; midwestern phase, reaching our range in WVa; *R. sullivantii* Boynton & Beadle—S; *R. speciosa* var. *sullivantii* (Boynton & Beadle) Robins.—F. .... var. *sullivantii* (Boynton & Beadle) Cronq.

Rays mostly 1–2.5 (3) cm long.

Cauline leaves mostly ovate or subcordate, sharply toothed to sometimes entire, abruptly contracted to a narrowly or scarcely winged petiole; woodlands; Appalachian phase; *R. chapmanii* Boynton & Beadle—S; *R. umbrosa* Boynton & Beadle—F, S. .... var. *umbrosa* (Boynton & Beadle) Cronq.

Cauline leaves mostly narrower than ovate and sessile or merely narrowed to a winged petiole or petioliform base, usually entire or merely denticulate; moist, low ground in shaded or open places; southeastern phase, widespread at lower elevations in our range; *R. acuminata* Boynton & Beadle—F, S; *R. foliosa* Boynton & Beadle—S; *R. palustris* Eggert—F, S; *R. spathulata* Michx.—F, S; *R. tenax* Boynton & Beadle—F, S; *R. truncata* Small—S. ..... var. *fulgida*.

4. R. auriculata (Perdue) Kral. Robust, rhizomatous and stoloniferous perennial, commonly 2–3 m tall, inconspicuously hairy or subglabrous. Leaves large and numerous, the lowest ones up to 6.5 dm long overall, with oblong to oblanceolate or lance-ovate blade half to fully as long as the petiole, the others sessile or nearly so and gradually reduced upward, elliptic to ovate or (especially the lower) pandurate, auriculate-clasping. Heads commonly long-pedunculate, the hemispheric or ovoid disk dark purple or brown, 1–1.8 cm wide; rays bright yellow, ca. 2 cm long; receptacular bracts ca. 6 mm long, oblong-cuneate, ciliolate on the margins and sparsely short-hairy on the back. Achenes 4–4.5 mm long, longer than the disk corollas. Pappus an evident, unevenly 4–6-toothed crown, ca. 2 mm long. Summer. Moist low ground; se Ala (GC), n occasionally to St. Clair Co., Ala (RV).

5. R. triloba L. Short-lived perennial, 5–15 dm tall, moderately hirsute or strigose to subglabrous. Leaves thin, sharply toothed to subentire, the basal ones broadly ovate or subcordate and long-petiolate, the cauline mostly narrower and short-petiolate or sessile, usually some of the larger ones deeply trilobed or pinnatifid. Heads terminating the branches, the disk hemispheric or ovoid, dark purple; rays yellow or (especially toward the base) orange; receptacular bracts glabrous, abruptly narrowed to a short but distinct awn tip often shortly surpassing the disk corollas. (n = 19) Summer. Woods and moist soil; widespread in e US. Three vars.:

Heads relatively large, the rays mostly 2–3 cm long and the disk 1.5–2 cm wide; leaves of var. *triloba*; high altitudes in BR of w NC and e Tenn; *R. rupestris*

Chickering—S. .... var. *rupestris* (Chickering) A. Gray.
Heads smaller, the rays mostly 1–2 (2.5) cm long and the disk 1–1.5 cm wide.
Some of the leaves generally pinnately few-lobed, the lobes often obtuse; leaves consistently small, the blade up to 5 cm long (to 8 cm in the mt. plants); plants relatively small and delicate; GC of w Fla, and apparently also in BR of NC; *R. beadlei* Small, the NC form—S; *R. triloba* var. *beadlei* (Small) Fern.—G, but defined to include pinnate-leaved individuals of var. *triloba*; *R. pinnatiloba* (T. & G.) Beadle—S. .... var. *pinnatiloba* T. & G.
Some of the leaves generally trilobed, rarely any of them pinnately lobed, the lobes generally acute; leaves often (not always) well over 5 cm long; widespread in e US, s to Ga, Miss, and La, but not at the altitudes of var. *rupestris* .... var. *triloba*.

6. R. mollis Elliott. Annual (reputedly), biennial, or occasionally short-lived perennial, 4–10 dm tall, soft-hairy throughout, the stem densely and conspicuously pilose-woolly toward the base. Leaves principally cauline, rather numerous, toothed or subentire, sessile or broadly short-petiolate (basal rosettes, when present, more petiolate), often clasping, relatively small, seldom more than 10 cm long overall and 2 (2.5) cm wide. Heads 1–several, terminating the branches, the disk dark, hemispheric to subconic, 1.2–1.8 cm wide; rays 2–3.5 cm long, yellow, often with orange base; receptacular bracts shortly viscidulous-canescent distally. Pappus none. (n = 19) Summer. Dry, sandy soil; CP of Ala, s Ga, and n Fla.

7. R. subtomentosa Pursh. Perennial from a stout rhizome. Stem mostly 6–20 dm tall, glabrous below, more or less densely short-hairy above. Leaves firm, short-hairy on both sides, especially beneath, ovate to sometimes lance-elliptic, petiolate, serrate, generally some of the larger ones deeply trilobed. Heads several, the disk dark purple or brown, mostly 0.8–1.6 cm wide, hemispheric or short-ovoid, not elongating; rays yellow, 2–4 cm long; receptacular bracts distally viscidulous-canescent. (n = 19) Summer. Prairies and low ground; prairie species, s and e occasionally to IP of w Tenn and GC of La.

8. R. heliopsidis T. & G. Perennial from a woody rhizome, 6–10 dm tall. Stem spreading-villous to subglabrous. Leaves relatively thin, subglabrous or sparsely hairy, serrate or subentire, the basal very long-petiolate, the cauline progressively less so, the principal ones with ovate to elliptic-ovate or lance-ovate blade mostly 6–10 cm long and 3–5 cm wide. Heads several, the disk dark, hemispheric or ovoid, 1–1.5 cm wide, rays yellow, 1.5–2.5 cm long; receptacular bracts distally viscidulous-canescent. Summer. Rare and local from se Va to SC, Ga, and ne Ala, variously on AC, PP, BR, and RV.

9. R. grandiflora (Sweet) DC. Fibrous-rooted perennial from a more or less woody caudex or stout rhizome, 5–12 dm tall. Leaves thick and firm, pubescent on both sides, few, toothed or subentire, large, the principal ones with ovate to lance-ovate or elliptic blade commonly 7–25 cm long and 3–11 cm wide, the basal long-petiolate, the others progressively less so, the upper reduced. Heads solitary or few, long-pedunculate, the disk dark purple or brown, hemispheric to ovoid-subconic, 1.5–2.5 cm wide and up to 2.5 cm high; rays orange-yellow, drooping, mostly 3–5 cm long; receptacular bracts distally viscidulous-canescent. (n = 19) Summer. Prairies, open places, and dry woods. Two well-marked vars.:

Leaves hirsute, the hairs mostly well over 0.5 mm long; stem evidently spreading-hairy toward the base; chiefly OZ and OU, Ark and nw La to Mo, Kans, and Tex, and isolated in RV of nw Ga .... var. *grandiflora*.
Leaves scabrous-puberulent, the hairs mostly about 0.5 mm long or less; lower part

of stem glabrous or nearly so. ME and GC, Ark to La and Tex; R. *alismaefolia* T. & G.—S. .... var. *alismaefolia* (T. & G.) Cronq.

10. R. graminifolia (T. & G.) Boynton & Beadle. Fibrous-rooted perennial, 5–8 dm tall, thinly to moderately strigose or hirsute-strigose, or the leaves often essentially glabrous. Leaves grasslike, elongate and narrow, entire, obscurely veined except for the strong midrib, basally disposed, the larger ones with the blade 10–20 cm long and less than 1 cm wide. Heads solitary, the disk 1–1.5 cm wide and high, deep maroon; rays 1–1.5 cm long, orange-red to more often dark red; receptacular bracts glabrous or sparsely strigose. (n = 19) Spring–summer. Wet low ground; Gulf, Bay, Calhoun, Liberty and Franklin Cos. in w Fla.

11. R. mohrii A. Gray. Fibrous-rooted perennial, 5–11 dm tall, essentially glabrous throughout. Leaves firm, grasslike, elongate and narrow, strongly 3 (–7)-nerved, basally disposed, the larger ones with the blade 10–20 cm long and 1 cm wide or less. Heads 3–10 on elongate, slender, arcuate peduncles, the disk dark purple or reddish-brown, 1–1.5 cm wide and high; rays yellow, 1.5–3.5 cm long; receptacular bracts glabrous or minutely glandular-pruinose distally. (n = 18) Late spring–summer. Wet low ground; CP of s Ga and n Fla.

12. R. nitida Nutt. Fibrous-rooted perennial, mostly 5–13 dm tall, glabrous but not glaucous. Basal leaves long-petiolate, with elliptic or narrower blade 2–6.5 cm wide and up to 25 cm long, up to half as wide as long, tending to have one or two pairs of lateral veins enlarged and elongate, the cauline leaves progressively reduced. Heads solitary or few, long-pedunculate, the disk light reddish-brown, hemispheric to ovoid-conic at first, often more elongate later; rays yellow, drooping, 2–5 cm long; receptacular bracts distally viscidulous-canescent. (n = 18) Late spring–summer. Moist, low ground, interruptedly from Fla to Tex on CP. Two morphologically overlapping but geographically disjunct vars., the more western one approaching (but distinct from) *R. maxima*:

Leaves relatively narrow, the basal ones with the blade up to about 1/3 as wide as long; disk up to 3 cm long at maturity; achenes mostly 3–5.5 mm long; ne Fla and e Ga, with outliers in Manatee Co., Fla, and Macon Co., Ala. *R. glabra* DC.—S. .... var. *nitida*.

Leaves broader, the basal ones with the blade up to about 1/2 as wide as long; disk up to 4.5 cm long at maturity; achenes mostly 5–7.5 mm long; c La to se Tex .... var. *texana* Perdue.

13. R. maxima Nutt. Fibrous-rooted perennial, mostly 10–25 dm tall, glabrous and glaucous. Leaves firm, large, the larger ones with the blade 7–16 cm wide and up to 25 cm long, 2/5–3/5 as wide as long, the basal ones long-petiolate, elliptic or elliptic-ovate, the cauline few, elliptic to obovate, narrowed to a clasping, broadly petioliform base, or the upper sessile. Heads solitary or few, long-pedunculate; disk light reddish-brown, 1.5–3 cm wide, ovoid-cylindric at first, in fruit usually elongating to 4–8 cm; rays yellow, drooping, 3–6 cm long; receptacular bracts distally viscidulous-canescent. (n = 18) Summer. Moist, low ground; OZ, OU, ME, and GC, from Ark and La to Mo, Okla, and Tex.

14. R. laciniata L. Perennial from a woody base, 5–30 dm tall. Stem glabrous, often glaucous. Leaves large, petiolate, coarsely toothed or laciniate, some or most of them pinnatifid or sometimes merely trilobed, subglabrous, or sometimes hairy especially beneath. Heads 1–many; rays yellow, drooping, 3–6 cm long; disk yellow or grayish, 1–2 cm wide, hemispheric at first; receptacular bracts distally viscidulous-canescent. (2n = 36, 54, 72, 102+; plants often apomictic) Summer. Moist places; widespread in e US, w to the Rocky Mts.; three confluent varieties in our range:

Heads relatively small and several or numerous on slender, flexuous peduncles, the disk mostly 1–1.5 cm wide, not elongating, the rays mostly ca. 5 or 8; leaves usually pinnatifid as in var. *laciniata*, varying to sometimes merely trifid as in var. *humilis*; plants mostly 1–2 m tall; more often sexual than apomictic; CP and PP, Va to Fla and La; *R. heterophylla* T. & G.—S, a Fla form with the leaves densely soft-hairy beneath, sparsely hairy or scabrous above .... var. *digitata* (Miller) Fiori.

Heads larger, the disk mostly 1.5–2 cm wide, the rays mostly ca. 8 or ca. 13.

Plants tall, 1.5–3 m when well developed, with several or many heads, the disk generally elongating in age (to 2–3 cm); leaves evidently pinnatifid or even subbipinnatifid; more often apomictic than sexual; widespread, but not at highest elevations, and not on CP .... var. *laciniata*.

Plants relatively small, 0.5–1.5 m tall, with only 1–6 heads, the disk not elongating; leaves mostly merely 3-cleft or undivided, occasionally some of them pinnatifid, more often sexual than apomictic; high elev. in s Appalachians, BR to CU, Va and NC to Ky .... var. *humilis* A. Gray.

2. ECHINACEA Moench CONEFLOWER

Perennial herbs, more or less leafy-stemmed (in numbers 3 and 4) or with more basally disposed leaves (in numbers 1 and 2). Leaves alternate, more or less evidently 3 (5)-nerved, at least the lower ones conspicuously petiolate. Heads naked-pedunculate, radiate, the rays mostly about 13 (21) relatively large, anthocyanic to white or yellow, neutral or with a vestigial style, often drooping; involucral bracts in 2–4 subequal or slightly imbricate series, with firm base and spreading or reflexed tip, the inner sometimes passing into those of the receptacle, but not sharply divisible into two types; receptacle high-conic, its bracts partly enfolding the achenes, firm, with stout, spinescent tip conspicuously surpassing the disk corollas; disk corollas slightly thickened at base, not narrowed into a slender tube; style branches flattened, without well-marked stigmatic lines, with slender, acuminate, hairy appendage. Achenes quadrangular, glabrous or sparsely hairy on the angles; pappus a short, toothed crown. (x = 11) (*Brauneria* Neck.; *Rudbeckia*, in part)

1 Leaf blades narrow, mostly 5–20 times as long as wide, entire, gradually tapering to the petiole; plants strongly taprooted.
  2 Rays yellow (unique in the genus); herbage strigose to subglabrous .... 1. *E. paradoxa*.
  2 Rays anthocyanic to white; herbage spreading-hirsute .... 2. *E. pallida*.
1 Leaf blades wider, mostly 1.5–5 times as long as wide, more abruptly contracted to the petiole, very often toothed; plants tending to be fibrous-rooted.
  3 Leaves more or less pubescent on both sides .... 3. *E. purpurea*.
  3 Leaves glabrous on both sides, or somewhat scabrous or short-hairy above ... 4. *E. laevigata*.

1. E. paradoxa (Norton) Britton. Stems clustered on a stout taproot, 3–8 dm tall, unbranched; herbage strigose to subglabrous. Leaves entire, basally disposed, elongate and narrow, the blade up to 20 cm long and 2.5 (4) cm wide, tapering to the petiole. Disk 2–3 cm wide; rays yellow, 3–7 cm long, mostly drooping. (n = 11) Summer. Prairies and open, rocky places; OZ of s Mo and adj. n Ark. *E. atrorubens* var. *paradoxa* (Norton) Cronq.—G.

2. E. pallida (Nutt.) Nutt. Stems clustered on a strong taproot, usually simple;

herbage coarsely spreading-hirsute. Leaves entire or nearly so, basally disposed, elongate and narrow, the blade up to 20 × 4 cm, mostly 5–20 times as long as wide (or the basal a little wider), tapering to the petiole. Disk 1.5–3 cm wide; rays pink, varying to purple or white. (n = 11, 22) Spring–midsummer. Dry, open places; species mainly of the prairies and plains, in our range common in Ark and w La, and extending e irregularly to Va, NC, and Ga. Var. *pallida*, the more eastern segment of the species, the principal phase in our range, is mostly tetraploid and robust, 4–10 dm tall, with drooping rays mostly 4–8 cm long, and very often with white pollen. (*E. sanguinea* Nutt., a diploid, southwestern form with yellow pollen, but with the habit and rays of var. *pallida*; perhaps better treated as a distinct var.) Var. *angustifolia* (DC.) Cronq., the more western segment of the species, is diploid and smaller, mostly 1–5 dm tall, with spreading to drooping rays mostly 2–4 cm long, and yellow pollen; an eastern outlier of var. *angustifolia*, in the cedar barrens of c Tenn, has been called *E. tennesseensis* (Beadle) Small—S. *E. angustifolia* DC.—F.

3. E. purpurea (L.) Moench. Stems solitary or few from a coarsely fibrous-rooted crown, caudex, or short, stout rhizome, hirsute to glabrous, mostly 6–18 dm tall, simple or often few-branched above, more leafy than in numbers 1 and 2. Leaves more or less hairy on both sides, toothed or less commonly entire, the principal ones with broadly lanceolate to elliptic or broadly ovate blade up to about 15 × 10 cm, mostly 1.5–5 times as long as wide, rather abruptly contracted (or even rounded or subcordate) to the petiole. Disk 1.5–3.5 cm wide; rays reddish-purple to occasionally pale pink, 3–8 cm long, drooping. (n = 11) Summer. Woods and prairies, generally in moister sites than numbers 1 and 2: chiefly Ozarkian and midwestern species, s to c La, and e irregularly to Ky, Tenn, Ga, and less commonly to Va and NC.

4. E. laevigata (Boynton & Beadle) Blake. Much like no. 3, and perhaps only a geographical var. of it, but glabrous and more or less glaucous, or the leaves sometimes scabrous or short-hairy above; stems consistently simple, 6–12 dm tall; rays up to 8 cm long; plants tending to have a short, quickly deliquescent taproot or vertical caudex. (n = 11) Summer. Woods and fields; PP to RV, locally from se Pa to n Ga. *E. purpurea* var. *laevigata* (Boynton & Beadle) Cronq.—G.

3. RATIBIDA Raf. CONEFLOWER

Perennial herbs. Leaves alternate, in most spp. (including ours) pinnatifid or bipinnatifid. Heads radiate, the rays mostly 3–13, neutral, relatively broad (often more than 1 cm), yellow or sometimes partly or wholly brown-purple; involucre a single series of green, subherbaceous, linear or lance-linear bracts; receptacle columnar, chaffy, its bracts subtending rays as well as disk flowers, more or less clasping the achenes, the tip densely velutinous and incurved; disk corollas numerous, short, cylindric, scarcely narrowed at the base; style branches flattened, with ovate to subulate, hairy appendage. Achenes compressed at right angles to the involucral bracts, often also evidently quadrangular, glabrous except for the sometimes ciliate margins; pappus coroniform, with 1 or 2 prolonged, awnlike teeth, or of teeth only, or absent. (x = 14, 16) (Genus formerly often submerged in *Rudbeckia*.) (*Lepachys*). Richards, E. L. 1968. A monograph of the genus *Ratibida*. Rhodora 70: 348–393.

1 Disk ellipsoid-globular, 1–1.6 times as long as thick; plants fibrous-rooted from a woody rhizome or caudex .... 1. *R. pinnata*.

1 Disk columnar, 2–4.5 times as long as thick; plants taprooted.
  2 Stem generally leafy to beyond the middle, the leaves not very finely dissected, generally many or most of the ultimate segments 1.5 cm long or more; style appendages very short and blunt, about 0.25 mm long or less .... 2. *R. columnifera*.
  2 Stem leafy only below or up to about the middle, the leaves more finely dissected, seldom more than a few if any of the ultimate segments as much as 1.5 cm long; style appendages more elongate and narrow, 0.4 mm long or more .... 3. *R. peduncularis*.

1. R. pinnata (Vent.) Barnhart. Fibrous-rooted from a stout, woody rhizome or sometimes a short caudex, 4–12 dm tall, more or less hirsute, or the stem strigose above. Lower leaves long-petioled, the upper short-petioled or sessile; segments lanceolate, acute, coarsely toothed or entire. Heads usually several, naked-pedunculate; disk ellipsoid-globular, 1–2 cm long, 1–1.6 times as long as thick, much shorter than the rays, these pale yellow, (2.5) 3–6 cm long, spreading or often reflexed; style appendages elongate, acuminate. Achenes smooth; pappus none. (n = 14) Summer. Prairies, old fields, and dry woods, often on limestone; sp. characteristically of c and sc US, extending e occasionally to Ky, Tenn, Ala, w Fla, Ga, and (intro.) SC.

2. R. columnifera (Nutt.) Woot. & Standl. Taprooted, with clustered stems 3–12 dm tall, strigose or partly hirsute, generally leafy to above the middle. Leaves pinnatifid or partly bipinnatifid, the ultimate segments linear or lanceolate, entire or nearly so, relatively few and often very unequal, generally many or all of them 1.5 cm long or more. Heads (1) several or many, naked-pedunculate, the disk columnar, 1.5–4.5 cm long, 2–4.5 times as long as thick; rays yellow or [f. *pulcherrima* (DC.) Fern.] partly or wholly brown-purple, 1–3.5 cm long, spreading or reflexed; style appendages very short and blunt, about 0.25 mm long or less. Achenes with the inner margin fringed-ciliate to nearly smooth, usually slightly winged; pappus an evident awn tooth on the inner angle of the achene, and often also a shorter one on the outer angle. (n = 13, 14, 17–19) Summer. Prairies and other dry, open places, as along roadsides and railroads; sp. chiefly of the prairies and plains of c US, probably native in Ark and n and w La, and occasionally intro. eastward, as in Ala and NC. *R. columnaris*—S.

3. R. peduncularis (T. & G.) Barnhart. Taprooted, strongly hirsute-strigose, with clustered or solitary, often basally curved stems 2–11 dm tall, leafy only toward the base or up to the middle. Leaves pinnatifid to more often bipinnatifid, the ultimate segments more crowded and often more numerous than in *R. columnifera*, linear to broadly oblong, seldom as much as 1.5 cm long. Heads 1–several, naked pedunculate, the disk columnar, 1.5–4.5 cm long, 2–4.5 times as long as thick; rays yellow or partly or wholly brown-purple, less than 1.5 cm long; style appendages relatively elongate and slender, commonly 0.4 mm long or more. Achenes with the inner margin conspicuously pectinate-fimbriate; pappus of 1 or 2 teeth on the angles of the achene. (n = 14) Spring–fall. Sandy soil in dry woods and disturbed sites; GC of La, Tex, and adj Mex. Ours is var. *peduncularis*, as described here.

4. DRACOPIS Cass. CONEFLOWER

1. D. amplexicaulis (Vahl) Cass. Glabrous and glaucous annual, 3–7 dm tall. Leaves alternate, entire or occasionally toothed, acute or acuminate, cordate-clasping, up to about 11 cm long and 5 cm wide, the lower mostly elliptic or oblong, the upper ovate. Heads few or solitary, radiate, the rays about 6–10, neutral, yellow, or partly orange or purple, 1–3 cm long; spreading or eventually reflexed; disk elongating to 1.5–3 cm in fruit, 10–14 mm wide; involucre a single series of lance-linear, green, somewhat herbaceous bracts; receptacle columnar, its bracts subtending rays as well as disk flowers, ciliate-margined near the tip and somewhat clasping the achenes; disk flowers perfect and fertile, the corolla narrowed to a distinct tube at the base; anthers subtruncate at the base; style branches slightly flattened, with elongate, acuminate, hairy appendage. Achenes of the disk glabrous, several-nerved, minutely cross-rugose, obscurely quadrangular or subterete, the pappus obsolete; ray achenes abortive, compressed-quadrangular, villous, their pappus a short crown. (n = 16) Late spring–midsummer. Open places, esp. in clay soils; GC and ME of Ala, Miss, La, and Ark, w and n to Tex, Mo, and Kans. *Rudbeckia amplexicaulis* Vahl—F.

5. **ECLIPTA** L., nom. conserv.

1. E. alba (L.) Hassk. Weak or spreading, branching strigose annual, often rooting at the nodes; leaves opposite, mostly lanceolate or lance-elliptic to lance-linear, narrowed to a sessile or shortly petiolar base, serrulate, commonly 2–10 cm × 4–25 mm. Heads on axillary and terminal peduncles (these 1–5 cm long), white or whitish, radiate, the rays pistillate and fertile, rather numerous, slender, only about 1 mm long; involucral bracts 1–2 seriate, herbaceous or herbaceous-tipped, subequal, or the inner narrower and shorter; receptacle flat or slightly convex, its bracts slender and fragile, somewhat bristlelike, the central sometimes wanting; disk flowers numerous, perfect, and fertile, the corolla mostly 4-toothed; style branches flattened, with short, obtuse, hairy appendage. Achenes thick, roughened, 2–2.5 mm long, the marginal often triquetrous and with 1 surface apposed to an involucral bract, the others generally quadrangular and somewhat compressed along radii of the head. Pappus a nearly obsolete crown well removed from the margins of the truncate-topped achene. (n = 11) Spring–fall, or all year southward. A weed in moist bottomlands and muddy places, native to the New World, now widespread in tropical and warm-temperate regions, and found throughout our range, more commonly on CP and PP than in the mts. *Verbesina alba* L.—S.

6. **MELANTHERA** Rohr.

Perennial herbs, sometimes basally shrubby. Leaves opposite, toothed, often also 3-lobed. Heads terminating the stem (and branches) or loosely cymose, discoid, the flowers all tubular and perfect; involucre of 2–3 subequal or imbricate series, the bracts firm but green-tipped or green nearly throughout; receptacle convex, chaffy, its bracts concave and clasping the achenes, tapering or abruptly contracted to a firm point or mucro; corolla white; anthers black with white appendage; style branches flattened, with introrsely marginal stigmatic lines and a short, commonly lance-triangular, papillate or hairy appendage. Achenes truncate, with a raised, ciliolate margin, the marginal ones often triquetrous, the others quadrangular and radially somewhat compressed. Pappus of 2–several fragile, caducous awns set well in from the margin of the achene. Typical forms of our 3 spp. are very different in appearance, but seeming intermediates are not rare.

1 Leaves narrow, mostly 4–12 times as long as wide and not more than 1.5 (2) cm wide .... 3. *M. angustifolia.*
1 Leaves wider, either less than 4 times as long as wide, or more than 2 cm wide, or very often both.
  2 Leaves small, up to ca. 4 × 2.5 cm; plants relatively low and slender-stemmed, mostly 2–6 dm tall, often decumbent .... 2. *M. parvifolia.*
  2 Leaves larger, the better developed ones mostly 5–15 × (2) 2.5–10 cm; plants taller and coarser, erect, mostly 5–20 dm tall .... 1. *M. nivea.*

1. M. nivea (L.) Small. Coarse, erect, scabrous or shortly rough-hairy plants, commonly 5–20 dm tall. Leaves evidently petiolate, with narrowly to broadly ovate or triangular, obviously triplinerved, often trilobed or hastate blade (sometimes with a long, narrow terminal lobe and flaring lateral lobes), the better developed ones mostly 5–15 × (2) 2.5–10 cm. Heads 1–2 cm wide. Achenes 2.5–3 mm long, warty. Pappus awns 2–3 mm long. (n = 15) Summer–fall, or all year southward. In a wide range of habitats, from moist woods to roadsides to sea beaches; CP from SC to Fla and La; widespread in trop Am. Plants with the involucral bracts relatively broad and blunt, and with the receptacular bracts only shortly mucronate or simply acute, have been distinguished as *M. aspera* (Jacq.) Small from typical *M. nivea*, with narrower, more pointed, often looser involucral bracts and gradually acuminate or shortly awn-tipped receptacular bracts, but the populational significance of the distinction is doubtful. *M. deltoidea* Michx.—S; *M. hastata* Michx.—R, S.

2. M. parvifolia Small. Differing from no. 1 in its low habit and small leaves. Plants mostly 2–6 dm tall, often decumbent, frequently much branched, slender-stemmed, the stems commonly only 1–2 mm thick. Leaves short-petiolate or subsessile, the blade up to ca. 4 × 2.5 cm. All year. Pinelands, hammocks, and disturbed sites; trop Fla. *M. radiata* Small—S.

3. M. angustifolia A. Rich. Habitally like no. 2, or coarser and up to ca. 1 m tall. Leaves mostly merely penniveined, narrow, commonly narrowly oblong to oblanceolate, lanceolate, or nearly linear, shortly or scarcely petiolate, 4–12 times as long as wide, up to 12 × 1.5 (2) cm. All year. Pinelands, hammocks, and disturbed sites; trop Fla; WI, Mex, and Guatemala. *M. ligulata* Small—S.

7. TETRAGONOTHECA L. SQUAREHEAD

Taprooted perennial herbs with clustered stems. Leaves opposite (seldom ternate), more or less toothed, the lower reduced. Heads few or solitary, radiate, the rays pistillate and fertile, light yellow; involucre of 4 (5) relatively large and leafy bracts, these united at the base; receptacle conic, especially in fruit, its bracts subtending rays as well as disk flowers, thin, partly clasping the achenes; disk flowers perfect and fertile, the corolla tube expanded at the base and covering the top of the achene; style branches flattened, with rather long, hairy appendage. Achenes quadrangular or subterete, scarcely compressed. Pappus of several short scales, or none.

1 Pappus essentially none; leaves not connate-perfoliate .... 1. *T. helianthoides.*
1 Pappus of several scales 1–2 mm long; upper foliage leaves connate-perfoliate .. 2. *T. ludoviciana.*

1. T. helianthoides L. Plants 3–8 dm tall, viscid-villous, especially the stem, the leaves also atomiferous-glandular. Leaves large and thin, up to 20 cm long and 10 cm wide, ovate to elliptic or rhombic, narrowed to a sometimes petioliform, often clasping (but not connate-perfoliate) base, coarsely toothed, with generally less than 15 teeth per side. Outer involucral bracts broadly ovate, 2–3 cm long, ciliate, otherwise subglabrous; rays 6–12, 1.5–3 cm long, often 1 cm wide. Achenes subterete or obscurely quadrangular, without pappus. (n = 17) Mid-spring–early summer, and sometimes again in autumn. Dry, open woods, often in sandy soil; all provinces, Va to Tenn, s to Miss and c Fla.

2. T. ludoviciana (T. & G.) A. Gray. Like *T. helianthoides* in habit, but averaging taller and more robust, up to 12 dm tall, and with more numerous heads, glabrous or nearly so, except the finely tomentulose inner margins and tips of the involucral bracts, these smaller, less than 2 cm long. Leaves more closely, deeply, and saliently toothed, commonly with 12–30 teeth to a side, the upper leaves evidently connate-perfoliate, the lower more often abruptly narrowed to a winged petiole. Achenes more evidently quadrangular, and with an evident pappus of several scales 1–2 mm long. (n = 17) Late spring. Sandy woods; GC of La, w to Tex.

8. BORRICHIA Adans. SEA-OXEYE

More or less succulent, maritime shrubs. Leaves rather small, opposite, entire or toothed. Heads radiate, the rays yellow, pistillate and fertile; involucral bracts in 2 or 3 slightly unequal series, the outer herbaceous to somewhat coriaceous, the inner often more chartaceous and sometimes passing into those of the receptacle; receptacle slightly convex, chaffy, its bracts clasping the achenes; disk flowers perfect and fertile; style branches flattened, with elongate, hairy appendage. Achenes glabrous, quadrangular, or those of the rays 3-angled; pappus a dentate crown up to about 1 mm long.

1 Bracts of the receptacle rigid, with an evident spine tip 1–3 mm long .... 1. *B. frutescens*.
1 Bracts of the receptacle thinner and softer, not at all spine-tipped .... 2. *B. arborescens*.

1. B. frutescens (L.) DC. Rhizomatous, colonial, sparingly to freely branched shrub 1.5–12 dm tall, strongly canescent throughout, or the stems often subglabrous. Leaves oblanceolate or spatulate to sometimes nearly elliptic, entire or spinulose-denticulate, mucronate, 2–7 cm long and 3–25 mm wide. Heads solitary or several at the branch tips, the disk 1–2 cm wide; involucral bracts canescent, firm, the outer ovate and somewhat spreading, the inner less hairy, more appressed, often some of them shortly spine-tipped; rays several, commonly about 13, up to about 1 cm long; receptacular bracts rigid, with an evident spine tip 1–3 mm long, the body tending to become more thickened and fleshy in fruit. (n = 14) All year. Seacoast, especially in salt marshes; Va to Fla, Tex and Mex; Bermuda.

2. B. arborescens (L.) DC. Much like *B. frutescens*, sometimes canescent as in that species, but more often largely or wholly glabrate; involucral bracts broader, more membranous, and more appressed, the middle and inner sometimes with loose or spreading tip that may be mucronate; receptacular bracts softer, thinner, and drier, not at all spine-tipped, commonly somewhat spoon-shaped, with a slightly enlarged, concave-convex distal portion above a narrower basal portion that does not become fleshy-thickened. (n = 14) All year. Maritime habitats. West Indian sp., extending into s Fla, where it sometimes grows with *B. frutescens* and hybridizes with it.

9. **WEDELIA** Jacq., nom. conserv.

Perennial herbs (ours) or shrubs. Leaves opposite. Heads terminating the branches or pedunculate from the upper axils, yellow, the rays pistillate and fertile; involucre 2–3-seriate, the outer bracts commonly more or less herbaceous, the inner often drier and more chaffy; receptacle flat or convex, chaffy, its bracts concave and clasping; disk flowers perfect and fertile; style branches flattened, with ventromarginal stigmatic lines and an acutish, externally hairy appendage. Achenes double-convex or compressed-quadrangular (or those of the rays triangular), more or less narrowed to the summit, or shouldered so that the morphologically marginal pappus appears to be set well in from the lateral margins. Pappus a short crown or ring of scales, sometimes also with 1 or 2 deciduous awns, or obsolete. (*Pascalia, Stemmodontia*)

In addition to the following spp., the West Indian sp. *W. calycina* L. C. Rich. has been reported (S) as an introduction on CP of Ala, but I have seen no specimens. It is a shrub or half-shrub up to 2 m tall, with lanceolate to ovate, scabrous leaves 1.5–3 times as long as wide.

1 Leaves mostly (1.5) 2–5.5 cm wide and 1–3 (4) times as long as wide, toothed and often trilobed; creeping herb with erect flowering branches (or peduncles) up to 3 (4) dm tall .... 1. *W. trilobata*.
1 Leaves mostly 0.5–2 cm wide and 5–10 times as long as wide, entire or few-toothed; erect, branching herb 3–10 dm tall .... 2. *W. glauca*.

1. W. trilobata (L.) A. S. Hitchc. Creeping or shallowly rhizomatous herb with more or less erect, mostly simple flowering branches 1–3 (4) dm tall, or with axillary peduncles from the prostrate stem; herbage coarsely strigose to somewhat spreading-hirsute, or subglabrous. Leaves somewhat succulent, subsessile, cuneate from the middle or below, triplinerved, irregularly toothed and often with a pair of lateral lobes, mostly 4–9 × (1.5) 2–2.5 cm and 1–3 (4) times as long as wide. Heads on peduncles 3–10 cm long, campanulate-hemispheric; involucre about 1 cm high, its outer bracts commonly 2.5–4 mm wide; disk 7–11 mm wide; rays commonly ca. 8 or ca. 13, 6–15 mm long and 4–5 mm wide. Achenes 4–5 mm long, tuberculate. Pappus a persistent crown. (n = ca. 20, ca. 27, ca. 28) All year. Disturbed habitats; WI, escaped from cult. in peninsular Fla, and established in tropical Fla.

2. W. glauca (Ortega) O. Hoffm. Rhizomatous, leafy-stemmed, branched perennial herb 3–10 dm tall, finely strigose, or the stems glabrous. Leaves trinerved, lance-linear, entire or irregularly few-toothed, the main ones mostly 5–15 cm long and 0.5–2 cm wide, 5–10 times as long as wide, tapering to a narrow or shortly petiolar base. Heads short-pedunculate at the branch tips, hemispheric; involucre 7–9 mm high, its outer bracts mostly 1.5–2 mm wide, acute; disk 10–15 mm wide; rays ca. (13) 21, 7–10 mm long and 3–7 mm wide. Achenes 4–5 mm long, tuberculate, broadly truncate-topped. Pappus an irregular short crown. (n = 33) Summer. Disturbed habitats; native of tropical and subtropical S Am, casually intro. in coastal La and Fla, and perhaps not persistent. *Pascalia glauca* Ortega—S. Reputedly poisonous to livestock.

10. **VIGUIERA** HBK.

1. V. porteri (A. Gray) Blake. CONFEDERATE DAISY. Annual, 4–10 dm tall, branched above, thinly strigose, especially upward, and sometimes also sparsely hispid. Leaves tending to be hispid-ciliate toward the base, otherwise nearly glabrous beneath, more scabrous above, the lower or lower and middle ones opposite, the others alternate, all simple and entire or nearly so, narrow, the lower ones mostly 5–10 cm long and 2–10 mm wide, shortly or scarcely petiolate. Heads several or many, radiate, the disk 7–15 mm wide, becoming elevated; involucral bracts more or less herbaceous, loose and narrow; rays mostly about 8, yellow, neutral, 1–2 cm long; receptacle conic, chaffy throughout, its bracts clasping the achenes; disk flowers yellow, perfect and fertile; anthers sagittate at the base; style branches flattened, externally hairy above, without well-marked stigmatic lines, the appendage minutely short-hairy. Achenes quadrangular, moderately compressed at right angles to the involucral bracts; pappus none. (n = 17) Late summer, fall. Thin soil about granite flatrocks and monadnocks on PP of Ga and e Ala, and extending into RV in St. Clair Co., Ala; an outlying station in Alexander Co., NC.

11. TITHONIA Desf.

1. T. diversifolia (Hemsley) A. Gray. SHRUB SUNFLOWER. Soft shrubs 2–5 m tall, villous and atomiferous-glandular to subglabrate. Leaves alternate (at least above), many or all of them palmately 3- to 5-lobed as well as serrate, the blade generally 8–20 cm long and nearly as wide. Heads large, borne on notably fistulous peduncles, *Helianthus*-like; involucral bracts more or less membranous, imbricate, broadly rounded above; rays 7–21, yellow, 3–6 cm long; disk 2.5–4 cm wide. Pappus of 2 awns and several shorter, basally connate scales. (n = 17) All year. Disturbed sites; native of Mexico, occasionally cult., and escaped in c and s Fla and in many warm regions.

12. HELIANTHUS L. SUNFLOWER

Coarse annual or more commonly perennial herbs, with various sorts of underground systems, but not producing horizontal tubers except in *H. tuberosus*. Leaves simple, at least the lowermost ones opposite. Heads borne on normal, nonfistulous peduncles, radiate (rarely discoid) the rays relatively large, yellow, neutral; involucral bracts subequal or evidently imbricate, generally green and more or less herbaceous, at least distally; receptacle flat to convex or low-conic, chaffy throughout, its bracts clasping the achenes. Disk flowers perfect and fertile; anthers entire or minutely sagittate at the base; style branches flattened, externally (and sometimes distally internally) hispidulous, the short or elongate appendage hispidulous on both sides, the marginal or introrsely submarginal stigmatic lines poorly developed. Achenes thick, moderately compressed at right angles to the involucral bracts, with 2 evident and usually 2 obscure angles, generally glabrous or nearly so; pappus of 2 readily deciduous awns with enlarged, thin, paleaceous base (or paleaceous throughout), rarely with some additional short scales. (x = 17) The tendency for the upper or middle and upper leaves to be alternate is variably expressed according to the species and the individual. Robust individuals are more likely to have more of the leaves alternate; small specimens often have wholly opposite leaves. Hybrids abound. Several names thought to be based on hybrids are not provided for in the key and descriptions. These are as follows: *H. doronicoides* Lam., thought to be a hybrid of *H. giganteus* and *H. mollis*;

H. glaucus Small, thought to be a hybrid of *H. divaricatus* and *H. microcephalus*; and *H. verticillatus* Small, thought to be a hybrid of *H. angustifolius* with either *H. eggertii* or *H. grosseserratus*. Heiser, C. B. Jr. 1969. The North American sunflowers (Helianthus). Mem. Torrey Club 22(3).

1 Plants annual, or occasionally short-lived, taprooted perennial.
  2 Principal leaf blades (0.8) 1–3 (3.5) times as long as wide, often more than 4 cm wide, cordate or deltoid or abruptly contracted to a petiole seldom less than 1.5 cm long.
    3 Herbage (including phyllaries) densely, softly, and conspicuously white-hairy . . . . . 1. *H. argophyllus*.
    3 Herbage more coarsely and less densely hairy, or glabrous.
      4 Central receptacular bracts conspicuously white-bearded at the tip; involucral bracts lanceolate or lance-ovate, not long-hairy, scarcely or not ciliate . . . . . 2. *H. petiolaris*.
      4 Central receptacular bracts inconspicuously short-hairy; involucral bracts various.
        5 Involucral bracts lanceolate, up to 3 mm wide, tapering above, scabrous or hispidulous-scabrous to subglabrous, not long-hairy; plants mostly less than 1 m tall . . . . . 3. *H. debilis*.
        5 Involucral bracts broader, chiefly ovate or ovate-oblong and mostly 4 mm wide or more, abruptly contracted above the middle, usually ciliate and with some long hairs on the back; plants 1–3 m tall when well developed . . . . . 4. *H. annuus*.
  2 Principal leaf blades (3.5) 4–10 times as long as wide, not more than 4 cm wide, tapering to a petiole seldom more than 1 (1.2) cm long . . . . . 5. *H. agrestis*.
1 Plants fibrous-rooted to tuberous-rooted perennials from rhizomes, tubers, or crown buds.
  6 Disk yellow.
    7 Leaves more or less well distributed along the stem, only gradually reduced upward, the basal or lower cauline ones generally not notably larger than those next above.
      8 Stem evidently hairy below as well as in the inflorescence.
        9 Involucral bracts appressed, firm, evidently imbricate, shorter than the disk, (2) 2.5–4 mm wide; larger leaves (1.3) 2–6 cm wide . . . . . 28. *H. laetiflorus*.
        9 Involucral bracts looser (at least at the tip), sometimes spreading or reflexed, generally softer and seldom much imbricate, often equaling or surpassing the disk, often less than 2.5 mm wide; leaves various.
          10 Involucral bracts conspicuously reflexed (at least the upper half) and generally much elongate; heads relatively large, the disk (1.5) 2–3 cm wide; leaf blades short-petiolate or tapering to a narrow, more or less sessile base, usually many of them alternate . . . . . 6. *H. resinosus*.
          10 Involucral bracts merely loose or somewhat spreading, not conspicuously reflexed; heads and leaves various.
            11 Herbage densely and rather softly short-hairy throughout; leaves sessile or nearly so, broad-based and more or less cordate-clasping, 1.4–3 (4) times as long as wide; rays 16–35 . . . . . 7. *H. mollis*.
            11 Herbage more coarsely or sparsely hairy or scabrous, or in part glabrous; leaves and rays various.
              12 Leaves relatively large, the better developed ones generally with the blade 4–12 cm wide on a petiole (1.5) 2–8 cm long; rhizomes commonly tuber-bearing . . . . . 8. *H. tuberosus*.
              12 Leaves smaller, or shorter-petiolate, or both; rhizomes without tubers, although the roots are sometimes tuberous thickened (a rare form of *H. angustifolius* with a yellow disk would be sought here).
                13 Plants with short rhizomes and more or less tuberous-thickened roots.
                  14 Heads small, the disk 1–1.5 cm wide; involucral bracts shorter than the disk . . . . . 9. *H. schweinitzii*.
                  14 Heads larger, the disk mostly 1.5–2.5 cm wide; involucral bracts equaling or surpassing the disk.
                    15 Stem spreading-hirsute; involucral bracts conspicuously ciliate with long, loose hairs; leaves commonly trip-linerved at the base, flat . . . . . 10. *H. giganteus*.

15 Stem rather loosely white-strigose; involucral bracts finely canescent-strigose, seldom ciliate; leaves not triplinerved, usually some folded .... 11. *H. maximilianii.*

13 Plants with more or less elongate rhizomes (only 5–10 cm in *H. simulans*) and fibrous roots.

16 Leaves mostly or all opposite, abruptly contracted to often broadly rounded or subcordate at the base, on a petiole 5–15 (20) mm long .... 12. *H. hirsutus.*

16 Leaves in well-developed plants mostly (or at least many of them) alternate, sessile or nearly so, the petiole if any not more than ca. 5 mm long.

17 Plants robust, mostly 1.5–2.5 m tall; rhizomes short (5–10 cm) and stout; leaves not undulate; sporadic, unstable cultigen .... 25. *H. simulans.*

17 Plants smaller and more slender, mostly 0.4–1.5 (2) m tall; rhizomes slender and elongate; leaves tending to be undulate-margined; native sp. .... 26. *H. floridanus.*

8 Stem essentially glabrous below the inflorescence, often also glaucous.

18 Heads middle-sized to large, the disk 1.5–3.5 cm wide, the rays (8) 10–25; rhizomes short to elongate.

19 Petiole short (seldom as much as 1 cm long) or none.

20 Leaves glaucous beneath, somewhat tapering toward the base .... 13. *H. eggertii.*

20 Leaves somewhat hairy to subglabrous beneath, but not glaucous.

21 Leaves widest near the truncate or broadly rounded base ...... 14. *H. divaricatus.*

21 Leaves tapering to a narrow base (occasional forms of) .... 10. *H. giganteus.*

19 Petiole longer, commonly 1–4 cm long on well-developed leaves.

22 Leaves relatively narrow, usually at least 3 times as long as wide, the middle and upper ones seldom over 4 cm wide.

23 Lower leaf surface pale, more or less glaucous, only sparsely or not at all hairy (occasional forms of) .... 17. *H. strumosus.*

23 Lower leaf surface green and generally evidently hairy (sometimes sparsely so), not glaucous.

24 Upper surface of leaves distinctly scabrous; lower surface rather coarsely (and often sparsely) long-hairy, many or all of the hairs on the order of 1 mm long or more (occasional forms of) .... 10. *H. giganteus.*

24 Upper surface of leaves usually scarcely scabrous, the short hairs tending to be more or less appressed; lower surface more finely and shortly hairy than in no. 10, most or all of the hairs well under 1 mm long .... 15. *H. grosseserratus.*

22 Leaves wider, seldom over 3 times as long as wide, often over 4 cm wide.

25 Involucral bracts very loose, evidently surpassing the disk; leaves thin, generally strongly serrate .... 16. *H. decapetalus.*

25 Involucral bracts only moderately loose, slightly or not at all surpassing the disk; leaves fairly firm, inconspicuously serrulate or subentire .... 17. *H. strumosus.*

18 Heads small, the disk mostly 0.5–1.5 cm wide, the rays 5–8 (10) in number; rhizomes short or none.

26 Lower surface of leaves resin-dotted and very often also evidently hairy; leaves evidently petiolate .... 18. *H. microcephalus.*

26 Lower surface of leaves glabrous or nearly so, not resin-dotted; leaves petiolate to subsessile.

27 Leaves rather thin, abruptly narrowed or broadly rounded to a petiole generally 1–4 cm long .... 19. *H. glaucophyllus.*

27 Leaves firm, narrowed to a subsessile base or short petiole less than 1 cm long .... 20. *H. laevigatus.*

7 Leaves basally disposed, the basal or lower cauline ones well developed and persistent, the others more or less strongly reduced and often few and distant.

28 Leaves relatively wide, the larger ones with the blade mostly 2–8 cm wide and 1.5–5 times as long as wide, well distinguished from the petiole; rhizomes well developed; leaves scabrous or hirsute to strigose or occasionally glabrous .... 21. *H. occidentalis.*

28 Leaves narrow, the larger ones with the blade less than 2 cm wide and well over 5 times as long as wide, often not well set off from the petiole; rhizomes poorly developed or wanting; leaves glabrous.
29 Heads mostly several; involucral bracts mostly 1–2.5 mm wide .... 22. *H. longifolius*.
29 Heads mostly solitary; involucral bracts mostly 3–6 mm wide .... 23. *H. carnosus*.
6 Disk red or purple.
30 Cauline leaves more or less numerous and not conspicuously reduced upward, sessile or nearly so, linear to lance-ovate, in well-developed plants mostly alternate; involucral bracts narrow, 1–2.5 (3) mm wide.
31 Rhizomes wanting, or very short and poorly developed; leaves mostly 10–30 times as long as wide, rarely more than 1 cm wide .... 24. *H. angustifolius*.
31 Rhizomes well developed; leaves mostly 2.5–10 (15) times as long as wide, often more than 1 cm wide.
32 Plants robust, mostly 1.5–2.5 m tall; rhizomes short (5–10 cm) and stout; leaves not undulate; sporadic, unstable cultigen .... 25. *H. simulans*.
32 Plants smaller and more slender, mostly 0.5–1.5 (2) m tall; rhizomes slender and elongate; leaves tending to be undulate-margined; native sp. .... 26. *H. floridanus*.
30 Cauline leaves otherwise, often more or less reduced, or wider and more petiolate, and often mostly or all opposite; involucral bracts wider, (2) 2.5–5 mm wide.
33 Cauline leaves well developed, (2) 2.5–8 times as long as wide, the middle ones not much if at all smaller than the lower; accessory pappus scales generally present; rhizomes well developed .... 27. *H. rigidus*.
33 Cauline leaves either obviously reduced (the middle ones evidently smaller than the basal or lower cauline ones), or not more than twice as long as wide; accessory pappus scales wanting; rhizomes wanting or very short.
34 At least the lower cauline leaves well developed, more or less resembling the basal ones; heads several.
35 Stem relatively more leafy, often to above the middle; leaf blades relatively wide, generally only 1–1.7 (2) times as long as wide, on short petioles that are seldom more than 1/3 as long as the blade and seldom strongly wing-flared upward .... 29. *H. silphioides*.
35 Stem less leafy, the leaves evidently basally disposed; leaf blades narrower, mostly (1.3) 1.7–2.5 (3) times as long as wide, the petiole often more than 1/3 as long as the blade and tending to be wing-flared upward .... 30. *H. atrorubens*.
34 Cauline leaves all very much smaller than the tufted basal ones, or nearly wanting; heads 1–3 (5).
36 Rays well developed, mostly 1.5–3.5 cm long; principal leaf blades (1.6) 2–several times as long as wide .... 31. *H. heterophyllus*.
36 Rays none, or rarely present and up to ca. 1 cm long; principal leaf blades 1–1.5 times as long as wide .... 32. *H. radula*.

1. H. argophyllus T. & G. Taprooted annual, 1–3 m tall, the herbage (including phyllaries) densely, softly, and conspicuously white-hairy. Leaves entire or shallowly serrate, the principal ones rather narrowly deltoid-ovate to broadly cordate, 10–25 cm long, at least half as wide; petiole 2–10 cm long or more. Heads on more or less leafy-bracteate peduncles, the disk 2–3 cm wide, generally red-purple; involucral bracts ovate or lance-ovate, abruptly long-attenuate, mostly 4–7 mm wide; rays (13) 15–25, mostly 2–4.5 cm long; receptacular bracts glabrous or lightly villous. (n = 17) Summer–fall. Sandy soil; native in e Tex, and adventive in Fla.

2. H. petiolaris Nutt. Taprooted annual, 0.4–1 (1.5) m tall; stems green to red (not mottled), hirsute or strigose, simple or often branched from below the middle. Leaves subglabrous to more often scabrous or shortly hispid-hirsute, entire or inconspicuously toothed, the principal ones lance-ovate to more often rather narrowly deltoid, varying to subcordate, mostly 4–13 cm long and 1.5–6 cm wide, 1–3 (3.5) times as long as wide, the petiole mostly 1.5–7 cm long. Heads naked-pedunculate, the disk 1–2.5 wide, generally red-purple; involucral bracts lanceolate or lance-ovate, tapering above, mostly 3–5 (6) mm wide, shortly scabrous-hispid, seldom at all ciliate or with any long hairs; rays mostly 15–30, 2–3 cm long; central receptacular bracts conspicuously white-bearded at the tip, the others inconspicuously scabrous. (n = 17) Summer. Disturbed, often low ground; native to the Great Plains, and occasional in our range as a weed. Ours is var. *petiolaris*.

3. H. debilis Nutt. Taprooted annual or occasionally short-lived perennial, seldom 1 m tall. Leaves scabrous to rough-hairy or seldom subglabrous, entire or toothed, the principal ones lance-ovate or broader, basally broadly cuneate to more often rounded, truncate or somewhat cordate, mostly 3–10 cm long and 1.5–9 cm wide, 1–2 (3) times as long as wide, the petiole mostly 1.5–7 cm long. Heads naked-pedunculate, the disk 1–2.5 cm wide, generally red-purple; involucral bracts lanceolate, up to 3 mm wide, tapering above, scabrous or hispidulous-scabrous to subglabrous, not long-hairy; rays mostly 11–21, 1–3 cm long; receptacular bracts not bearded. (n = 17) Nearly all year. Sandy beaches, and disturbed soil inland; Tex to Fla, and intro. in Atlantic coastal states northward. Two subspecies, each with two vars. in our range:

Stem erect, often conspicuously red-brown mottled .... ssp. *cucumerifolius* (T. & G.) Heiser.

  Peduncles elongate, generally 10–30 cm long; rays mostly 1.5–3 cm long; leaves entire to variously toothed; mostly inland, often in disturbed sites (*H. c.* T. & G.—S) .... var. *cucumerifolius* (T. & G.) A. Gray.

  Peduncles shorter, 5–15 (20) cm long; rays seldom more than 2 cm long; leaves often strongly and irregularly serrate; plants averaging more densely bush-branched than the previous var.; chiefly on sandy beaches, sometimes inland in pine woods; Gulf Coast of Fla, w to Miss .... var. *tardiflorus* (Heiser) Cronq.

Stem more or less prostrate, with ascending or erect flowering branches, not strongly mottled; sandy beaches along the Fla coast .... ssp. *debilis*.

  Stem conspicuously spreading-white-hirsute; leaves averaging more deeply and irregularly serrate than in the next var.; Gulf Coast of Fla (*H. v.* E. E. Wats.—S) .... var. *vestitus* (E. E. Wats.) Cronq.

  Stem scabrous or shortly scabrous-hispid to occasionally subglabrous; Atlantic Coast of Fla .... var. *debilis*.

4. H. annuus L. COMMON SUNFLOWER Coarse, taprooted annual, (0.5) 1–3 m tall, unbranched and single-headed (in cultivated forms) to more often branched and several- to many-headed, the herbage rough-hairy. Leaves chiefly alternate (except the lowermost), mostly toothed, ovate or broader, often (especially the lower) cordate, commonly 10–40 cm long and 5–25 cm wide, long-petiolate. Disk generally red-purple, seldom less than 3 cm wide (to 30 cm in cultivated forms); involucral bracts mostly ovate to ovate-oblong and rather abruptly contracted above the middle, mostly 4 mm wide or more, ciliate on the margins and commonly more or less hispid or hirsute on the back; rays mostly 17 or more, 2.5 cm long or more; receptacular bracts inconspicuously hairy at the tip. (n = 17) Summer. A weed in disturbed sites, especially in moist, low ground, throughout the US, and extending into Can and Mex.

5. H. agrestis Pollard. Taprooted annual, 0.5–1.5 (2) m tall; stems glabrous or nearly so, glaucous. Leaves glabrous or scabrous, irregularly dentate or entire, lance-elliptic or narrower, the principal ones 5–15 cm long and 0.6–3 (4) cm wide, (3.5) 4–10 times as long as wide, tapering to a short (to 1.2 cm) petiole that is often conspicuously hispid-ciliate. Heads naked-pedunculate, the disk 1–2 cm wide, generally red-purple; involucral bracts narrow, elongate, loose-tipped; rays from about 8 to about 13, 1–2 cm long; receptacular bracts glabrous. (n = 17) Summer–late fall. Mucky, wet soil; Fla and s Ga.

6. H. resinosus Small. Stout perennial, 1–3 m tall, from a short rhizome; stem coarsely spreading-hairy. Leaves numerous, scabrous to hispid above, hirsute to velvety-subtomentose (often also resin-dotted) beneath, narrowly lanceolate to broadly ovate, 5–25 cm long and 2–8 cm wide, entire or serrate, subsessile or with a short, wing-flared petiole, usually many of them alternate. Heads 1–several on short, rigid peduncles; disk yellow, (1.5) 2–3 cm wide; involucral bracts hispid-scabrous to long-villous, narrow, with an elongate, slender tip, generally all except sometimes the inner ones with the upper half (or more) reflexed; rays mostly 8–15, 2–3 cm long. (n = 51) Summer. Moist or dry woods and open ground; CP to Mt provinces, NC to Ga, w Fla, and Miss. *H. tomentosus* Michx.—sensu R, S, not Michx.

7. H. mollis Lam. Rhizomatous perennial, 5–10 (12) dm tall, densely and rather softly short-hairy throughout. Leaves sessile, subcordate, ascending, broadly lanceolate to broadly ovate or oblong, 6–15 cm long and 2–8 cm wide, 1.4–3 (4) times as long as wide, serrulate or entire. Heads few or solitary; disk yellow, 2–3 cm wide; involucral bracts slightly imbricate, lanceolate, acuminate, often finely glandular as well as densely white-hairy, the upper part loose or spreading; rays mostly 16–35, 1.5–3.5 cm long. (n = 17) Summer. Prairies and other dry, open places; a chiefly Ozarkian and midwestern sp., s to s La, e to Ky, e Tenn, and n Ga, and (intr) AC of Md, Va, and NC.

8. H. tuberosus L. JERUSALEM ARTICHOKE. Perennial with well-developed, commonly tuber-bearing rhizomes; stems stout, (0.7) 1–3 m tall, more or less pubescent with mostly spreading hairs. Leaves numerous, scabrous above, short-hairy beneath, broadly lanceolate to broadly ovate, the better-developed ones mostly 10–25 cm long and 4–12 cm wide, serrate, abruptly contracted or somewhat tapering to the well-developed, more or less winged petiole, this 2–8 cm long. Heads several or generally numerous in a corymbiform inflorescence, the disk yellow, 1.5–2.5 cm wide; involucral bracts generally rather dark, especially near the base, narrowly lanceolate, acuminate or subattenuate, loose especially above the middle, often hispidulous; rays 10–20, 2–4 cm long. (n = 51) Late summer–fall. Moist soil and waste places; widespread weedy sp of e and c US and adj Can, cult. for its edible tubers and readily escaping, the pre-Columbian range uncertain, but now throughout our area except possibly c and s Fla.

9. H. schweinitzii T. & G. Perennial (0.6) 1–2 m tall from a short rhizome with clustered, tuberous roots; stems strigose or hirsute-strigose. Leaves scabrous above, resin-dotted and loosely soft-white-hairy beneath, entire or nearly so, lanceolate to lance-linear, up to 18 cm long and 2.5 cm wide, tapering to a subpetiolar base or short petiole to 1.5 cm long. Heads several; disk yellow, 1–1.5 cm wide; involucral bracts slender, long-pointed, loosely or partly squarrose, shorter than the disk, ciliate-margined at least below, hairy on the back. (n = 51) Summer. Irregularly and infrequently on CP and PP in NC and SC.

10. H. giganteus L. Perennial with crown buds, short rhizomes, and thickened, often fleshy roots; stems 1–3 m tall, coarsely (and often rather sparsely) spreading-hairy to occasionally subglabrous. Leaves flat, strongly scabrous above, usually hirsute beneath, more or less triplinerved at the base, strongly toothed to subentire, lanceolate, acuminate, 8–20 cm long and 1–3.5 cm wide, tapering to the short petiole or petiolar base, the upper generally alternate. Heads several or many in an open, corymbiform inflorescence; disk yellow, 1.5–2.5 cm wide; involucral bracts narrow, thin, green (or dark below), acuminate or attenuate, loose, often conspicuously surpassing the disk, strongly hirsute-ciliate and often hairy on the back; rays 10–20, 1.5–3 cm long. (n = 17) Late summer–fall. Swamps and other wet places; widespread in ne US and adj Can, s to NC, n SC, n Ga, e Tenn, and even s Miss. *H. alienus* E. E. Wats.—S; *H. validus* E. E. Wats.—S.

11. H. maximilianii Schrader. Perennial with crown buds, short rhizomes, and thickened, often fleshy roots; stems usually several, 0.5–3 m tall, conspicuously pubescent, especially upward, with mostly short, white, appressed or subappressed hairs. Leaves strongly scabrous on both sides, lanceolate, acuminate, commonly 7–20 cm long and 1–3 cm wide, entire or slightly toothed, firm, pinnately veined, not at all triplinerved, gradually narrowed to the short, winged petiole or petioliform base, generally at least some of them conduplicate and often falcate, the upper often mainly alternate. Heads generally several, the inflorescence tending to be elongate and racemiform; involucral bracts narrow, firm, attenuate or subcaudate, loose, often much exceeding the disk, canescent with short white hairs and sometimes basally ciliate; disk yellow, 1.5–2.5 cm wide; rays mostly 10–25, 1.5–4 cm long. (n = 17) A characteristic species of the prairies and plains of c US, now widely established in similar habitats and waste places farther e, in our range s only to Tenn and NC.

12. H. hirsutus Raf. Perennial, 0.6–2 m tall, from elongate rhizomes; stems spreading-hairy. Leaves all or mostly opposite, hirsute on both sides or scabrous above, narrowly lanceolate to ovate, mostly 7–16 cm long and (1.5) 2–6 cm wide, serrate to entire, triplinerved at the abruptly contracted to often broadly rounded or subcordate base, generally ascending on short petioles 5–15 (20) mm long. Heads 1–several on short, stout peduncles; disk yellow, (1.2) 1.5–2 (2.5) cm wide; involucral bracts conspicuously ciliate and often also hairy on the back, slender, long-pointed, often with loose or reflexed tip; rays 10–15, 1.5–3.5 cm long. (n = 34) Late summer–fall. Dry, wooded or open places; widespread from Pa to n Fla and w to Minn and Tex, but only seldom on AC. *H. stenophyllus* (T. & G.) E. E. Wats.—S.

13. H. eggertii Small. Fibrous-rooted perennial 1–2 m tall from elongate rhizomes; stem glabrous and glaucous. Leaves glabrous or nearly so above, glabrous and strongly glaucous beneath, lanceolate, mostly 9–15 cm long and 1.5–3.5 cm wide, entire or serrulate, tapering to a sessile or subsessile base. Heads rather few, on elongate peduncles often 1–1.5 dm long; disk yellow, 1.2–2 cm wide; involucral bracts firm, lance-acuminate, ciliolate, about equaling the disk; rays 10–14, up to about 2 cm long. (n = 51) Summer. Rocky hills and barrens; c Tenn to ec Ky.

14. H. divaricatus L. Fibrous-rooted perennial from elongate rhizomes, 0.5–1.5 m tall; stem glabrous below the inflorescence, often glaucous. Leaves scabrous above, loosely hirsute or hispidulous beneath, at least on the main veins, narrowly lanceolate to broadly lance-ovate, 5–18 cm long and 1–5 (8) cm wide, broadest near the truncate or broadly rounded base, serrate to more often subentire, sessile or rarely with a short petiole up to 5 mm long. Heads 1–several at the tips of cymose stiff branches; disk yellow, 1–1.5 cm wide; involucral bracts lance-acuminate or -attenuate, ciliolate, rather loose, often with reflexed tip; rays 8–15, 1.5–3 cm long. (n = 17) Summer. Dry woods and other open places; widespread in e US and adj Can, s to c Fla, Miss, and La.

15. H. grosse-serratus Martens. Coarsely fibrous-rooted, rhizomatous perennial, 1–4 m tall; stems strigose in the inflorescence, otherwise glabrous and often glaucous. Leaves strigose or somewhat scabrous-strigose on both sides, or more hirtellous or puberulent especially on the lower surface, lanceolate, the middle and upper generally 10–20 cm long and 1.5–4 cm wide, the lower often larger, tapering to an often winged petiole 1–4 (5.5) cm long. Heads relatively large, generally several or many in an often corymbiform inflorescence, the disk yellow, 1.5–2.5 cm wide; involucral bracts lance-linear, loose, attenuate, surpassing the disk, more or less ciliate at least near the base, often short-hairy on the back; rays 10–20, 2–4.5 cm long. (n = 17) Summer–fall. Bottomlands, damp prairies, and other moist places; midwestern species, known at scattered stations in our area, especially in Ky, Tenn, and n La, mainly or wholly as an introduction.

16. H. decapetalus L. Perennial from slender rhizomes, 0.5–1.5 (2) m tall; stem short-hairy in the inflorescence, otherwise glabrous. Leaves thin, pale beneath, moderately scabrous to subglabrous, broadly lanceolate to ovate, mostly 8–20 cm long and 3–8 cm wide, long-acuminate, serrate, generally sharply so, more or less abruptly contracted near the base and decurrent onto the 1.5–6 cm petiole. Heads generally several, the disk 1–2 cm wide, yellow; involucral bracts very loose, thin, green, conspicuously ciliate, occasionally hispidulous on the back, attenuate-acuminate, at least some of them usually conspicuously surpassing the disk; rays 8–15, 1.5–3.5 cm long. (n = 17, 34) Late summer–fall. Woods and along streams; widespread in ne US, s to AC in NC and adj SC, PP in Ga, and IP and mt. provinces of Tenn. *H. trachelifolius* Mill.—F.

17. H. strumosus L. Rhizomatous perennial, 1–2 m tall; stem glabrous below the inflorescence or with a few long hairs, often glaucous. Leaves relatively thick and firm, scabrous-hispid above, green and moderately short-hairy to more often glaucous and subglabrous beneath, mostly broadly lanceolate to ovate and 8–20 cm long by 2.5–10 cm wide, but sometimes narrower and even lance-linear, long-acuminate, shallowly toothed or subentire, more or less abruptly contracted or sometimes broadly rounded at the base, commonly shortly decurrent onto the 6–30 mm petiole. Heads 1–several, the disk 1.2–2.5 cm wide, yellow; involucral bracts subequal, lanceolate, somewhat loose, especially the long acuminate tips, which commonly equal or slightly surpass the disk; rays 8–15, 1.5–4 cm long. (Alloploid with n = 34, 51) Summer. Woods and open places; widespread in e US, s to La and Fla. *H. montanus* E. E. Wats.—S; *H. saxicola* Small—S.

18. H. microcephalus T. & G. Fibrous-rooted perennial with crown buds and a short (seldom more than 5 cm) rhizome; stems (0.7) 1–2 m tall, glabrous and generally glaucous. Leaves scabrous above, resinous-dotted and usually also more or less densely and loosely short-hairy beneath, sometimes also glaucous, lanceolate or lance-ovate, mostly 7–15 (20) cm long and (1) 2–5 (6) cm wide, toothed or entire, gradually tapering distally, more or less abruptly narrowed to the 1–3 cm petiole. Heads few to many on long, slender peduncles, small, the disk yellow, 0.5–1 cm wide; involucral bracts few, lanceolate, acuminate or attenuate, ciliolate, otherwise glabrous or nearly so; rays 5–8, 1–1.5 cm long. (n = 17, 34) Summer. Woods and brushlands; NJ to nw Fla, w to s Mich, e Ark, and se La, but seldom on AC. *H. smithii* Heiser. A hybrid with *H. divaricatus* has been called *H. glaucus* Small.

19. H. glaucophyllus D. M. Smith. Fibrous-rooted perennial 1–2 m tall with a short rhizome and numerous crown buds; stem glabrous, usually glaucous. Leaves glabrous or scabrous above, glabrous and glaucous beneath, without resin dots, evidently serrate, lanceolate or lance-ovate, mostly 8–20 cm long and 1–6 cm wide, abruptly narrowed or broadly rounded to a petiole 1–4 cm long. Heads few to many in a leafy-bracteate inflorescence, slender-pedunculate, rather small, the disk yellow, 0.5–1.5 (2) cm wide; involucral bracts ciliolate, otherwise glabrous or minutely puberulent, the outer loose-tipped; rays 5–8, 1–2 (2.5) cm long. (n = 17) Summer. Moist woods above 2500 ft elevation; BR of NC and adj Tenn.

20. H. laevigatus T. & G. Fibrous-rooted perennial 1–2 m tall from a crown or short rhizome; stem essentially glabrous, often glaucous. Leaves firm, essentially glabrous, serrulate to entire, lanceolate to sometimes lance-elliptic or lance-ovate, mostly 6–18 cm long and 1–4 cm wide, narrowed to a subsessile base or short petiole less than 1 cm long. Heads 1–several, rather small, the disk yellow, 1–1.5 cm wide; involucral bracts often ciliolate, otherwise glabrous; rays 5–10, 1–2 cm long. (n = 34) Summer. Mainly on shale barrens; RV, BR, and PP of Va and adj WVa to w NC and n SC. *H. reindutus* (Steele) E. E. Wats.—S.

21. H. occidentalis Riddell. Rhizomatous and often stoloniferous perennial, 0.5–1.5 m tall. Leaves basally disposed, the lower ones much the largest, with ovate to lance-elliptic or lanceolate blade mostly 6–15 cm long and 2–8 cm wide, 1.5–5 times as long as wide, sharply set off from the elongate petiole, the others more or less reduced and distant. Heads (1) 3–several; disk 1–1.5 cm wide, yellow; involucral bracts ciliolate, more or less imbricate, lanceolate or lance-ovate, (1.5) 2–3 mm wide, at least the inner with loose, slender tip; rays mostly 10–15, (1) 1.5–3 cm long. (n = 17) Summer. Dry, often sandy soil; Md and DC to Minn, s to Ga (CP), w Fla, and Tex. Three vars.:

Leaves scabrous to hirsute, entire or nearly so; widespread, but rare or absent in the range of the other vars. .... var. *occidentalis*.
Leaves strigose (the hairs closely appressed) to glabrous.
  Leaves more or less evidently toothed; Garland Co., Ark (OU) to sw La and se Tex .... var. *plantagineus* T. & G.
  Leaves entire or obscurely toothed, often not so quickly reduced upward as in the other vars., the basal and lowermost cauline ones often deciduous, some of the lower cauline ones (but still well above the base) often fairly well developed; Appalachian region from DC, Md, and WVa to n Ga [*H. dowellianus* (T. & G.) M. A. Curtis—S.] .... var. *dowellianus* T. & G.

22. H. longifolius Pursh. Glabrous perennial with several stems 0.5–1.2 m tall from a crown or short rhizome or caudex. Leaves basally disposed, narrow, the basal and lower cauline ones with linear-elliptic to lance-linear or linear blade 10–25 cm long and 0.7–1.2 (2) cm wide, tapering to a slender, petioliform base or long petiole, or the radical leaves sometimes short and broad, to 3 cm wide, but the lower cauline still elongate and narrow; cauline leaves progressively reduced upward, the uppermost alternate and bractlike. Heads several (commonly 4–10) in an open-corymbiform inflorescence; disk yellow, 1–1.5 cm wide; involucral bracts imbricate or not, linear to lance-oblong, 1–2.5 mm wide, often some or all with loose, slender tip that may surpass the disk; rays mostly 8–10, 1.5–2.5 cm long. (n = 17) Summer. Rocky glades; mt. provinces and PP from sw NC to ne Ala, s to CP of wc Ga and ec Ala.

23. H. carnosus Small. Much like *H. longifolius;* stem usually solitary from one side of a crown, to 0.6 m tall, the radical leaves from the other side, much larger than the obviously reduced cauline ones, up to 20 cm long and 1.5 cm wide. Head mostly solitary; involucral bracts broader, commonly 3–6 mm wide; rays up to 15. (n = 17) Summer–late fall. Moist or wet sandy soil in ne Fla.

24. H. angustifolius L. Fibrous-rooted perennial with crown buds, nearly or quite without rhizomes; stem solitary, 0.5–1.5 (2) m tall, more or less hairy, especially below. Cauline leaves numerous, sessile or nearly so, dark green and scabrous above, pale beneath with fine, loose, sometimes deciduous hairs, and often also atomiferous-glandular, slender and elongate, mostly 5–15 (20) cm long and 2–10 (15) mm wide, 10–30 times as long as wide, revolute-margined, mostly alternate (except toward the base) in well-developed plants; broader, tufted, petiolate basal leaves up to 7 × 2 cm sometimes present. Heads commonly rather few; disk red-purple, rarely yellow, mostly (1) 1.5–2 cm wide; involucral bracts lanceolate or lance-linear, 1.3–2 mm wide, inconspicuously hairy or subglabrous, at least the inner with loose, narrow tip seldom surpassing the disk; rays (8) 10–15 (21), 1.5–3 cm long. (n = 17) Summer–late fall. Moist or wet low ground, open fields, and roadsides, hybridizing freely with *H. floridanus*; SE, and adj states.

H. salicifolius A. Dietr., with equally narrow leaves and red-purple disk, occurs in w Mo, Kans, Okla, and Tex, and may be expected in nw Ark. It differs from *H. angustifolius* in its well-developed rhizomes, glabrous, often glaucous stem, glabrous or subglabrous leaves, and more elongate, slender involucral bracts, the inner generally surpassing the disk.

25. H. simulans E. E. Wats. Like *H. floridanus*, but more robust, 1.5–2.5 m tall; rhizomes short (5–10 cm long) and stout (up to 1 cm thick); leaves not undulate, sometimes wider; rays 12–23, 1.5–3 cm long. (n = 17) Fall. CP: sporadically from w Fla to s La. Possibly a hybrid cultigen, derived from *H. angustifolius* and *H. maximilianus*.

26. H. floridanus A. Gray. Perennial, 0.4–1.5 (2) m tall, from long rhizomes; stems hispid-scabrous or hispidulous, especially below, often merely strigose above. Leaves numerous, the lower opposite, the upper generally alternate, scabrous above, paler and more softly (sometimes sparsely) short-hairy and glandular beneath, the principal ones sessile or nearly so (petiole less than 5 mm long), lance-ovate to lance-linear, 4–20 cm long and up to about 2.5 cm wide, 2.5–10 (15) times as long as wide, the margins tending to be undulate and often also revolute. Heads 1–several, the disk red-purple or less often yellow, mostly 1–2 cm wide; involucral bracts firm, short-hairy or subglabrous, lanceolate, 1.5–3 mm wide, at least the inner with loose to spreading reflexed tip; rays mostly about (8) 13, 2–4 cm long. (n = 17) Summer–late fall. Moist, low ground, often contaminated by introgression from the closely related *H. angustifolius*; CP: SC to c Fla and se La.

27. H. rigidus (Cass.) Desf. Rhizomatous perennial, mostly 1–2 m tall; stem sparsely to densely scabrous-hispid. Leaves all or nearly all opposite, scabrous or shortly hispid on both sides, mostly 9–15 pairs below the inflorescence, lanceolate to rather narrowly ovate, mostly 8–15 (25) cm long and 1.5–6 cm wide, (2) 2.5–8 times as long as wide, entire or serrate, tapering to a short petiole or petiolar base, the upper often reduced and distant. Heads several or solitary, the disk red-purple, very rarely yellow, 1.5–2.5 (3) cm wide; involucral bracts evidently imbricate, broad, firm, appressed, mostly ovate or broadly lanceolate and (2.5) 3–5 mm wide, sharply acute to obtuse, conspicuously ciliolate, otherwise generally glabrous, not exceeding the disk; rays mostly 10–21, 1.5–3 (3.5) cm long. Pappus nearly always with some short scales in addition to the two longer awns. (n = 51) Summer. Roadsides and prairies; midwestern and Great Plains species, entering our range in nw Ark. Ours is var. *rigidus*. (Under *H. laetiflorus* Pers.—F, G.)

28. H. laetiflorus Pers. Rhizomatous perennial, much like *H. rigidus*, but with a yellow disk; leaves often larger and with much longer (to 5 cm) petiole; involucral bracts averaging a little narrower, generally (2) 2.5–4 mm wide, less imbricate, more pointed, and occasionally sparsely short-hairy on the back. (n = 51) Late summer–fall. Roadsides and other disturbed sites, mainly or wholly as an escape from cult., often (always?) more or less sterile; scattered in e and midw US, s to SC, Ga, and Tenn. "Species" thought to consist of hybrids and hybrid progeny of *H. rigidus* and *H.*

tuberosus, perhaps mainly a cultigen. [Broadly interpreted to include *H. rigidus* (Cass.) Desf.—F, G.]

29. H. silphioides Nutt. Similar to *H. atrorubens*, and hybridizing extensively with it from La to Tenn. Stems commonly several, up to 3 m tall, relatively more leafy, often to above the middle. Leaf blades relatively broad, generally only 1–1.7 (2) times as long as wide, on short, generally inconspicuously hairy petioles that are seldom more than 1/3 as long as the blade and seldom conspicuously wing-flared upward. Involucral bracts not always ciliolate. (n = 17) Summer–fall. Basically Ozarkian sp, from Ark and s Mo to c La, s Ill, and w Ky, and e through most of Tenn. Perhaps better called *H. atrorubens* var. *pubescens* Kuntze, as in G.

30. H. atrorubens L. Fibrous-rooted perennial from a very short, stout rhizome or crown, 0.5–2 m tall; stems mostly solitary, usually conspicuously spreading-hairy at least below. Leaves nearly all opposite, 3–8 pairs below the inflorescence, commonly hirsute or hispid on both sides, especially on the main veins beneath, lance-ovate to sometimes broadly ovate, mostly (1.3) 1.7–2.5 (3) times as long as wide, abruptly contracted to the petiole, the largest ones near the base, these commonly 6–20 (25) cm long, toothed; petioles tending to be conspicuously wing-flared upward, often conspicuously spreading-hairy, the lower generally from 1/3 to fully as long as the blade. Heads several on long, naked peduncles in a corymbiform inflorescence; disk red-purple, mostly 1–1.5 (2) cm wide; involucral bracts evidently imbricate, broad, firm, appressed, mostly oblong or elliptic, rounded to acutish, sometimes with an abrupt, very short acumination, 2.5–4 (5) mm wide, ciliolate, otherwise glabrous or nearly so; rays 10–15, 1–3 cm long. Pappus without accessory scales. (n = 17) Summer–fall. Open woods; basically App and Atl sp., from Va and e Ky to c Ga and Ala, w to w Tenn and se La.

31. H. heterophyllus Nutt. Fibrous-rooted perennial from a crown or very short rhizome, the herbage generally rough-hairy at least below; stems mostly solitary, stiffly erect, 0.4–1.2 m tall. Basal leaves tufted and persistent, entire or nearly so, mostly 6–30 cm long (petiole included) and 1–4 cm wide, with lanceolate to more or less elliptic or nearly linear blade (1.6) 2–several times as long as wide and tapering to the petiole; cauline leaves few and distant, only the lowest pair sometimes as large as the basal ones, the upper bractlike and often alternate. Heads 1–3 (5), typically solitary on a long, scapiform peduncle; disk red-purple, 1–2 (2.5) cm wide; involucral bracts greenish, firm, imbricate, 1.5–3.5 (5) mm wide, more or less ciliolate-margined, sometimes also scabrous on the back, the outer merely acute, the inner with longer, loose, slender tip, but hardly surpassing the disk; rays 10–20, mostly 1.5–3.5 cm long. (n = 17) Late summer–fall. Moist or wet pinelands; CP from NC to se La, excl peninsular Fla, but rare in Ga.

32. H. radula (Pursh) T. & G. Fibrous-rooted perennial from a short, erect crown; stems (1–) 2–several, erect above a decumbent base, 0.3–1 m tall, generally spreading-hairy below. Basal leaves rosulate, pilose-hirsute, entire or nearly so, with short-petiolate or subsessile, orbicular or elliptic-orbicular to broadly obovate or subrhombic blade 4–11 cm long and 1–1.6 times as long as wide; cauline leaves few and much reduced, mostly near the base, the upper distant, alternate, and bractlike. Head solitary (2–3) on a long, scapelike peduncle; disk deep purple, 1.5–3 cm wide; involucral bracts firm, imbricate, glabrous to hispid, 2–5 mm wide, more or less acute or the inner acuminate, up to about equaling the disk; ray flowers none, or rarely present and with a short, purple or yellowish-purple ray up to ca. 1 cm long. (n = 17) Fall. Wet or dry pinelands, less often under oaks; CP from s SC to c Fla, w to se La.

13. PHOEBANTHUS Blake

Perennial herbs from rhizomes that are tuberous-thickened at intervals or produce horizontal tubers. Leaves numerous, simple, all cauline, narrow, entire, a few of the lowermost ones opposite, the others alternate. Head solitary and terminating the simple stem, or heads 2–6 and terminating the branches, radiate, the rays relatively large, yellow, neutral; involucral bracts in several subequal or slightly imbricate series, rather narrow, green and herbaceous at least distally, often loose or spreading; receptacle somewhat convex, chaffy throughout, its bracts firm, shortly trilobed or tricuspidate, folded and clasping the achenes. Disk flowers perfect and fertile; style of *Helianthus*. Achenes thick, somewhat compressed at right angles to the involucral bracts, with 2 prominent and generally 2 more obscure angles, varying to almost equably quadrangular. Pappus of 2 short but fairly firm and persistent awn scales from the 2 principal angles of the achene (these awn scales sometimes basally expanded and lacerate, or conversely one or both of them more or less reduced or even obsolete), plus an indefinite number of minute (sometimes scarcely evident), more or less coalescent bristles. (x = 17)

1 Leaves all very slender, mostly 1–2 mm wide .... 2. *P. tenuifolius*.
1 Leaves coarser, at least the larger ones mostly 3–5 mm wide .... 1. *P. grandiflorus*.

1. P. grandiflorus Blake. Plants mostly 5–12 dm tall, averaging coarser than in the next sp.; stem hirsute-puberulent to loosely strigose, often only sparsely so. Leaves linear-oblong or lance-linear, the larger ones commonly 3.5–5 cm × 3–5 mm, pustulate-scabrous above, sparsely hispid-scabrous and often atomiferous-glandular beneath. Involucral bracts averaging shorter, broader, and less spreading than in the next sp.; disk yellow, 1.5–2.5 cm wide; rays 10–15 (21), 2–4 cm long; receptacular bracts subequally cleft, or the central cusp a little the longer. (n = 34) Spring–summer. Sandy pine and oak woods in c peninsular Fla.

2. P. tenuifolius (T. & G.) Blake. Plants slender, mostly 4–10 dm tall. Leaves linear or linear-filiform, mostly 2–8 cm × 1–2 mm, scabrous-hispid and pustulate. Involucral bracts, especially the outer, with elongate, slender, loose or spreading tip; disk yellow or red-purple, 1.2–2.2 cm wide; rays (8) 10–15, 3–4.5 cm long; central cusp of the receptacular bracts more or less distinctly longer and narrower than the lateral ones. (n = 17) Spring–midsummer. Sandy pinelands near the Apalachicola River, Fla.

14. VERBESINA L. FLATSEED SUNFLOWER

Ours herbs. Leaves opposite or alternate, entire to toothed or sometimes pinnately cleft. Heads radiate or sometimes discoid, yellow or less often white, the rays pistillate (and often fertile) or neutral; involucre of subequal or slightly imbricate, often somewhat herbaceous bracts; receptacle chaffy, shortly conic to sometimes merely polsterform or nearly flat; disk flowers perfect and fertile; style branches flattened, with papillate or hairy, acute appendage. Achenes more or less strongly flattened at right angles to the involucral bracts, winged or less often wingless; pappus mostly of 2 short to well-developed awns, occasionally with a few shorter scales or minute awns. *Actinomeris, Phaethusa, Pterophyton, Ridan, Ximenesia*. Our spp. all sharply distinct.

1 Plants evidently perennial; roots fibrous or fleshy-fibrous.
 2 Involucral bracts well developed, appressed or somewhat loose, but not deflexed; achenes not much spreading, not at all reflexed; leaves opposite or alternate.
  3 Heads few (1–20), larger than in the next group, the disk commonly 7–20 mm wide at anthesis, the rays often more than 5; leaves sessile or subsessile; receptacle shortly conic (*Pterophyton*).
   4 Stem generally winged by the decurrent leaf bases.
    5 Stem more or less leafy to the inflorescence, the leaves alternate .... 1. *V. helianthoides*.
    5 Stem subnaked above; principal leaves opposite .... 2. *V. heterophylla*.
   4 Stem wingless, the leaves merely sessile.
    6 Rays yellow; heads 3–15 (20) .... 3. *V. aristata*.
    6 Rays none; heads solitary, rarely 2–5 .... 4. *V. chapmanii*.
  3 Heads numerous, mostly 20–100 or more, small, the disk about 3–7 mm wide at anthesis, the rays 1–5; principal leaves usually evidently petiolate, the petiole sometimes wing-margined; receptacle tiny, merely convex (*Phaethusa*).
   7 Rays yellow; leaves opposite .... 5. *V. occidentalis*.
   7 Rays white; leaves alternate .... 6. *V. virginica*.
 2 Involucral bracts few, narrow, inconspicuous, soon deflexed; achenes spreading in all directions, forming globose heads; leaves alternate (*Actinomeris, Ridan*).
  8 Heads radiate, yellow .... 7. *V. alternifolia*.
  8 Heads discoid, white .... 8. *V. walteri*.
1 Plants annual, taprooted (*Ximenesia*) .... 9. *V. encelioides*.

1. V. helianthoides Michx. Perennial from a short, stout rhizome, 5–12 dm tall, more or less hirsute throughout, especially the lower leaf surfaces, leafy throughout, the upper leaves only gradually reduced. Leaves alternate, mostly 6–15 cm long and 2–6 cm wide, lanceolate to narrowly ovate, sharply acute, serrate, sessile or broadly short-petiolate, the stem winged by their decurrent bases. Heads mostly 1–10 in a rather compact inflorescence, the disk 9–16 mm wide at anthesis; involucral bracts erect but rather loose; receptacle shortly conic; rays 8–15, pistillate or neutral, 1–3 cm long, yellow. Achenes winged or wingless, often a little spreading, but not at all reflexed. (n = 17) Late spring–early fall. Prairies and dry woods; Ozarkian and midw sp., s to GC of La, Miss, and w Ala, and e to Ky, c Tenn, and n Ala, with outliers in BR of NC and CP of Ga. *Pterophyton helianthoides* (Michx.) Alexander—S.

2. V. heterophylla (Chapman) A. Gray. Perennial from a slender rhizome, 4–8 dm tall, sparsely to moderately hispid-scabrous, leafy chiefly below the middle, the stem naked or sparsely bracteate above, the bracts often alternate. Principal leaves opposite, oblong to elliptic or elliptic-ovate, 3–8 cm long and 1–3 cm wide, obtuse or acutish, blunt-toothed, sessile, the stem winged by their decurrent bases. Heads mostly 1–6, on slender peduncles, the disk 8–12 mm wide at anthesis; involucral bracts few, firm, appressed or loose-tipped, often somewhat imbricate; receptacle shortly conic; rays mostly about (5) 8, 1–2 cm long, light yellow, pistillate or neutral. (n = 17) Summer. Dry pine barrens on AC of ne Fla. *Pterophyton heterophyllum* (Chapm.) Alexander—S.

3. V. aristata (L.) A. A. Heller. Fibrous-rooted perennial from a short, stout rhizome, 4–12 dm tall. Herbage hispid or hispid-scabrous. Leaves opposite or seldom alternate, oblong to rather narrowly ovate, 4–8 cm long and 1–3.5 cm wide, serrate, sessile or nearly so, but not decurrent. Heads 3–15 (20) on slender peduncles in a mostly open and nearly naked inflorescence, yellow; disk mostly 7–12 mm wide at anthesis; involucral bracts subherbaceous, shorter than the disk, somewhat loose-tipped; rays mostly (5) 7–10, 1.5–3 cm long; receptacular bracts distally rather pale, often yellowish, not indurated. Achenes with dark body and prominent, pale wings. (n = 17) Summer. Dry woods; AC (mainly GC) of Ga, n Fla, and Ala. *Pterophyton aristatum* (Ell.) Alexander—S.

4. V. chapmanii J. R. Coleman. Fibrous-rooted perennial from a crown or short, stout rhizome, 5–10 dm tall. Herbage scabrous-hispidulous. Leaves opposite or seldom alternate, oblong to lance-elliptic, 4–11 cm long and 1–3 cm wide, firm, toothed, sessile but not decurrent. Head solitary (2–5) on a terminal, subnaked peduncle, discoid, mostly 1–2 cm wide, the flowers conspicuously exserted beyond the firm, distally dark (often anthocyanic) receptacular bracts; involucral bracts firm, green or anthocyanic, shorter than the disk, somewhat loose-tipped. Achenes wholly dark, or the wings sometimes paler. (n = ca. 17) Late spring–midsummer. GC of w Fla. *Pterophyton pauciflorum* (Nutt.) Alexander—S; *V. warei* Gray, often misapplied here, properly a synonym of no. 2.

5. V. occidentalis (L.) Walter. Coarse, leafy-stemmed perennial 1–2 m tall from fibrous or fleshy-fibrous roots. Stem glabrous below, puberulent above. Leaves opposite, ovate, acute or acuminate, serrate, 7–17 cm long and 4–11 cm wide, strigillose to scabrous, the petioles more or less winged, decurrent on the stem. Heads small, numerous, 20–100 or more in a congested to rather open inflorescence, the disk 3–7 mm wide at anthesis; involucral bracts erect, the tips loose, often imbricate, commonly sparsely puberulent; rays 2–5, yellow, usually pistillate and fertile, 0.5–2 cm long; receptacle small, merely convex. Achenes wingless. (n = 17) Late summer–fall. Bottomlands, thickets, woods and waste places, all provinces, throughout southeast except peninsular Fla, and to Ohio, Mo, and Tex. *Phaethusa occidentalis* (L.) Small—S.

6. V. virginica L. Coarse, leafy-stemmed perennial up to 2 m tall, from fleshy-fibrous roots. Stem puberulent. Leaves alternate, ovate to lance-ovate or lance-elliptic, serrate to subentire (but see var. *laciniata*, below), mostly 9–20 cm long and 3–10 cm wide, scabrous to subglabrous above, velutinous or sometimes merely appressed-puberulent beneath, the petioles more or less winged, often decurrent on the stem. Heads small, numerous, 20–100 or more in a dense, generally more or less flat-topped inflorescence, the disk 3–7 mm wide at anthesis; receptacle small, nearly flat; involucral bracts appressed, evidently imbricate, puberulent; rays 1–5, pistillate and fertile, white, less than 1 cm long. Achenes erect, flattened, winged or wingless. (n = 16, 17) Late summer–fall. Bottomlands, thickets, woods, and waste places, all provinces, throughout southeast, and to Mo, Kans, and Tex. The var. *virginica*, as principally described above, is widespread, but not common in Fla. Most Fla plants, and a dribble up AC to SC, belong to the poorly defined var. *laciniata* (Poir.) Gray—R, with some or all leaves pinnately few-lobed or cleft and often larger and scarcely petiolate, up to 30 cm long and 16 cm wide. *Phaethusa laciniata* (Poir.) Small—S.

7. V. alternifolia (L.) Britton. Coarse, homely, leafy stemmed, fibrous-rooted perennial 1–3 m tall. Stem spreading-hirsute to subglabrous, usually winged by the decurrent leaf-bases. Leaves alternate, lanceolate or lance-elliptic to occasionally ovate, usually gradually narrowed to a petiolar base, sharply serrate or subentire, 10–25 cm long and 2–8 cm wide, scabrous-hirsute, especially above. Heads 10–100 or more in an open inflorescence, yellow, the disk 1–1.5 cm wide in flower; involucral bracts few, glabrous or subglabrous, rather small, narrow, soon deflexed, the disk flowers loosely spreading even before anthesis; rays 2–10, neutral, yellow, 1–3 cm long; receptacle polsterform. Achenes spreading in all directions, forming a globose head 8–15 mm thick, broadly winged or sometimes wingless. (n = 34) Late summer–fall. Thickets, woods, and bottomlands; widespread sp. of e and c US, irregularly distributed with us, but rare on CP (except Ark and Va) and in Fla only on the panhandle. *Ridan alternifolia* (L.) Britt.—S. *Actinomeris alternifolia* (L.) DC.—F.

8. V. walteri Shinners. Much like *V. alternifolia*, but the heads discoid, with white flowers. (n = 17) Late summer. Floodplains and other moist, low places on CP from SC to Ga and La, with outliers in PP of NC and OU of Ark. *Ridan paniculata* (Walt.) Small—S.

9. V. encelioides (Cav.) Benth. & Hook. Taprooted, branching annual 2–10 dm, the stem and lower leaf surfaces strigose-canescent, the upper leaf surfaces sometimes greener but still strigose. Leaves alternate (except the lower), the blade mostly ovate or deltoid, sometimes narrowly so, rather coarsely toothed, especially near the base, mostly 4–13 cm long and 2–10 cm wide, the petiole generally a bit shorter than the blade, commonly dilated to form a pair of stipulelike auricles at the base. Heads yellow, long-pedunculate in an open inflorescence, the disk 13–20 mm wide in flower; involucral bracts loose or a little spreading, scarcely imbricate, canescently strigose or hirsute, mostly 1–2 cm long; receptacle polsterform; rays 10–15, pistillate, 1–2.5 cm long. Achenes winged, a little spreading but not reflexed, the fruiting head hemispheric. (n = 17) Spring–fall. Native of Mex and sw US, casually intro. in our range, as in Ark, and from Fla to NC. Ours, as described here, is var. *encelioides*. *Ximenesia encelioides* Cav.—S.

15. SPILANTHES Jacq.

1. S. americana (Mutis) Hieron. Subglabrous to hirsute, fibrous-rooted perennial herb, the stems weak, often rooting at the lower nodes, 2–6 dm tall. Leaves opposite, toothed, ovate or lanceolate, 2.5–9 cm long and 0.5–3.5 cm wide, on petioles 0.5–2.5 cm long. Heads few, naked-pedunculate, radiate, the rays several, pistillate, yellow, 3-toothed, 3–10 mm long; involucral bracts in 1 or 2 series, subherbaceous and subequal; receptacle elongate, conic or subcylindric, its bracts clasping the disk achenes and eventually deciduous with them; disk 4–9 mm wide, elongating conspicuously in fruit, its flowers perfect and fertile; style branches flattened, truncate, exappendiculate. Achenes of the rays 3-angled or tangentially flattened, those of the disk radially flattened, commonly ciliate-margined. Pappus of 1 or 2 very short awns, or none. (n = 13, 25, 26, 39) Summer–fall. Moist woods, streambanks, and bottomlands, somewhat weedy; widespread in tropical Am, n to SC, w Tenn, Ill, and Mo, with us mostly on CP.

16. HELIOPSIS Pers. SUNFLOWER EVERLASTING

Perennial (ours) or sometimes annual herbs. Leaves opposite, petiolate, toothed or entire. Heads 1–several on terminal and axillary peduncles, radiate, the rays pistillate and fertile, yellow, persistent and becoming papery; involucral bracts herbaceous at least distally, in 1–3 subequal series, or the outer ones enlarged and leafy; receptacle convex to conic, chaffy throughout, its bracts concave and clasping, subtending the rays as well as the disk flowers; disk flowers perfect and fertile; style branches flattened, with short, hairy appendage. Achenes nearly equably quadrangular (or those of the rays triangular). Pappus none, or a short irregular crown or a few teeth. (x = 14) Fisher, Richard T. 1957. Taxonomy of the genus *Heliopsis* (Compositae). Ohio J. Sci. 57: 171–191.

1 Robust plants, mostly (4) 8–15 dm tall; principal leaves mostly 7–15 cm long; heads (1–) several, with (8) 10–16 rays (1.5) 2–4 cm long .... 1. *H. helianthoides*.
1 Small, slender plants, mostly 3–8 dm tall; principal leaves mostly 3–8 cm long; heads 1 (–few), with 6–10 (13) rays 1–2 cm long .... 2. *H. gracilis*.

1. H. helianthoides (L.) Sweet. Robust, fibrous-rooted perennial from a caudex or stout rhizome, (4) 8–15 dm tall. Principal leaves ovate or broadly lanceolate, often subtruncate at the base, serrate, mostly 7–15 cm long and 2.5–8 cm wide, on petioles 0.5–3.5 cm long. Heads (1–) several, naked-pedunculate, the disk 1–2.5 cm wide, commonly elongating to 1.5–2 cm in fruit; rays (8) 10–16, (1.5) 2–4 cm long. Achenes essentially glabrous. (n = 14) Midsummer–fall. Rich to dry woods, less often in fields and waste places; widespread in e and c US and adj Can, but seldom on CP southward, and absent from Fla. The common form over most of our range is var. *helianthoides*, with glabrous, often glaucous stem and rather thin leaves smooth on both sides or merely slightly scabrous above. Westward (as in Mo, Ark, and La) this gives way to var. *scabra* (Dunal.) Fern., with firmer leaves scabrous on both sides, the stem often scabrous as well. (*H. scabra* Dunal.)—S.

2. H. gracilis Nutt. Like a small, slender form of *H. helianthoides*, but forming a distinctive taxon. Stems mostly 3–8 dm tall, usually from a slender rhizome. Herbage glabrous to inconspicuously short-hairy or scabrous. Principal leaves lanceolate to deltoid-ovate, strongly to obscurely toothed or subentire, mostly 3–8 cm long and 1–3 cm wide. Heads solitary or sometimes few, relatively small, the disk 7–15 mm wide, scarcely more than 1.2 cm high even in fruit; rays 6–10 (13), 1–2 cm long. Spring–summer. Woods, especially of pine; largely or wholly on GC, from sw Ga and w Fla to Ala and La (where it apparently passes into no. 1). *H. minor* (Hook.) C. Mohr—S.

17. ZINNIA L., nom. conserv. ZINNIA

Annual (our spp.) or perennial herbs or low shrubs. Leaves opposite, sessile or short-petiolate, entire. Heads terminating the stem (and branches), radiate, the rays pistillate and fertile, persistent on the achenes and becoming papery; involucral bracts imbricate in several series, largely chartaceous or coriaceous, stramineous, with a darker band near the rounded erose ciliate tip; receptacle flat to conic (convex to elongate-conic in our spp.) its bracts folded and embracing the achenes; disk flowers perfect and fertile; style branches flattened, with ventromarginal stigmatic lines, in our spp. mostly truncate and penicillate but sometimes also with a short, conical appendage. Achenes compressed along the radii of the head. Pappus of 1–4 often unequal awns, or none. Torres, Andrew M. 1963. Taxonomy of *Zinnia*. Brittonia 15: 1–25.

1 Receptacular bracts with strongly differentiated, fimbriately cleft tip; achenes cartilaginous-winged, awnless .... 1. *Z. elegans.*
1 Receptacular bracts with slightly differentiated, merely erose or toothed tip; achenes wingless, those of the disk with a prominent awn on the inner angle .... 2. *Z. peruviana.*

1. **Z. elegans** Jacq. GARDEN ZINNIA Stout annual 1–20 dm tall, more or less hirsute to strigose (or partly scabrous) throughout. Leaves ovate to oblong, prominently 3 (5)-nerved, up to 10 × 6 cm. Heads hemispheric, many-flowered; rays mostly 8–21, or more numerous in the commonly cult. (and escaped) double forms, 1–3 cm long, red to variously anthocyanic, xanthic, or white; receptacle elongating, becoming elongate-conical, its bracts with prominent, strongly modified, colored (commonly anthocyanic), conspicuously fimbriate tip. Achenes compressed, ciliate on the cartilaginous margins, awnless, but often deeply emarginate at the summit. (n = 12) Summer–fall, or all year southward. Disturbed sites; native of Mex, commonly cult. and casually escaped throughout our range, and naturalized in parts of Fla.

2. **Z. peruviana** (L.) L. Like a smaller (to 1 m tall), more slender form of no. 1, but with the tips of the receptacular bracts only slightly differentiated, merely erose or toothed, and with the achenes compressed but often narrowly 3-angled, wingless, those of the disk with a prominent awn on the inner angle. Heads cylindric to campanulate, fewer-flowered, and commonly with less elongate receptacle; rays 5–15 (21), 0.8–2.5 cm long, typically maroon, varying to scarlet, yellow, or ochroleucous. (n = 12) Summer, or more or less all year. Disturbed sites; native to trop Am and adj sw US, casually intro. in our area from SC to Fla. *Z. pauciflora* L.—S.

18. CALYPTOCARPUS Less.

1. **C. vialis** Less. Slender, taprooted, apparently short-lived perennial. Herbage strigose. Stems several, 1–4 dm long, branched, prostrate and often creeping, or sometimes ascending. Leaves opposite, the blade lance-ovate to deltoid-ovate, toothed, mostly 1.5–6 cm × 1–4 cm, on a petiole 0.5–1.5 cm. Heads yellow, solitary in the axils, sessile or slender-pedunculate, radiate, the rays 3–8, pistillate and fertile, only 1.5–3 mm long; involucre campanulate, of 3–5 broad, herbaceous bracts 5–10 mm high; receptacle small, chaffy, its bracts flat or merely a bit concave; disk flowers perfect and fertile, the corolla 4-toothed; style branches flattened, with ventromarginal stigmatic lines and a slender, hirsutulous appendage. Achenes flattened (at least those of the rays flattened parallel to the involucral bracts), wingless or somewhat thick-winged distally, tuberculate or sometimes smooth, 3.5–5.5 mm long. Pappus a pair of stout, divergent marginal awns 1–2 (4) mm long, sometimes with some shorter intermediate ones. (n = 12, 36) All year. Tropical Am weed, occ. intro. in disturbed sites in our coastal states, as in SC, Fla, Ala, and La.

C. blepharolepis B. L. Robinson, based on a specimen purportedly from Ala, is *Sanvitalia ocymoides* DC., the specimen actually from Brownsville, Tex. *S. ocymoides* is not known in our range.

19. SYNEDRELLA Gaertn., nom. conserv.

1. **S. nodiflora** (L.) Gaertn. Freely branched to subsimple, erect or ascending annual, 1.5–10 dm tall. Herbage strigose. Leaves opposite, the blade ovate to elliptic, mostly 2–10 cm long, toothed, on a petiole up to 2.5 cm long. Heads yellow, mostly sessile or subsessile in small, axillary clusters, radiate, the rays 2–9, pistillate and fertile, only 1–2 mm long; involucre subcylindric to campanulate, 7–9 mm high, 2 or 3 outer bracts more or less herbaceous or herbaceous-tipped, the others passing into the bracts of the receptacle; receptacle small, chaffy, its bracts flat, with evident midvein but not folded; disk flowers 4–12, perfect and fertile, with 4-toothed corolla, the innermost sometimes bractless; style branches flattened toward the base, with short, introrsely marginal stigmatic lines and a long, slender, hirsutulous appendage. Achenes black, 4–5 mm long, dimorphic, those of the rays broad, strongly flattened parallel to the involucral bracts, smoothish, with conspicuous, pale, lacerate wing margins produced into a pair of flattened pappus awns; disk achenes narrower, wingless, muricate or tuberculate, with 2 or 3 pappus awns. (n = 18, 19, 20) All year. Pantropical weed of American origin, casually intro. in Fla.

20. COREOPSIS L. TICKSEED

Herbs (ours). Leaves opposite or less commonly alternate, entire to pinnatifid, ternate, trifoliolate, or dissected. Heads radiate, the rays conspicuous, neutral (ours), yellow (sometimes marked with red-brown toward the base) or less often lavender or pink to white, in our spp. (except *C. latifolia*) commonly 8; involucral bracts biseriate and dimorphic, all joined at the base, the outer narrower, usually shorter, and commonly more herbaceous than the generally membranous and striate inner; receptacle flat or somewhat convex, chaffy, with slender, thin, flat bracts; disk flowers perfect and fertile; style branches flattened, with short or elongate, subtruncate to subcaudate, hairy appendage, commonly without well-marked stigmatic lines. Achenes flattened parallel to the involucral bracts, winged or in a few spp. wingless. Pappus of 2 smooth or upwardly barbed short awns or teeth, or a minute crown, or obsolete. A large genus, poorly delimited from the even larger genus *Bidens*; species with flattened, wingless achenes and reduced pappus should be sought in both genera. Mueller, A. M. 1974. An evolutionary study of *Coreopsis* section Palmatae (Compositae). Ph.D. thesis, Univ. N. Carolina. Parker, H. 1972. A biosystematic study of section *Calliopsis* of *Coreopsis*. Ph.D. thesis, Univ. Arkansas. Smith, E. B. 1973. A biosytematic study of *Coreopsis saxicola* (Compositae). Brittonia 25: 200–208. Smith, E. B. 1975. A biosystematic survey of *Coreopsis* in eastern United States and Canada. Sida 6: 123–215.

1 Style appendages evidently acute, often cuspidately so; disk corollas mostly 5-toothed; achenes winged or wingless, the wings (except sometimes in *C. grandiflora*) entire or nearly so, not fimbriate-lacerate; plants nearly always perennial.

2 Leaves large (mostly 10–25 × 3.5–10 cm), toothed but not pinnatifid; achenes wingless . . . . 1. *C. latifolia*.

2 Leaves smaller (either shorter, or narrower, or both), entire or pinnatifid or ternate, but not toothed; achenes winged.

3 Receptacular bracts chaffy-flattened below, caudate-attenuate above; leaves simple or pinnatifid.

4 Plants with clustered stems, not stoloniferous; wings of the achenes thin, not strongly incurved.
5 Stems leafy almost to the top; peduncles mostly 0.5–2 dm long, less than half as long as the leafy part of the stem.
6 Leaves relatively broad, mostly (1) 1.5–4 cm wide, entire or often with 1 (2) pair of much smaller but similarly shaped pinnae at the base .... 2. *C. pubescens.*
6 Leaves pinnately parted into narrow segments mostly less than 1 cm wide .... 3. *C. grandiflora.*
5 Stems leafy chiefly toward the base, the long, naked peduncles about as long as or longer than the leafy part of the stem, seldom only half as long .... 4. *C. lanceolata.*
4 Plants spreading by elongate, slender stolons or superficial rhizomes; wings of the achenes finally strongly incurved and callous-thickened .. 5. *C. auriculata.*
3 Receptacular bracts linear or clavate, blunt to acutish; leaves mostly ternately cleft or compound, or sometimes entire.
7 Leaves divided to the base, essentially trifoliolate, or rarely simple and entire.
8 Leaflets or leaf segments (or entire leaves) relatively broad, mostly 5–30 mm wide, generally entire.
9 Leaves evidently petiolate, with 3 leaflets, or the upper (rarely all) entire .... 6. *C. tripteris.*
9 Leaves essentially sessile, the leaflets commonly appearing as whorled leaves .... 7. *C. major.*
8 Leaf segments relatively narrow, seldom more than ca. 5 (9) mm wide; generally at least some of the leaves with the central segment again cleft.
10 Disk yellow, sometimes turning brown in age or in drying; plants with well-developed creeping rhizomes.
11 Leaf segments broadly linear, mostly (1.5) 2–5 (–9) mm wide ..... 8. *C. delphinifolia.*
11 Leaf segments narrowly linear or filiform, mostly 0.3–1 mm wide 9. *C. verticillata.*
10 Disk dark red-purple; plants without creeping rhizomes .... 10. *C. pulchra.*
7 Leaves trifid distinctly above the base, the segments mostly 2–7 mm wide 11. *C. palmata.*
1 Style appendages short and blunt, broader than long, or nearly obsolete; disk corollas mostly 4-toothed (except in *C. basalis*); achenes winged or wingless, the wings often fimbriate-lacerate; plants annual, biennial, or perennial.
12 Achenes winged or wingless, the wings entire or nearly so, not fimbriate-lacerate; leaves pinnatifid or entire.
13 Rays yellow, or marked with red-brown at the base; annual to short-lived perennial, not rhizomatous.
14 Disk corollas 5-toothed; outer involucral bracts more or less elongate, half to fully as long as the inner .... 12. *C. basalis.*
14 Disk corollas 4-toothed; outer involucral bracts shorter, up to ca. half as long as the inner.
15 Pappus of definite short awns mostly (0.3–) 0.5–1 (–1.4) mm long; rays usually wholly yellow, seldom with a small basal red-brown spot .... 13. *C. leavenworthii.*
15 Pappus obsolete or nearly so, the awns, if present, hardly more than 0.2 mm long; rays usually marked with red-brown at the base, seldom wholly yellow .... 14. *C. tinctoria.*
13 Rays pink to sometimes white; rhizomatous perennial .... 15. *C. rosea.*
12 Achenes with deeply fimbriate-lacerate or deeply scalloped wings; leaves entire, or in *C. gladiata* seldom with a single pair of lateral lobes.
16 Rays yellow; leaves flat (though often fleshy-thickened), not juncoid.
17 Leaves relatively narrow and basally disposed, those 1–2 dm above the base with the blade mostly 5–10 times as long as wide (often hardly distinguished from the petiole) and seldom much more than 1 cm wide; plants glabrous .... 16. *C. gladiata.*
17 Leaves broader in shape and not so clearly (or not at all) basally disposed, those 1–2 dm above the base commonly with the blade 1–4 cm wide and 1.5–5 times as long as wide; plants glabrous, or often with the leaves hirsute on both surfaces, the stem sometimes also hirsute.
18 Leaves alternate; petioles not ciliate toward the base .... 17. *C. helianthoides.*
18 Leaves opposite; petioles evidently ciliate toward the base .... 18. *C. integrifolia.*
16 Rays anthocyanic; leaves terete, juncoid .... 19. *C. nudata.*

1. C. latifolia Michx. Rhizomatous perennial 7–15 dm tall, glabrous or somewhat hairy, leafy throughout. Leaves short-petiolate or subsessile, veiny, lance-ovate or lance-elliptic to ovate to elliptic, mostly 10–25 × 4–10 cm, acuminate, ciliate and sharply serrate. Heads several or many, narrow; outer involucral bracts spreading, linear-oblong, half to nearly or fully as long as the inner; rays 4–5, yellow, 1–2 cm long; receptacular bracts linear, obtuse; disk flowers few (ca. 10–18), yellow, 5-toothed; style appendages cuspidately acute. Achenes 6–8 mm long, lance-oblong, wingless, truncate and naked at the narrow summit. (n = 13) Late summer. Rare and local in rich, moist woods at upper elev. in BR of NC and adj. SC and n Ga.

2. C. pubescens Elliott. Clustered perennial 6–12 dm tall from a fibrous-rooted crown or caudex, shortly spreading-hairy or occasionally glabrous, leafy throughout. Leaves relatively broad, mostly elliptic or ovate, short-petiolate, 4–10 × (1–) 1.5–4 cm, entire or very often with 1 (2) pair of much smaller but similarly shaped pinnae at the base. Heads few to several on naked peduncles 0.5–2 dm long; outer involucral bracts narrowly lance-triangular, generally at least half as long as the inner; rays yellow, 1.–2.5 cm long; receptacular bracts flat, chaffy below, caudate-attenuate above; disk flowers yellow, 5-toothed; style appendages lance-triangular, acute. Achenes 2.5–3 mm long, with spreading, thin wings. Pappus commonly of 2 short, chaffy teeth. (n = 13, 14) Summer. Chiefly in woods, especially in sandy soil; Va to n Fla, w to Mo, Okla, Ark, and La.

3. C. grandiflora Hogg. Similar to *C. lanceolata*, but often taller (to 1 m). Stems leafy nearly to the summit, the 1–many slender peduncles 0.5–2 dm long, seldom more than half as long as the leafy part of the stem. Leaves mostly pinnatifid into linear-filiform to narrowly lanceolate segments (or the lowermost entire), the lateral lobes rarely more than 5 mm wide, the terminal one often a little wider (to 1 cm). Heads averaging a little smaller. (n = 13, 14, 15; i.e., 2n = 26 + 0–4 B). Rather dry, often sandy places; Ark (OU, OZ) to Mo, Kans, Okla, and Tex; n of CP in Ala and Ga; scattered elsewhere (perhaps only intro.) as in La, Miss, NC and SC. Two morphologically and ecologically distinctive but wholly confluent vars: Var. *grandiflora*, as principally described above, has the wings of the achene entire; its petioles are generally ciliate, and the stems are usually green; it occurs throughout the range of the species, sometimes even in the specialized habitat of the next var., and blooms mainly in the spring. The var. *saxicola* (Alexander) E. B. Smith differs most notably in having the wings of the achenes deeply fimbriate-lacerate; the plants are usually essentially glabrous, even the petioles nonciliate; the stem (except in the Stone Mt. population) is commonly deep purple; in the Stone Mt. population the leaf segments average a little wider, sometimes more than 1 cm, and are sometimes a little hairy on the surface, perhaps reflecting hybridization with *C. pubescens*; var. *saxicola* occurs in pockets of thin soil on rocks of PP of Ga and Ala, and OU of Ark, most notably on Stone Mt. and the granite flatrocks of Ga, and blooms mainly in the summer. *C. saxicola* Alexander—S.

4. C. lanceolata L. Clustered perennial 2–7 dm tall from a short caudex, glabrous or spreading-villous. Stems leafy below, elongate and naked above. Leaves spatulate to linear or lance-linear, simple and entire or with 1 or 2 pairs of small lateral lobes, the lower long-petiolate, mostly 5–20 cm long overall and 0.5–2 cm wide, the others reduced and sessile or nearly so. Heads few or solitary on long (mostly 2–4 dm) naked peduncles, the stout peduncles usually equaling or longer than the leafy part of the stem, seldom only half as long; outer involucral bracts lanceolate to oblong-ovate, more or less scarious-margined, 5–10 mm long; rays yellow, 1.5–3 cm long; disk yellow, 1–2 cm wide, the corollas 5-toothed; receptacular bracts flat and chaffy below, caudate-attenuate above; style appendages cuspidately acute. Achenes 2–3 (–4) mm long, with thin, flat wings. Pappus of 2 short, chaffy teeth. (n = 12, 13, 24) Spring. Dry, often sandy places, especially in disturbed soil; probably originally of c and sw US, but widely escaped from cult. and established throughout most of our range s to La and w and n Fla. *C. crassifolia* Ait.—S; *C. heterogyna* Fern.—F.

C. nuecensis A. A. Heller, a taprooted annual sp. of Tex with the rays marked with maroon near the base, rarely escapes from cult., but apparently does not persist. It is much like *C. auriculata* in aspect (but without the stolons), but has achenes more nearly like those of *C. lanceolata*.

5. C. auriculata L. Perennial with slender, naked stolons or stoloniform rhizomes. Stems erect or ascending, spreading-hairy or glabrate, 1–6 dm tall, leafy below. Leaves petiolate, the blade ovate to broadly elliptic or suborbicular, up to 8 × 3.5 cm, often with a pair of much smaller lateral lobes at the base, glabrous or appressed hairy. Heads solitary or few on long, naked peduncles; outer involucral bracts lance-oblong or ovate-oblong, often narrowly thin-margined, hispid-ciliate or glabrous; rays yellow, 1.5–2.5 cm long; disk yellow, 7–15 mm wide, the corollas 5-toothed; receptacular bracts narrowly linear, attenuate upward; style appendages cuspidately acute. Achenes 2–3 mm long, the rather narrow wings finally cartilaginous and involute. Pappus of 2 minute, deciduous scales. (n = 12) Spring. Woods; Va (AC) to Ky (IP), s to CP of Ga, Ala, Miss, and La.

6. C. tripteris L. Single-stemmed perennial mostly 1–3 m tall from a short, stout rhizome, usually glabrous and somewhat glaucous, but occasionally short-hairy. Leaves mainly cauline, numerous, with evident petiole 0.5–4 cm long, mostly trifoliolate (or the upper entire), the leaflets rather firm, lanceolate or narrowly elliptic, evidently penniveined, 5–10 (–13) cm × 6–25 (–30) mm, the terminal one often again divided. Heads several or many in an open inflorescence; outer involucral bracts linear-oblong, scarcely half as long as the inner; rays yellow, 1–2.5 cm long; disk yellow, becoming purplish or deep red, its corollas 5-toothed; receptacular bracts narrowly linear or linear-clavate; style appendages cuspidately acute. Achenes obovate, 4–7 mm long. Pappus of a few minute, erect bristles, sometimes also with 2 short, upwardly barbed awns. (n = 13) Late summer–fall. Mostly in moist or wet low places and in woods; Mass and s Ont. to Wis, s to n Fla, Miss, La, and Tex.

7. C. major Walter. Much like *C. tripteris*, but smaller, 5–10 dm tall, vigorously rhizomatous but the stems commonly tufted, the herbage often short-hairy. Leaves sessile, trifoliolate (rarely simple and entire), the leaflets mostly lance-elliptic to nearly ovate and 3–8 × 1–3 cm (narrower in var. *rigida*), rather inconspicuously veined; outer involucral bracts longer and often more foliaceous, sometimes equaling the inner. (n = 13, 39, 52) Summer. Mostly in dry, open woods; s Pa and s O to n Fla, s Miss, and se La. Var. *major*, as principally described above, occurs throughout the range of the sp. (n = 13, 39, 52). Var. *rigida* (Nutt.) Boynton, with firmer, scarcely veiny, notably narrower leaflets mostly 5–10 (12) mm wide, often occupying more exposed or less favorable sites, is frequent in SC (AC, PP, BR) and occasionally extends w to nw Ga and n to Va; an outlying station in RV of Tenn. Var. thought to have originated through introgression of *C. delphinifolia* into octoploid forms of *C. major* var. *major*; extreme forms approach *C. delphinifolia*. (n = 52).

8. C. delphinifolia Lam. Vigorously rhizomatous perennial 3–9 dm tall, commonly hispidulous at the nodes, otherwise glabrous or nearly so. Stem leafy, the leaves sessile, ternately parted to the base into linear or broadly linear segments mostly (1.5) 2–5 (9) mm wide, at least some of the leaves generally with the central segment again trifid near or somewhat below the middle or asymmetrically bifid. Heads several or rather numerous, the disk ca. 1 cm wide or less, yellow, often turning brown in drying; outer involucral bracts more or less herbaceous, linear, 3–7 mm long; rays 1.5–3 cm long; receptacular bracts linear-clavate, obtuse or acutish; disk corollas 5-toothed; style appendages cuspidately acute. Achenes 4.5–6 mm long. Pappus a pair of minute, fimbriate scales. Species thought to be an allopolyploid of nos. 7 and 9. (n = 52) Summer. Pine woods, savannas, and moist swales on AC, barely encroaching on PP; SC and Ga, and reputedly to Ala; disjunct in BR of Tenn.

9. C. verticillata L. Perennial, 3–9 dm tall, from slender, yellow, more or less elongate rhizomes; plants often hispidulous at the nodes, otherwise essentially glabrous. Stem leafy, the leaves sessile, ternately parted to the base, at least the central segment pinnatifid or again ternate, the ultimate segments linear-filiform, 0.3–1 mm wide, some of the secondary segments generally arising well above the base of the primary ones, often above the middle. Heads on very slender and rather short peduncles, the disk 5–10 mm wide, yellow to dark purple; outer involucral bracts oblong or linear-oblong, herbaceous, 3–7 mm long; rays 1–2.5 cm long; receptacular bracts linear-clavate, acutish; disk corollas dark reddish-purple distally, 5-toothed; style appendages cuspidately acute. Achenes 3–5 mm long, narrowly winged. Pappus obsolete. (n = 13, 26, 39) Summer. Dry, open woods; AC of MD and DC, s to AC of NC and PP of SC, w to BR of Va and NC; cult. and escaped elsewhere.

10. C. pulchra F. E. Boynton. Much like *C. verticillata* in aspect, but the aerial stems clustered on a compact set of underground, very short and cormose-thickened, densely rooting stems, the plants not otherwise rhizomatous; leaves softer and the segments somewhat wider, commonly 1.4–2.4 mm wide, the primary segments branched only below the middle, often only near the base; disk dark red-purple. (n = 13) Late summer. Sandstone outcrops on Lookout Mt. (CU) in ne Ala and nw Ga.

11. C. palmata Nutt. Rhizomatous perennial 4–9 dm tall, glabrous except for the scabro-ciliate leaf margins and sometimes hispidulous nodes. Leaves wholly cauline, rather numerous, firm, narrow, 3–8 cm long, essentially sessile, deeply trifid near or somewhat below the middle, the lobes linear-oblong, 2–7 mm wide, the central one sometimes again lobed. Heads few or solitary, short-pedunculate, the disk 8–15 mm wide, yellow; outer involucral bracts linear-oblong, nearly or fully as long as the inner; rays 1.5–3 cm long; receptacular bracts linear-clavate, acutish; disk corollas 5-toothed; style appendages cuspidately acute. Achenes 5–6.5 mm long, narrowly winged. Pappus obsolete or of 2 callous teeth. (n = 13) Summer. Prairies and open woods; Ark (OU, OZ) and n La to Mich, Man, e Kans, and e Tex.

12. C. basalis (Otto & Dietr.) Blake. Taprooted annual or biennial, single-stemmed and erect, or several-stemmed and basally decumbent, 1–6 dm tall, loosely villous-hirsute to subglabrous. Leaves petiolate or subsessile, pinnatifid or pinnately compound, the middle and lower ones mostly with elliptic to ovate or subrhombic terminal segment 1–3 × 0.5–1.5 cm and 1–3 pairs of smaller but similarly shaped lateral segments (these sometimes again cleft), or sometimes most of them simple and entire; upper leaves often with narrower segments. Heads terminating the branches, on naked peduncles mostly 0.5–1.5 dm long, the disk dull red to dark purple, ca. 1 cm wide or less; outer involucral bracts lance-linear, subherbaceous, half to fully as long as the inner; rays yellow, usually marked with red-brown at the base, 1–2.5 cm long; receptacular bracts chaffy flattened below, tapering to an often long and slender but generally somewhat clavate tip; disk corollas 5-toothed; style appendages short and blunt. Achenes 1–2 mm long, with thickened, incurved marginal ribs, but hardly winged. Pappus of 2 minute scales, or obsolete. (n = 13) Spring. Open, sandy places; native to Tex and sw La, but locally intro., sometimes as an escape from cult., on CP e to Fla and n to NC. *C. drummondii* (D. Don) T. & G.—S.

13. C. leavenworthii T. & G. Taprooted annual or short-lived, more fibrous-rooted, usually single-stemmed perennial, mostly (3–) 5–15 dm tall, essentially glabrous. Leaves slender, sometimes all entire and linear to oblanceolate, more often the larger ones irregularly pinnatifid into a few narrow segments less than 1 cm wide. Heads (1–) several–many, terminating the slender branches, the disk dark purple or maroon, ca. 1 cm wide or less; outer involucral bracts lance-oblong to narrowly ovate, up to half as long as the inner; rays 1–2 cm long, yellow, seldom with a small, basal, red-brown spot; receptacular bracts chaffy-flattened below, tapering to a long, slender, but often distally slightly expanded tip; disk corollas 4-toothed; style appendages short and blunt. Achenes 2–3.5 mm long, broadly winged. Pappus awns (0.3–) 0.5–1 (–1.4) mm long. (n = 12) All year, but mostly spring. Mostly in moist, low, sandy, open places; throughout Fla. *C. lewtonii* Small—S.

14. C. tinctoria Nutt. Similar to *C. leavenworthii*; rays usually bicolored, with a reddish-brown basal blotch, seldom wholly yellow; achenes wingless or narrowly to broadly winged; pappus obsolete, or of minute awns up to ca. 0.2 mm long. (n = 12, 24) Summer. Moist, low places and disturbed sites; primarily c and s Great Plains and Tex, e to Ark and La, but widely cult., escaped, and irregularly established elsewhere. *C. cardaminaefolia* (DC.) T. & G.—R, S; *C. stenophylla* F. E. Boynton—S.

15. C. rosea Nutt. Glabrous, rhizomatous perennial 2–6 dm tall. Leaves 2–5 cm long, entire or occasionally some irregularly lobed. Heads (1–) several or many, rather shortly pedunculate at the branch tips, the disk yellow, ca. 1 cm wide or less; outer involucral bracts lance-triangular, mostly less than half as long as the inner, eventually spreading; rays pink (white), ca. 1 cm long; receptacular bracts linear, acute to obtuse; disk corollas 4-toothed; style appendages short and blunt. Achenes narrow, wingless, ca. 2 mm long. Pappus a minute cup, or nearly obsolete. (n = 13) Late summer. Wet, often sandy or acid soil, or in shallow water; AC of Del, Ga, Md and SC, n in coastal states to Mass; NS.

16. C. gladiata Walter. Short-lived, fibrous-rooted, glabrous perennial mostly 5–12 dm tall. Leaves fleshy-firm, entire (or the lower occasionally with a pair of slender lateral lobes), basally disposed, the lower usually persistent, oblanceolate or spatulate, conspicuously long-petiolate, up to 30 × 3.5 cm, or much smaller; leaves 1–2 dm above the base with the blade mostly 5–10 times as long as wide (often hardly distinguished from the petiole) and seldom much more than 1 cm wide, the middle ones linear, the upper mere tiny bracts. Heads several, the disk 6–18 mm wide, dark purplish; outer involucral bracts broadly lance-triangular to ovate, up to half as long as the inner, membranous and often striate-veiny; rays yellow, mostly 1–2.5 cm long; receptacular bracts linear, acute; disk corollas 4-toothed; style appendages short and blunt. Achenes oblong, 2–4.5 mm long, the wings deeply and subpectinately lacerate. Pappus of 2 short, upwardly barbed awns. (n = 13, 26, 78) Spring–fall, or all year southward. Wet, often acid places; CP from s Va to s Fla, and w to Tex. Two vars., well marked in the extreme forms but wholly intergradient and with widely overlapping ranges: Var. *gladiata*, with the leaves not dark-dotted, all or nearly all of them alternate, the lower conspicuously elongate and mostly 1–3.5 cm wide, occurs from NC to s Fla; and w occ. to Miss. *C. falcata* F. E. Boynton—R, S; *C. longifolia* Small—S. Var. *linifolia* (Nutt.) Cronq., with the leaves minutely darkdotted in transmitted light in life, many or all the leaves opposite, the lower up to ca. 1 cm wide, not greatly elongate, occurs from se Va to n Fla, and w to Tex. *C. angustifolia* Ait.—R, S; *C. oniscicarpa* Fern.—F.

17. C. helianthoides Beadle. Short-lived, fibrous-rooted, commonly single-stemmed perennial mostly 3–10 dm tall. Leaves alternate, firm, cartilaginous-margined, glabrous or often hirsute on both sides (but the petiole nonciliate at least toward the base), not so clearly basally disposed as in *C. gladiata*, the lowermost ones often poorly developed but varying to large and elliptic (to 12 × 6 cm), those 1–2 dm above the base commonly with lanceolate to ovate or elliptic blade 4–15 × 1–4.5 cm, 2–5 times as long as wide, the upper or middle and upper ones reduced and distant. Heads (1–) several, essentially similar to those of *C. gladiata*, but the wings of the achenes more nearly scalloped-lacerate. Mostly fall. Peat bogs and other wet, low places; CP from c and w Fla to se Miss and n to NC, and disjunct in BR of NC.

18. C. integrifolia Poiret. Slender, fibrous-rooted, rhizomatous perennial 3–10 dm tall. Stem hirsute, especially below the nodes, or glabrous. Leaves opposite, largely or wholly cauline, with lance-ovate to elliptic or ovate, glabrous or hirsute blade mostly 2–6 × 1–2.5 cm, the rather short petioles hirsute-ciliate toward their connate bases. Heads few or solitary, essentially similar to those of *C. helianthoides*. (n = 26) Late summer. Moist, low places; SC (AC), Ga (CP, PP), and Fla panhandle.

19. C. nudata Nutt. Glabrous and glaucous, fibrous-rooted perennial from a short stout rhizome or crown, mostly 4–12 dm tall. Leaves somewhat basally disposed, alternate, linear-terete, junciform, the lower ones 1–4 dm long and 1–2 mm thick, the others few and progressively reduced. Heads 1–10, the disk 1–1.5 cm wide, yellow with dark anthers; outer involucral bracts lance-triangular to lance-ovate, membranous and striate, up to ca. half as long as the inner; rays 1.5–3 cm long, pale to dark lavender or pink; receptacular bracts broadly linear or linear-oblong, blunt or erose; disk corollas 4-toothed; style appendages short and blunt. Achenes 2.5–3.5 mm long, with subpectinately scalloped wings. Pappus a pair of upwardly barbed awn scales 1–1.5 mm long. (n = 13) Spring. Wet pinelands and shallow standing water; CP of Ga and n Fla to se La.

21. BIDENS L. BEGGAR TICKS

Herbs (ours), our spp. commonly weedy, often in wet places. Leaves opposite, simple to pinnately or ternately compound or dissected. Heads radiate to disciform or discoid, the rays when present sterile, generally neutral, yellow or less often white or pink; involucral bracts biseriate and dimorphic, the outer more or less herbaceous, often enlarged and leafy, the inner membranous, often striate; receptacle flat or a little convex, chaffy, its bracts narrow, flat or nearly so; disk flowers perfect and fertile; style branches flattened, with externally hairy, usually short appendage, commonly without well-marked stigmatic lines. Achenes flattened parallel to the involucral bracts, but not (or scarcely) winged, varying to sometimes almost regularly tetragonal. Pappus of mostly 2–4 awns or teeth, commonly retrorsely barbed, but sometimes antrorsely barbed or even barbless, or in a few spp. obsolete.

1 Achenes linear, not widened upward, sometimes narrowed above; rays white or ochroleucous or none.
  2 Leaves simple or pinnately 3–5-parted or compound, with rather finely toothed blade or leaflets; rays evident and white, or very short, or none .... 12. *B. pilosa*.
  2 Leaves mostly 2–3 times pinnately dissected, the ultimate segments entire or few-toothed; rays very short and inconspicuous .... 13. *B. bipinnata*.
1 Achenes otherwise, either distinctly broader, or cuneate and tapering to the base; rays yellow or none.
  3 Leaves simple, though sometimes rather deeply 3 (–7)-cleft.
    4 Leaves (except sometimes the lower) sessile by a broad or narrow base; heads often nodding in age; achenes narrowly cuneate, mostly 2.5–4 times as long as wide.
      5 Rays up to ca. 1.5 cm long, or often wanting; receptacular bracts yellowish at the tip .... 1. *B. cernua*.
      5 Rays 1.5–3 cm long; receptacular bracts reddish at the tip .... 2. *B. laevis*.
    4 Leaves with a more or less distinct (sometimes winged) petiole 1–4 cm long; heads erect; achenes various.
      6 Heads discoid or very shortly radiate, the rays, if present, less than 5 mm long.
        7 Heads narrow, with mostly 8–30 flowers; estuarine sp. .... 4. *B. bidentoides*.
        7 Heads broader, at least the terminal one with 30–150 flowers; widespread .... 3. *B. tripartita*.
      6 Heads with well developed rays mostly 1–2 cm long .... 5. *B. mitis*.
  3 Leaves once to thrice pinnately divided or compound (or trifoliolate), the terminal segment or leaflet often stalked.
    8 Heads with well-developed rays ca. 1 cm long or more.
      9 Achenes relatively small, mostly 2.5–4.5 (5) mm long, scarcely ciliate; pappus of 2 short teeth 0.5–1 mm long, or obsolete; outer involucral bracts mostly 7–10 .... 5. *B. mitis*.
      9 Achenes larger, mostly 5–9 mm long, ciliate on the margins; pappus and involucral bracts various.
        10 Achenes relatively broad, mostly obovate or elliptic-obovate or broadly cuneate, 1.5–2 (2.5) times as long as wide.
          11 Outer involucral bracts mostly ca. 8 (–10), shorter than the inner ... 6. *B. aristosa*.
          11 Outer involucral bracts mostly 12–21, commonly longer than the inner .... 7. *B. polylepis*.
        10 Achenes narrowly cuneate-oblong (or the inner cuneate-linear), mostly 2.5–4 times as long as wide .... 8. *B. coronata*.
    8 Heads discoid, or with very short rays less than 5 mm long.
      12 Outer involucral bracts 5–16 or more, ciliate at least near the base.
        13 Outer involucral bracts mostly 10–16 (21), typically ca. 13 .... 9. *B. vulgata*.
        13 Outer involucral bracts mostly 5–10, typically ca. 8 .... 10. *B. frondosa*.
      12 Outer involucral bracts 3–5, not ciliate .... 11. *B. discoidea*.

1. B. cernua L. Annual, 1–10 dm tall, glabrous or the stem scabrous-hispid. Leaves sessile (the lower sometimes tapering to a narrow base), often connate, simple, lance-linear to lance-ovate or lance-elliptic, coarsely serrate to subentire, mostly 4–20 × 0.5–4.5 cm. Heads hemispheric, many-flowered, the disk mostly 12–25 mm wide, commonly nodding in age; outer involucral bracts 5–8, lance-linear, unequal, usually rather leafy and spreading or reflexed, commonly surpassing the disk; rays (6–) 8, yellow, up to ca. 1.5 cm long, or wanting; receptacular bracts usually yellowish toward the tip. Achenes narrowly cuneate, compressed-quadrangular, mostly 5–8 mm long and 2.5–4 times as long as wide, the margins tending to be thickened, cartilaginous and pale at maturity, as also the finally convex summit. Pappus of 4 retrorsely barbed awns. (n = 12, 24) Summer–fall. Low, wet places, sometimes in shallow water; widespread in N Temp regions, in our range mostly avoiding CP.

2. B. laevis (L.) B. S. P. Much like no. 1. Plant wholly glabrous, sometimes perennial. Heads not so consistently nodding; outer involucral bracts scarcely leafy, seldom surpassing the disk; rays 1.5–3 cm long; receptacular bracts reddish toward the tip. Achenes 3–4 angled, or often flat. Pappus of 2–4 retrorsely barbed awns. (n = 11, 12) Summer–fall, or all year southward. Low, wet places, sometimes in shallow water; NH to Fla, Tex, and tropical Am, mainly coastal, but occasionally inland, as in Powell Co., Ky. *B. nashii* Small—S.

3. B. tripartita L. Annual, glabrous or nearly so, 1–20 dm tall. Leaves simple, serrate, sometimes rather deeply 3 (–7)-cleft, 3–15 cm long, up to 4 cm wide, on a more or less evident, sometimes winged petiole up to 3 cm long. Heads erect, discoid or with rays up to ca. 4 mm long, broadly campanulate to hemispheric, at least the terminal one generally with more than 30 flowers, the disk 8–20 mm wide; outer involucral bracts 4–9, herbaceous, often enlarged and leafy. Achenes cuneate or obovate-cuneate, flat or compressed-quadrangular, glabrous or short-hairy, often tuberculate, commonly with a sharp median rib on at least one face, 4–8 mm long. Pappus of 2–4 retrorsely or seldom antrorsely barbed awns. (n = 12, 24, 36, based on Old World material) Summer–fall. A more or less cosmopolitan weed of waste places, in our range seldom s of NC and Tenn. *B. comosa* (A. Gray) Wieg.—F, S; *B. connata* Muhl.—F, S. European plants more often have tripartite leaves than our plants, but are otherwise hardly to be distinguished.

4. B. bidentoides (Nutt.) Britton. Much like no. 3, but the heads narrower, with mostly 8–30 flowers. Achenes narrowly linear-cuneate, appressed-hispidulous, 6–13 × 1 mm, the median ribs obscure or none. Pappus of 2 (4) retrorsely barbed awns. Summer–fall. Estuaries; Md to NY. *B. mariana* Blake—F.

5. B. mitis (Michx.) Sherff. Annual, 2–10 dm tall, glabrous or nearly so. Leaves 4–12 cm long including the 0.3–3 cm petiole, highly variable in form, pinnately compound to more often pinnately 3–7-parted with the terminal segment the largest, varying to simple, ovate, and merely toothed, or even narrow and subentire. Heads rather small, the disk ca. 1 cm wide or less; outer involucral bracts 7–10, linear or linear-spatulate, 5–10 mm long, mostly short-ciliate, glabrous on the back, or the involucre sparsely hispid at the base; rays ca. 8, yellow, 1–2 cm long. Achenes rather broadly cuneate, 2.5–4.5 (5) mm long, glabrous or nearly so, scarcely ciliate. Pappus of 2 sharp, antrorsely barbed teeth 0.5–1 mm long, or obsolete. (n = 12) Summer–fall, or all year southward. Swamps and other wet places; CP from Md to Fla and Tex.

6. B. aristosa (Michx.) Britton. Annual or biennial, 3–15 dm tall, glabrous or slightly hairy. Leaves pinnately or bipinnately divided or compound, 5–15 cm long including the 1–3 cm petiole, the segments lanceolate or lance-linear, acuminate, incised-serrate or pinnatifid. Heads radiate, the rays ca. 8, 1–2.5 cm long, the disk 8–15 mm wide; outer involucral bracts mostly ca. 8 (–10), linear, 5–12 mm long, mostly shorter than the inner, the back glabrous or finely short-hairy, the margins smooth or moderately ciliate. Achenes flat, 5–7 mm long, mostly 1.5–2 (2.5) times as long as wide, strigose, the margins antrorsely ciliate and commonly narrowly and interruptedly thickish-winged. Pappus of 2 (–4) antrorsely or retrorsely barbed awns, or none. Summer–fall. Wet places; widespread in e and c US, and nearly throughout our range except Fla and Ga; thought to be only intro. in e part of range.

7. B. polylepis Blake. Much like no. 6, but the outer involucral bracts mostly 12–21, 10–25 mm long, mostly surpassing the inner, conspicuously hispid-ciliate and often also coarsely short-hairy on the back. Pappus more commonly short or obsolete than in no. 6. (n = 12) Summer–fall. Wet places; sp. of Ozark, plains, and prairie region, in our range native to Ark and La, and casually intro. e to Del, Va, and NC. *B. involucrata* (Nutt.) Britt.—S, a preoccupied name.

8. B. coronata (L.) Britton. Glabrous annual or biennial, 3–15 dm tall. Leaves up to 15 cm long, pinnately parted, with mostly 3–7 lance-linear to linear, incised-dentate to entire, acute or acuminate segments; petiole 0.3–1.5 cm long. Heads radiate, with ca. 8 rays 1–2.5 cm long, the disk 8–15 mm wide; involucre glabrous or nearly so, the 6–8 outer bracts linear or linear-spatulate, sometimes ciliate-margined, very rarely exceeding the disk. Achenes flat or nearly so, narrowly cuneate-oblong or the inner cuneate-linear, mostly 5–9 mm long and 2.5–4 times as long as wide, dark, smooth or hairy, the margins antrorsely ciliate. Pappus of 2 short, erectly setose strong awns or awn scales, generally 1–2 mm long. Summer–fall. Wet places; widespread in ne US, s to NC, Ky, and Neb. Reports from farther south need verification.

9. B. vulgata Greene. Much like no. 10, averaging a little more robust and larger-headed, the herbage glabrous to densely villous-puberulent. Outer involucral bracts in well-developed heads 10–16 (21), typically 13, averaging a little more leafy; disk yellow. Achenes sometimes as much as 12 mm long, more commonly olivaceous or somewhat yellowish, sometimes darker as in *B. frondosa*. Summer–fall. Wet or occasionally dry waste places; widespread in n US and adj Can, in our range s to NC (chiefly mt. provinces) and Ky.

10. B. frondosa L. Annual, generally glabrous or nearly so, 2–12 dm tall. Leaves on petioles 1–6 cm long, pinnately compound, with 3–5 lanceolate, acuminate serrate leaflets up to 10 × 3 cm, at least the terminal one petiolulate. Heads campanulate to hemispheric, or narrower in depauperate plants, discoid or nearly so, the disk up to 1 cm wide in flower; outer involucral bracts 5–10, typically 8, green and more or less leafy, usually conspicuously surpassing the orange disk, evidently ciliate on the margins, at least toward the base. Achenes flat, narrowly cuneate, strongly 1-nerved on each face, commonly dark brown or blackish, subglabrous or appressed-hairy, 5–10 mm long. Pappus of 2 retrorsely or sometimes antrorsely barbed awns. (n = 12, 24, 36) Summer–fall. Waste places, especially in wet soil; widespread in n US and adj Can, s less commonly nearly throughout our range.

11. B. discoidea (T. & G.) Britton. Glabrous annual 3–18 dm tall. Leaves thin, trifoliolate, on petioles 1–6 cm long, the leaflets (or at least the terminal one) petiolulate, lanceolate to lance-ovate, serrate, acuminate, the terminal one the largest, up to 10 × 4 cm. Heads numerous, small, discoid, the disk 3–10 mm wide; outer involucral bracts 3–5, linear-spatulate, leafy, much surpassing the disk, scarcely or not at all ciliate-margined. Achenes narrowly cuneate, short-hairy, 3–6 mm long. Pappus of 2 antrorsely setose awns to 2 mm long, or nearly obsolete. (n = 12) Summer–fall. Wet places; widespread in e US, and found occasionally nearly throughout our range.

12. B. pilosa L. Glabrous or variously hairy annual 1–20 dm tall. Leaves on petioles 1–6 cm long, the blade simple or more often pinnately 3–5-parted or compound, with rather finely toothed, lance-ovate to elliptic or ovate leaflets, the terminal one the largest, up to 8 × 4 cm. Heads radiate to disciform or discoid, the rays when present commonly ca. 5, mostly white, up to ca. 1.5 cm long; outer involucral bracts mostly ca. 8, linear-spatulate, shorter than the inner. Achenes linear, compressed-quadrangular to flat, the inner ones up to ca. 1.5 cm long, much longer than the outer. Pappus of 2–3 yellowish, retrorsely barbed awns. (n = 12, 18, 24, 36) Summer–fall, or all year southward. Pantropical weed of disturbed sites, common in Fla, and extending n occ. to CP of SC, Miss and La.

13. B. bipinnata L. Glabrous or minutely hairy annual 3–17 dm tall. Leaves 4–20 cm long including the 2–5 cm petiole, mostly 2–3 times pinnately dissected, the ultimate segments tending to be rounded. Heads narrow, disciform, the disk only 4–6 mm wide at anthesis, the short, ochroleucous rays not surpassing the disk; outer involucral bracts 7–10, linear, more or less acute, not evidently expanded upward, shorter than the inner. Achenes linear, tetragonal, narrowed above, often sparsely hairy, 10–13 mm long, or some of the outer ones shorter. Pappus of 3–4 yellowish, retrorsely barbed awns. (n = 12, 36) Summer–fall. Moist or wet places; widespread weed of e American origin, found nearly throughout our range.

22. COSMOS L. COSMOS

Herbs, ours annual. Leaves opposite, pinnately dissected (ours) to simple. Heads radiate, the rays neutral, white or pink to red or yellow, commonly ca. 8 in our spp.; involucral bracts biseriate and dimorphic, the outer subherbaceous, the inner membranous or almost hyaline; receptacle flat, chaffy, its bracts plane; disk flowers perfect and fertile; style branches slender, flattened, with short, hairy appendage. Achenes quadrangular, not much compressed, linear, beaked. Pappus of 2–8 retrorsely or rarely antrorsely barbed awns, or none.

1 Rays anthocyanic or white.
2 Ultimate leaf segments very narrow, ca. 1 mm wide or less; rays large, commonly 1.5–4 cm long .... 1. *C. bipinnatus*.
2 Ultimate leaf segments broader, mostly (2) 3–10 mm wide; rays smaller, up to ca. 1.5 cm long .... 2. *C. caudatus*.
1 Rays intensely orange-yellow or orange-red .... 3. *C. sulphureus*.

1. C. bipinnatus Cav. Annual, 6–20 dm tall, glabrous or minutely scabrous. Leaves sessile or short-petiolate, 6–11 cm long, the ultimate segments linear or linear-filiform, ca. 1 mm wide or less. Heads rather numerous, the disk yellow, 1–1.5 cm wide; rays anthocyanic to sometimes white, 1.5–4 cm long, often half as wide. Achenes 7–16 mm long, the body longer than the beak. Pappus of 2–3 short awns, or none. (n = 12) Late summer–fall. Native of Mex and adjacent US, commonly cult. and casually escaped into disturbed sites in our range.

2. C. caudatus HBK. Much like *C. bipinnatus*, often somewhat hairy. Leaves more petiolate (petiole 1–7 cm long), often larger (blade sometimes 2 dm long), less finely dissected, the ultimate segments mostly (2) 3–10 mm wide; rays up to ca. 1.5 cm long, sometimes much reduced. Achenes 1.2–3.5 cm long, the beak nearly or fully as long as the body. (n = 24) Late summer–fall. Native of tropical Am, cult. and reputedly escaped in s Fla.

3. C. sulphureus Cav. Vegetatively much like *C. caudatus*; ultimate leaf segments broadly linear to lanceolate, mostly (1.5) 2 mm wide or more. Heads several or numerous, long-pedunculate, the disk yellow, ca. 1 cm wide; rays intensely orange-yellow or orange-red, 2–3 cm long. Achenes 1.5–3 cm long, the beak mostly shorter than the body. Pappus of 2–3 short awns, or none. (n = 12) Late summer–fall. Native of tropical Am, cult. and casually escaped in our range.

23. THELESPERMA Less.

1. T. filifolium (Hook.) A. Gray. Taprooted, short-lived, mostly annual or biennial herb, commonly 2–6 dm tall, glabrous throughout. Leaves opposite (or the upper alternate), up to about 7 cm long, once to thrice pinnatifid, the ultimate segments linear or filiform. Heads solitary on long, naked peduncles ending the branches, radiate, the rays few (commonly 8) and broad, neutral, yellow, mostly 1.5–2 cm long. Involucral bracts biseriate and dimorphic, the outer loose, linear or lance-linear, mostly 4–10 (12) mm long, the inner broad, membranous and striate, with broad hyaline margins, united for about 1/4 to 1/2 their length; receptacle flat, chaffy throughout, its bracts broad, thin, subhyaline, flat or slightly convex and lightly clasping the achenes; disk dark red or purplish, mostly 8–20 mm wide, its flowers perfect and fertile; style branches slender, flattened, with ventromarginal stigmatic lines and an evident, slender, hispidulous appendage ca. 1 mm long. Achenes somewhat compressed parallel to the involucral bracts, but rather thick, often coarsely warty. Pappus of 2 short, stout, retrorsely barbed awn scales, or these nearly obsolete. (n = 8, 9) Spring. Prairies, glades, and open, rocky woods, often on calcareous substrate; Tex and NM to se Wyo, e to SD, sw Mo, and w Ark (OU, OZ); disjunct (possibly not native) in the Black Belt of Miss. Ours is var. *filifolium*. *T. trifidum* (Poir.) Britt.—F, misapplied.

24. GALINSOGA Ruiz & Pavon

Annual herbs. Leaves opposite. Heads small, radiate, the rays few, short, broad, white or seldom pink, only slightly surpassing the disk, pistillate and fertile; involucral bracts few, relatively broad, greenish at least in part, several-nerved, generally each subtending a ray, and tending to be joined at the base with the two adjacent receptacular bracts; a few shorter and narrower outer involucral bracts often also present; receptacle small, conic, chaffy, its bracts membranous, rather narrow, nearly flat; disk flowers perfect and fertile, yellow; style branches flattened, with short, minutely hairy appendage. Achenes quadrangular, scarcely compressed, or (especially the outer) somewhat flattened parallel to the bracts. Pappus of several or many (in ours commonly 15–20) scales, often fimbriate or awn-tipped, that of the rays more or less reduced or none. Canne, J. M. 1977. 'A revision of the genus *Galinsoga* (Compositae: Heliantheae). Rhodora 79: 319–389.

1 Rays with well-developed pappus about equaling the tube; pappus scales of the disk flowers tapering to a short awn tip .... 1. *G. quadriradiata*.
1 Rays without pappus, or nearly so; pappus scales of the disk not awn-tipped ... 2. *G. parviflora*.

1. G. quadriradiata Ruiz & Pavon. Freely branching, somewhat hairy annual, 2–7 dm tall, the hairs of the stem generally fairly coarse and spreading. Leaves petiolate, with ovate (often broadly so), rather coarsely toothed blade mostly 2.5–7 × 1.5–5 cm. Heads rather numerous (in well-developed plants), in open, leafy cymes, slender-pedunculate, the peduncles and often also the involucres spreading-villous with gland-tipped hairs; involucral bracts ca. 3–4 mm long; disk mostly 3–6 mm wide; rays mostly (3–) 5 (6), white, rarely pink, strongly 3-toothed, 2–3 mm long and nearly as wide. Achenes black, hispidulous with appressed or spreading hairs. Pappus of the disk of slender, fimbriate scales tapering to a short but definite awn tip, often shorter than the corolla, that of the rays of short but well-developed fimbriate scales, about equaling the tube. (n = 8, 16) Spring–fall, or all year southward. Native of C and S Am, now a cosmopolitan weed, and found throughout our range. *G. ciliata* (Raf.) Blake—F, G, R, S; *G. bicolorata* St. John & White—F, G; *G. caracasana* (DC.) Schultz-Bip.—F, G.

2. G. parviflora Cav. Similar to no. 1, but less hairy, the stem glabrous or sparsely pubescent with appressed or sometimes spreading hairs, the peduncles appressed-hairy, or finely villous with spreading, gland-tipped hairs. Leaves ovate or lance-ovate, mostly less coarsely toothed. Achenes sparsely appressed-hairy or glabrous. Pappus scales of the disk flowers conspicuously fimbriate, generally blunt, not awn-tipped, nearly or quite as long as the corolla; rays nearly or quite without pappus. (n = 8) Spring–fall. Waste places, but much less common with us than no. 1; native from sw US to S Am, but now a cosmopolitan weed.

25. TRIDAX L.

1. T. procumbens L. Hirsute, taprooted perennial herb with curved-ascending to trailing stems 1.5–4 dm long. Leaves opposite, short-petiolate, the blade ovate to lanceolate, 2–7 × 1–4 cm, irregularly toothed and sometimes shallowly few-lobed. Peduncles terminal and from the upper axils, (0.5) 1–3 dm long; heads campanulate-hemispheric, radiate, the ray flowers mostly 3–8, pistillate and fertile, with ochroleucous or nearly white, short, broad, prominently (2-) 3-toothed ligule 2.5–4 mm long; involucre 7–10 mm high, 2–3 seriate, the outer bracts herbaceous and acuminate, the others membranous, striate, and often broad and blunt; receptacle small, chaffy, its bracts hyaline-scarious, with prominent midvein but not strongly folded; disk yellow or ochroleucous, 7–15 mm wide, its flowers perfect and fertile; style branches slender, flattened, with inconspicuously ventromarginal stigmatic lines and a slender, externally hispidulous appendage. Achenes thick, densely hairy, 2 mm long. Pappus of 20 long, stout, conspicuously plumose bristles. (n = 18) All year. Pine woods and disturbed habitats; pantropical weed of Am origin, intro. in trop. Fla. Powell, A. M. 1965. Taxonomy of Tridax (Compositae) Brittonia 17: 47–96.

26. MARSHALLIA Schreber, nom. conserv. BARBARA'S BUTTONS

Fibrous-rooted perennial herbs. Leaves alternate (or all basal), entire. Heads 1–several, discoid, the flowers all tubular and perfect, with slender, elongate, hairy, conspicuously lobed, white or more often anthocyanic corolla; involucre of 1 or 2 series of narrow, subequal, more or less herbaceous bracts; receptacle conic or convex, chaffy, its bracts linear to linear-spatulate, commonly herbaceous distally; style branches flattened, elongate, with short, minutely hairy appendage. Achenes 5-angled, 10-ribbed, the ribs commonly hairy. Pappus of 5 (6) short, scarious or hyaline scales. Channell, R. B. 1957. A revisional study of the genus Marshallia (Compositae). Contr. Gray Herb. 181: 41–132.

1 Leaves relatively broad, the principal ones with the blade mostly 2.5–7 (10) times as long as wide, often more than 1 cm wide.

2 Involucral bracts acute; receptacular bracts linear or nearly so, acute.

3 Leaves equably distributed along the stem, or the lower reduced; internodes 10–25 .... 1. *M. trinervia.*
3 Leaves inequably distributed, the lower ones the larger; internodes 5–12.
4 Heads 2–10 .... 2. *M. mohrii.*
4 Heads strictly solitary .... 3. *M. grandiflora.*
2 Involucral bracts obtuse; receptacular bracts linear-spatulate, obtuse .... 4. *M. obovata.*
1 Leaves relatively narrow, the principal ones with the blade mostly 7–20 times as long as wide (some basal ones sometimes wider, but the cauline ones then numerous, well developed, and narrow), and seldom more than about 1 (1.5) cm wide.
5 Involucral bracts merely acute; flowers white or occasionally pale lavender; hairs of the peduncle not anthocyanic.
6 Heads relatively large, commonly 2.5–3.5 cm wide at anthesis, the involucre (6) 8–11 mm high; pappus mostly 2–2.5 mm long and as long as or longer than the achene .... 5. *M. caespitosa.*
6 Heads small, commonly 1.5–2 cm wide at anthesis, the involucre 4–6 mm high; pappus mostly 1–1.5 (2) mm long and shorter than the achene .... 6. *M. ramosa.*
5 Involucral bracts strongly acuminate-attenuate or subulate-tipped; flowers purple to pale lavender, rarely white; hairs of the peduncle purple, or with purple cross-walls.
7 Basal leaves relatively thin and soft, more or less spreading, strongly differentiated from the cauline ones, the blade only obscurely trinerved, very often less than 7 times as long as wide; old leaf bases not persistent .... 7. *M. tenuifolia.*
7 Basal leaves firm, ascending or erect, not strongly differentiated from the cauline ones (except in size and in length of petiole), the blade evidently trinerved, at least 7 times as long as wide; fibrous base of old leaves persistent and conspicuous at base of plant .... 8. *M. graminifolia.*

1. M. trinervia (Walter) Trelease. Plants glabrous, more or less colonial from short-creeping rhizomes, the stems arising singly, 4–8 dm tall. Leaves thin, strongly trinerved, mostly 5–10 × 1–3.5 cm, equably distributed, the lower oblanceolate or spatulate, petiolate, obtuse or rounded, often deciduous before anthesis, those above fully as large, lance-ovate or lance-elliptic to ovate, acuminate or acute, sessile or nearly so; internodes 10–25. Heads mostly solitary (few) on a terminal peduncle 1–2 dm long, 2–3 cm wide at anthesis; involucre (7) 8–10 (12) mm high, its bracts and those of the receptacle acute. Pappus scales minutely serrulate. Spring–midsummer. Woods, streambanks, and cliffs, often on calcareous clay. Ala (all), Miss (GC) and Tenn (IP), and irregularly to nw and wc Ga, sw NC, e SC, Va, se La, and perhaps Ark.

2. M. mohrii Beadle & Boynton. Plants mostly single-stemmed from a crown or short caudex, 3–7 dm tall, often puberulent under the heads, otherwise glabrous, the stem branched above. Leaves rather firm, 3-nerved, somewhat basally disposed, the lower ones evidently petiolate, with rather narrowly elliptic to spatulate blade 6–10 cm long and 1–2 cm wide, the others sessile and more or less reduced upward; internodes on the main axis mostly 5–10. Heads 2–6 (10) on naked or subnaked peduncles commonly 0.5–1 dm long terminating the stem and branches, mostly 1.5–2.5 cm wide at anthesis; involucre 5–9 mm high, its bracts and those of the receptacle acute. Pappus scales with irregular, broken margins, as well as minutely serrulate. Summer. Moist woods and meadows; rare and local in CU and RV of Ala and nw Ga.

3. M. grandiflora Beadle & Boynton. Plants single-stemmed or clustered from a short caudex, 2–9 dm tall, puberulent under the heads, otherwise glabrous. Leaves rather firm, 3-nerved, somewhat basally disposed, the lower ones evidently petiolate, with broadly oblanceolate or spatulate to more or less elliptic blade 5–15 cm long and 1–3 cm wide, the others sessile and more or less reduced upward; internodes commonly 5–12. Head solitary on a terminal peduncle mostly 1–3 dm long, mostly 2–3.5 cm wide at anthesis; involucre 7–12 mm high, its bracts and those of the receptacle acute. Pappus scales minutely serrulate. Summer. Moist to wet places in woods, meadows, and along streams; s App region, from sw Pa irregularly to NC and Tenn, on PP to AP.

4. M. obovata (Walter) Beadle & Boynton. Plants more or less colonial, but individually mostly single-stemmed from a short caudex, mostly 1.5–6 dm tall, pubescent under the heads, otherwise glabrous, abruptly naked-pedunculate, the peduncle at least as long as the leafy part of the stem. Leaves 3-nerved, the basal ones often tufted, these and (when present) the lowermost cauline leaves evidently petiolate, with oblanceolate or spatulate to elliptic, obtuse or rounded blade commonly 3–9 cm long and (0.5) 0.8–2 cm wide, the others less petiolate or sessile and often more acute, but mostly not much smaller. Heads solitary, mostly 2–3.5 cm wide at anthesis; involucre 6–12 mm high, its bracts relatively broad, obtuse; receptacular bracts linear-spatulate, obtuse (sometimes minutely mucronulate). Spring. Meadows and open woods; s Va to Ga, Ala, and w Fla. Var. *obovata,* with the leafy part of the stem 1/4 to fully as long as the peduncle, and with the pappus scales mostly 1–1.5 mm long, occurs chiefly on PP, extending into the mt. provinces in Ga and Ala, and onto CP in sw Ga, se Ala, and w Fla. (*M. obovata* var. *platyphylla* M. A. Curtis—F) Var. *scaposa* Channell, scapose or with the leafy part of the stem less than 1/4 as long as the peduncle, and with the pappus scales mostly 1.5–2.5 mm long, occurs chiefly on CP, especially inner CP. (*M. obovata,* sens. strict., of F.)

5. M. caespitosa Nutt. Plants mostly 2–5 dm tall, the stems clustered on a short caudex or crown, evidently pubescent at least under the heads. Leaves glabrous, firm, mostly or all clustered at or near the base, or some scattered along the stem, relatively narrow, the principal ones broadly linear to linear-oblanceolate or linear-elliptic, tapering gradually to a more or less petiolar base, 5–15 cm long (petiole included) and 3–8 (10) mm wide, or some of the very basal ones shorter and relatively broader, occasionally as much as 15 mm wide. Heads white or occasionally pale lavender, relatively large, commonly 2.5–3.5 cm wide at anthesis; involucre (6) 8–11 mm high, its bracts obtuse to more often acute, often shortly mucronate-pointed, sometimes narrowly wing-margined below, but not shouldered; receptacular bracts merely acute or inconspicuously mucronate. Pappus mostly 2–2.5 mm long, as long as or longer than the achene. Spring. Prairies, dry meadows, and pinelands; Tex and Okla to La, Ark, and sw Mo. Two vars.:

Plants scapose or subscapose, the cauline leaves, if any, hardly surpassing those of the basal tuft; heads solitary; La and Tex to Okla and sw Mo, not in Ark ...... var. *caespitosa*.

Plants more or less leafy-stemmed, the leafy part of the stem usually at least as long as the peduncles; heads solitary to more often several; chiefly of Tex, but with isolated stations in Ark (OU) and Mo (OZ) .... var. *signata* Beadle & Boynton

6. M. ramosa Beadle & Boynton. Plants single-stemmed, or the stems more often clustered on a short caudex, 2.5–6 dm tall, pubescent under the heads, otherwise glabrous, evidently leafy-stemmed, but with the largest leaves borne near the base. Leaves rather firm, 3-nerved, narrow and elongate, the principal ones mostly 8–20 cm long overall and 3–10 mm wide, the blade tapering gradually to the petiolar base; middle and upper leaves scattered, smaller, and sessile. Heads (2) 4–12 (20), white or occasionally pale lavender, relatively small, mostly 1.5–2 cm wide at anthesis; involucre 4–6 mm high, its bracts obtuse to acute and commonly mucronulate, often winged below, sometimes shouldered as in the next 2 spp., receptacular bracts obtuse and mucronulate to almost subulate-tipped. Pappus mostly 1–1.5 (2) mm long and shorter than the achene. Spring. Savannas and pinelands; local on the Altamaha Grit formation of Dodge, Johnson, Tatnall, and Telfair Cos. on inner CP of Ga, s to Coffee Co., and into Washington Co., Fla.

7. **M. tenuifolia** Raf. Stems mostly solitary or few from a short caudex or crown, commonly 3–10 dm tall, simple or more often branched near or below the middle, pubescent above (the hairs purple or with purple cross-walls), glabrous or glabrate below. Basal leaves rather soft, spreading, oblanceolate, obtuse or rounded to acutish, inconspicuously 3-nerved, 2–10 cm long overall, and 3–12 mm wide, gradually tapering to the petiolar base, the blade mostly 3–10 times as long as wide. Cauline leaves numerous, sharply differentiated from the basal ones, erect or closely ascending, mostly linear and sessile or nearly so, gradually reduced upward, the stem (and branches) ending in a naked or sparsely bracteate peduncle 1–3 dm long. Heads purple to pale lavender, rarely white, mostly 2–3 cm wide at anthesis; involucre 4–6 mm high, its bracts short in relation to the size of the head, glandular-punctate, often sparsely hairy, frequently wing-margined below the middle (the white-hyaline wing often forming an evident shoulder), tending to be anthocyanic above, tapering to a slender, more or less subulate point; receptacular bracts punctate and pointed like those of the involucre. Pappus mostly 1–2 mm long, commonly shorter than the achene. Summer. Variously in wet or dry meadows, pine forests, and bogs, and with some ecotypic differentiation; CP from Ga to extratropical Fla and Tex. [*M. graminifolia* (Walt.) Small, sensu Small] Perhaps better treated as a var. of no. 8.

8. **M. graminifolia** (Walter) Small. Much like *M. tenuifolia*, differing as indicated in the key; leaves up to 25 × 1.5 cm; involucre mostly 6–8 mm high. (n = 9) Summer. Moist or wet, low ground; NC and SC, and e Ga, largely on CP. (*M. lacinarioides* Small and *M. williamsonii* Small—S.)

27. SILPHIUM L. ROSIN-WEED

Perennial herbs. Leaves variously opposite, alternate, or whorled, entire or toothed to once or twice pinnatifid or palmatifid, but not with articulated leaflets. Heads medium-sized to large, subhemispheric, radiate, the rays yellow, pistillate and fertile, their ovaries imbricate in 2–3 series; involucral bracts subequal or imbricate in 2–several series, firm, herbaceous or partly membranous-chartaceous; receptacle flat, chaffy, its bracts, or sometimes in part the inner bracts of the involucre, subtending the rays as well as the disk flowers; disk flowers sterile, with undivided style. Ray achenes glabrous, strongly flattened parallel to the involucral bracts, wing-margined. Pappus none, or of 2 awns confluent with the wings of the achene. (n = 7).

1 Leaves, or their petiolar bases, connate-perfoliate; stem square .... 1. *S. perfoliatum*.
1 Leaves sessile or petiolate, not connate-perfoliate; stem more or less round.
  2 Plants fibrous-rooted from a short rhizome, caudex, or crown; leaves entire to coarsely toothed, from mainly cauline to somewhat basally disposed, but the stem still more or less leafy.
    3 Leaves well distributed along the stem and only gradually reduced upward; basal leaves sometimes large and persistent, but often smaller and deciduous.

4 Heads relatively large and many-flowered, commonly with 16–35 rays; leaves sessile or nearly so, broad-based and often clasping, commonly 2–4 times as long as wide.
5 Stem velvety to scabrous, with most of the hairs ca. 0.5 mm long or less, varying to largely or wholly glabrous .... 2. *S. integrifolium*.
5 Stem hispid or hispid-scabrous, many or all of the hairs ca. 1 mm long or more .... 3. *S. radula*.
4 Heads mostly smaller and fewer-flowered, mostly with ca. 6–15 (21) rays; leaves various, often petiolate, or tapering to the base, but sometimes as in the foregoing group.
6 Leaves, except the uppermost, mostly cordate to truncate or broadly rounded at the base, on petioles commonly 1.5–4.5 cm long; stem glabrous .... 4. *S. brachiatum*.
6 Leaves otherwise, many of them either narrowed to the base, or sessile, or both, the middle and upper ones seldom with a petiole more than 1.5 cm long (small-headed forms of *S. integrifolium* might be sought here).
7 Stem evidently spreading-hairy, many of the hairs ca. 1 mm long or more .... 5. *S. asteriscus*.
7 Stem essentially glabrous, often glaucous.
8 Receptacular bracts with some evident, gland-tipped, usually amber-colored hairs on the back toward the tip; leaves mostly opposite or alternate .... 6. *S. dentatum*.
8 Receptacular bracts not evidently glandular; leaves often ternate or quaternate, sometimes opposite, seldom alternate or scattered .. 7. *S. trifoliatum*.
3 Leaves somewhat basally disposed, the basal and/or lower cauline ones persistent and evidently larger and more petiolate than those above, which tend to be progressively reduced.
9 Herbage essentially glabrous .... 8. *S. confertifolium*.
9 Herbage evidently pubescent.
10 Stem conspicuously shaggy-hispid, many or most of the hairs more than 2 (to 5) mm long; Cumberland sp. .... 10. *S. mohrii*.
10 Stem more shortly and less copiously hispid, the hairs up to ca. 2 mm long; Coastal Plain sp. .... 9. *S. gracile*.
2 Plants distinctly taprooted; leaves variously entire or toothed to deeply cleft, always more or less basally disposed, the lower or basal ones notably larger than the progressively reduced middle and upper ones, or the stem subnaked above the base.
11 Stem evidently leafy, the cauline leaves only gradually reduced upward; leaves deeply pinnatifid or bipinnatifid .... 11. *S. laciniatum*.
11 Stem nearly naked, the principal leaves all at or near the base; leaves variously entire or toothed to pinnatifid or palmatifid and again cleft.
12 Heads large, with ca. 13 to ca. 21 rays, the involucre mostly 13–25 mm high, the disk 1.5–2.5 cm wide.
13 Principal leaves deeply pinnatifid to rarely entire, in the latter case lanceolate and long-tapering to the petiole .... 12. *S. pinnatifidum*.
13 Principal leaves merely toothed (or subentire), usually more or less cordate (seldom only truncate, or abruptly narrowed) at the base, always broader than lanceolate .... 13. *S. terebinthinaceum*.
12 Heads smaller, with ca. 5 or 8 (10) rays, the involucre 6–11 mm high, the disk 0.8–1.5 cm wide .... 14. *S. compositum*.

1. S. perfoliatum L. Stems coarse, 1–2.5 m tall, square. Leaves mostly opposite; leaves or their petiolar bases connate-perfoliate, the blade deltoid to ovate, coarsely toothed, up to 3 × 1.5 dm, scabrous, or hispidulous beneath. Heads several or rather numerous in an open inflorescence, the disk mostly 1.5–2.5 cm wide; rays 1.5–4 cm long. (n = 7) Summer. Moist woods and low ground; s Ont. to e ND, s to NC, Tenn, Miss, La, Ark, and Okla, and intro. into New England. Two geographically segregated, intergradient vars.:

Stem generally glabrous or nearly so, sometimes spreading-hairy toward the base. Most of the leaves, except for a few upper ones, generally more or less evidently wing-petiolate, the lower surface glabrous or short hairy, seldom any of the hairs as much as 1 mm long. Heads with relatively numerous (mostly 16–35, commonly ca. 21 or 34) rays; involucral bracts glabrous except for the ciliate margins. Widespread, in our range from Ky and Tenn to Miss (GC) and Ark (ME, OZ) .... var. *perfoliatum*.

Stem usually spreading-hispid, varying to largely glabrous. Many or most of the leaves sessile, only the lower of lowest ones evidently petiolate; lower leaf surface, at least along the midrib, relatively long-hairy, most of the hairs ca. 1–2 mm long. Heads with fewer (mostly ca. 8 or ca. 13) rays; involucre usually finely hairy, varying to glabrous. Scattered stations in mt. provinces and adj. PP of NC, Va, and WVa. *S. connatum* L.—F. .... var. *connatum* (L.) Cronq.

2. S. integrifolium Michx. Plants fibrous-rooted from a short rhizome or caudex. Stem 5–15 dm tall, velvety or scabrous, with most or all of the hairs ca. 0.5 mm long or less, varying to glabrous. Leaves firm, well distributed along the stem, mostly opposite, seldom ternate or alternate, entire or toothed, ovate or lance-ovate to elliptic, mostly 7–15 × 2–6.5 cm, 2–4 times as long as wide, scabrous above, variously scabrous or velvety to glabrous beneath, sessile and often clasping. Heads generally several or many, relatively large and many-flowered, with mostly 16–35 (commonly ca. 21 or 34) rays 2–5 cm long, the disk mostly 1.5–2.5 cm wide. (n = 7) Summer. Prairies and roadsides, less often in open woods; s Mich to e Neb, s to Ala, Miss, and Tex. Two well-marked but geographically overlapping and morphologically intergradient vars. Var. *integrifolium*, the more hairy phase, with the backs of the involucral bracts, the peduncles, and usually also the lower surfaces of the leaves and at least the upper part of the stem velvety or scabrous, is typical of the tall-grass prairie, from Ill and Mich to e Kans, s to c Ala, s Miss, c La, and ne Tex. Var. *laeve* T. & G., the less hairy phase, with the involucral bracts, peduncles, lower surfaces of the leaves and the stem glabrous or nearly so and often somewhat glaucous, is more western, typically in the mixed-grass prairie, in our range occurring in nw Ark and with an outlier in c Tenn. *S. speciosum* Nutt.—F, S.

3. S. radula Nutt. Much like the more densely hairy forms of *S. integrifolium*, but averaging more robust (commonly 1–2 m tall), often with alternate leaves, and with distinctly longer, coarser pubescence, many or all of the hairs of the stem ca. 1 mm long or more. (n = 7) Summer. Prairies, sometimes in open woods or invading old fields; Ozarkian sp., chiefly of Okla and Tex, with outlying stations in Ala and nw Ga (RV); reported from and to be expected in Ark. *S. asperrimum* Hook., misapplied by S and others. The use of the name *S. radula* for this species is based on the protologue and on my examination of the probable holotype at BM.

4. S. brachiatum Gattinger. Plants fibrous-rooted from a short rhizome or caudex. Stem mostly 10–15 dm tall, glabrous and glaucous. Leaves opposite, well distributed along the stem, the blade narrowly ovate to triangular, coarsely toothed, mostly 13–25 × 5–10 cm, scabrous above, scabrous or hispid beneath, truncate or broadly rounded at the base, or the lower cordate, on petioles 1.5–4.5 cm long. Heads rather numerous on slender peduncles in an open inflorescence, relatively small, the disk ca. 1 cm wide, with 4–6 (8) rays ca. 1–1.5 cm long; involucral bracts glabrous except for the ciliate margins. (n = 7) Summer. Woods; CU of Ky, Tenn, Ga and Ala.

5. S. asteriscus L. Plants fibrous-rooted from a short rhizome or caudex, mostly 5–12 dm tall. Herbage sparsely to densely spreading-hispid, or the leaves merely hispid-scabrous, many of the hairs of the stem ca. 1 mm long or more. Stem leafy, the leaves opposite or alternate, coarsely toothed or entire, mostly 6–15 × 1.5–5 cm, (2)

2.5–5 times as long as wide, the lower often rather large and evidently long-petiolate, the others variously sessile or short-petiolate, rounded or tapering at the base. Heads generally several or many in a more or less leafy-bracteate inflorescence, with mostly ca. 8 or ca. 13 (21) rays 1.5–3 cm long, the disk mostly 1–2 cm wide. (n = 7) Summer. Open woods, glades, and clearings. Va to n Fla, w to Mo and Tex. Three regionally differentiated but wholly intergradient vars.:

Pales (receptacular bracts) without evidently gland-tipped hairs, the tip of the pale relatively broad, rounded to obtuse or merely acute, not densely white-hairy; stem nearly or quite without under-pubescence; mt. provinces from Va s, and in NC and SC extending onto PP and seldom AC (*S. incisum* Greene—S) .... var. *asteriscus*.

Pales generally with some evident, amber-colored, gland-tipped hairs on the back near the tip, or these lacking in forms of var. *scabrum* with the tip densely white-hairy.

Pales with broad, rounded to obtuse or merely acute tip, not densely white-hairy; stem with or without a short underpubescence; leaves tending to taper to a short petiole or petiolar base; CP from n Fla to SC and Ala, and n onto PP in Ga and adj SC (*S. nodum* Small—S) .... var. *angustatum* A. Gray.

Pales with relatively slender, acute or acuminate tip that tends to be rather densely white-hairy; stem with a short, soft, crinkled, commonly rufous underpubescence (at least toward the top) as well as copiously spreading-hispid; involucre often more densely and copiously pubescent than in the other vars.; leaves tending to be broader-based than in var. *angustatum*, more often sessile or nearly so, and even clasping; OZ of Mo and Ark, e to mt. provinces (except BR) of Tenn, Ga, and Ala, and encroaching onto GC in Ark, Miss, and Ala. (*S. gatesii* Mohr—F, S; *S. dentatum* var. *gatesii* (Mohr) Ahles—R; *S. scaberrimum* Ell.—S) .... var. *scabrum* Nutt.

6. S. dentatum Elliott. Much like *S. asteriscus* var. *angustatum*, but with essentially glabrous, often glaucous stem. Leaves scabrous or scabrous-hispid above, hispid or hirsute at least along the main veins beneath, the cauline ones with the blade 8–25 × 1.5–10 cm, (2.3) 2.5–8 times as long as wide, tapering or sometimes more abruptly contracted to a shortly petiolar (seldom more broadly subsessile) base, the basal ones with more evident petiole often equaling the blade, this up to 30 × 12 cm. (n = 7) Summer. Dry woods; CP, PP, and BR of NC, SC, Ga, and Ala. *S. elliottii* Small—S. Perhaps better treated as *S. asteriscus* var. *laevicaule* DC.

7. S. trifoliatum L. Plants fibrous-rooted from a short rhizome or caudex. Stem robust, mostly (0.7) 1–2 m tall, glabrous, often glaucous. Leaves well distributed along the stem, variously whorled or opposite or seldom alternate, entire or merely toothed, petiolate or tapering to a subpetiolar base. Heads several to commonly more or less numerous in an open inflorescence, rather small, with ca. 8 or ca. 13 rays mostly 1.5–3 cm long, the disk ca. 1–1.5 (2) cm wide; involucral bracts glabrous except for the ciliate margins. (n = 7) Summer. Open woods, prairies, and disturbed open places; se Pa to Ohio and Ind, s to NC, Ga, Ala, and Miss. Two morphologically confluent, geographically partly differentiated vars.:

Leaves mostly ternate or quaternate, occasionally opposite, seldom partly alternate, scabrous above, hirsute at least on the midrib beneath, the blade lanceolate or lance-elliptic to rather narrowly lance-ovate, (2.5) 3–5 times as long as wide, 7–20 × 1.5–5 cm, tapering to a hirsute-ciliate petiole up to 1.5 cm long, or to a shortly subpetiolar base. Northern var., from se Pa to Ohio and Ind, s to Va (AC), NC (PP and mt. provinces), Tenn (mt. provinces and IP) and Ga (BR). *S. atropurpureum* Retz—F. .... var. *trifoliatum*.

Leaves prevailingly opposite, often wider than in var. *trifoliatum*, up to 7.5 cm wide and sometimes only twice as long as wide, tending to be less pubescent, often essentially glabrous, often more petiolate, the petiole sometimes up to 5 cm long.

Southern var., extending s to AC of NC, PP of Ga, and GC of Ala, and Miss, but broadly overlapping the range of var. *trifoliatum* in Tenn, Ky, and sw Va. *S. glabrum* Eggert—S; *S. laevigatum* Pursh—R. .... var. *latifolium* A. Gray.

8. S. confertifolium Small. Plants fibrous-rooted from a short rhizome, (2) 4–10 dm tall, essentially glabrous throughout except for the scabro-ciliate margins of the leaves and involucral bracts. Leaves firm, opposite, entire or nearly so, somewhat basally disposed, the lower ones well developed and persistent, with narrowly elliptic to lanceolate blade ca. 6–15 × 2–5 cm tapering to a short petiole, the others mostly sessile or nearly so and more or less strongly reduced upward. Heads (1) several to rather many in an open, corymbiform inflorescence, small to middle-sized for the genus, the disk 1–2 cm wide, with ca. 8 to ca. 13 rays mostly 1–1.5 cm long. Summer. Chalk prairies; local on and near GC in Ala (Choctaw, Greene, Hale, Jefferson, and St. Clair Cos.)

9. S. gracile A. Gray. Plants fibrous-rooted from a short rhizome or crown, 2–10 dm tall. Herbage sparsely to moderately hispid or scabrous-hispid, the hairs up to ca. 2 mm long. Leaves opposite or alternate, entire or toothed, the basal and lowermost cauline ones well developed and persistent, with oblanceolate to narrowly or broadly elliptic or ovate blade up to ca. 20 × 6 cm, tapering or rather abruptly contracted to a short petiole or petiolar base; other leaves generally fairly well developed but progressively reduced upward and mostly sessile. Heads 1–few, with ca. (8) 13 (21) rays 1.5–3 cm long, the disk mostly 1.5–2.5 cm wide. (n = 7) Spring–summer. Prairies and dry woods; GC from Fla to Tex. *S. simpsonii* Greene—S.

10. S. mohrii Small. Plants fibrous-rooted from a short rhizome or crown, 6–15 dm tall. Herbage copiously shaggy-hispid, many or most of the hairs of the stem more than 2 (to 5) mm long. Leaves alternate, opposite, or partly ternate, entire or toothed, the basal and lower cauline ones well developed and persistent, with broadly ovate to merely lanceolate, basally rounded or abruptly contracted blade 12–25 × 4–13 cm on a petiole 5–15 cm long; cauline leaves progressively reduced upward, becoming sessile, often ovate. Heads several or rather numerous, with ca. 13 rays 1.5–2 cm long, the disk mostly 1.5–2.5 cm wide. Summer. Glades and open woods; CU of Ala, Ga, and Tenn (where also on IP). *S. incisum* Greene—S.

11. S. laciniatum L. COMPASS PLANT. Plants taprooted, coarse, rough-hairy, 1–3 m tall. Leaves alternate, deeply pinnatifid or bipinnatifid, the lower very large, to 5 dm long, progressively reduced upward, the uppermost entire and well under 1 dm long. Heads in a narrow, sometimes racemiform inflorescence, large, the disk 2–3 cm wide, the rays mostly (13) 17–25 (34), 2–5 cm long; involucre 2–4 cm high, surpassing the disk, its bracts ovate, acuminate, squarrose, not much imbricate. (n = 7) Summer. Prairies; Ohio to Minn and SD, s to w Tenn (ME), w Ala (GC), Miss, La and Tex. The basal leaves tend to align themselves in a north–south direction. *S. laciniatum* var. *robinsonii* Perry—F, a form with the peduncles and involucre glandular as well as hairy.

12. S. pinnatifidum Elliott. Plants taprooted. Stem glabrous, 4–10 dm tall. Basal leaves large and persistent, the petiole 1–3 dm long, the blade commonly (1) 1.5–4 × 0.4–3 dm, deeply pinnatifid to rarely entire, (in that case lanceolate and tapering to the base), glabrous above except for the sometimes hirsute midrib, generally hirsute at least on the midrib beneath; cauline leaves alternate, few, much reduced and bractlike. Heads 1–several on long, naked peduncles, with ca. 21 rays 1–2 cm long, the disk 1.5–2.5 cm wide; involucre 1.5–2.5 cm high, the inner bracts with loose, broad tip. (n = 7) Summer. Cedar glades, prairies, and open woods; IP of Tenn and Ala, e to RV & CU of Ga. *S. terebinthinaceum* var. *pinnatifidum* (Ell.) A. Gray—F.

13. S. terebinthinaceum Jacq. Plants taprooted. Stem mostly 0.7–3 dm tall, essentially glabrous, nearly naked, its leaves, except those near the base, few and reduced to mere large bracts. Principal leaves large, long-petiolate, the blade narrowly to broadly ovate, oblong, or elliptic, usually cordate (seldom only truncate or abruptly narrowed) at the base, 1–5 dm long and 0.7–3 dm wide, glabrous or scabrous, sometimes hirsute on the midrib beneath. Heads several or rather numerous in an open-corymbiform inflorescence, relatively large, with ca. 13 to ca. 21 rays 2–3 cm long, the disk 1.5–2.5 cm wide; involucre 13–25 mm high, its bracts essentially glabrous, strongly imbricate, the outer broadly elliptic, the inner elongate, more oblong, loose and broad-tipped. (n = 7) Summer. Prairies; s Ont and Ohio to Minn, s to Tenn (RV), Ga (RV), Ala (RV), and Miss (ME), and reported by R from NC. *S. rumicifolium* Small—S.

14. S. compositum Michx. Much like no. 13, but with smaller and often more numerous heads, the disk 0.8–1.5 cm wide, the involucre 6–11 mm high, the rays 5–10, 1–2 cm long; leaves often wider than long, variously merely toothed to deeply pinnatifid or palmatifid and again lobed or cleft. (n = 7) Summer. Dry, often sandy places, esp in pine woods; throughout most of Va, NC, SC, and Ga to n Fla (CP), e Ala (CU to GC), and e Tenn (CU to BR). *S. lapsuum* Small—S; *S. orae* Small—S; *S. reniforme* Raf.—S; *S. compositum* var. *reniforme* (Raf.) T. & G.—F, R; *S. venosum* Small—S.

28. CHRYSOGONUM L.

1. C. virginianum L. Fibrous-rooted perennial, to 5 dm tall, often beginning to flower when very small. Leaves opposite, ovate to suborbicular, hairy, long-petiolate, crenate, 2.5–10 cm long and 1.5–6 cm wide, the middle and upper generally deltoid or cordate at the base. Heads few or solitary on slender terminal and axillary peduncles, radiate, the rays few (commonly 5), broad, uniseriate, yellow, pistillate and fertile, mostly 7–15 mm long; involucral bracts in 2 series of about 5, the outer herbaceous and surpassing the disk, villous like the stem, the inner more chartaceous and subtending the rays; receptacle flat, chaffy; disk 7–10 mm wide, its flowers sterile, with undivided style. Achenes more or less compressed parallel to the involucral bracts, wingless; each inner involucral bract grown to 2 or 3 adjacent receptacular bracts at the base, thus partly enclosing the achene, the group falling as a unit; pappus a short, half-cup-shaped crown. (n = 8, 16) Early spring–midsummer. Woods; from s Pa and se Ohio to Fla, Ala, and se Miss, chiefly on AC and PP, but occasionally inland to mt. provinces and AP. Two vars.:

Plants low, stoloniferous, tending to form loose mats with several stems not arising more than about 1 (1.5) dm above the ground; southern, from NC and Tenn to c Fla and ec Ala and reputedly La. *C. australe* Alexander—S. .... var. *australe* (Alexander) Ahles.

Plants taller (at least at maturity), few-stemmed or single-stemmed, not mat-forming, mostly 1.5–4 (5) dm tall; northern, from DC and Va to WVa and NC, and reputedly SC .... var. *virginianum*.

29. BERLANDIERA DC. GREEN EYES

Pubescent perennial herbs from a fleshy taproot. Leaves alternate (sometimes largely basal), crenate to sinuate-pinnatifid or lyrate-pinnatifid. Heads hemispheric, solitary and terminal to several in a terminal inflorescence, radiate, the rays 2–13 (most often 8), pistillate and fertile, yellow, with 9–12 prominent, green (in ours) to red or maroon, anastomosing veins beneath; involucral bracts about 3-seriate, relatively broad, herbaceous at least distally, the inner often enlarged and membranous-veiny in fruit; receptacle flat or nearly so, chaffy throughout; disk flowers sterile, with undivided style. Ray achenes strongly flattened parallel to the involucral bracts, wingless, generally hairy on the inner face; two adjacent receptacular bracts and disk flowers and the subtending bract adhering to each ripening achene and deciduous with it. Pappus a minute, incomplete crown, or a few short teeth or awns, or none. (n = 15) Pinkava, Donald J. 1967. Biosystematic study of Berlandiera (Compositae). Brittonia 19: 285–298.

The taxonomy of *Berlandiera* is complicated by hybridization and introgression. *B. pumila* hybridizes and introgresses to some extent with both *B. subacaulis* and *B. texana*. Both hybrids are more or less intermediate between the parents in habit and leaf form, but the *pumila-subacaulis* hybrids (called *B. humilis* Small) have the disk red as in *B. pumila*, and the *pumila-texana* hybrids [called *B. betonicifolia* (Hook.) Small] lack the distinctive pubescence of *B. pumila*.

1 Disk yellow; leaves sinuate-pinnatifid to shallowly lyrate-pinnatifid, tending to be basally disposed .... 1. *B. subacaulis*.
1 Disk red to maroon; leaves crenate, well distributed along the stem.
  2 Stem finely and closely tomentose or tomentose-puberulent, not very densely leafy, most of the internodes generally at least 3 cm long; middle leaves usually evidently petiolate .... 2. *B. pumila*.
  2 Stem loosely spreading-hairy (often woolly-villous), very densely leafy, most of the internodes generally less than 3 cm long; middle leaves mostly sessile or inconspicuously short-petiolate .... 3. *B. texana*.

1. B. subacaulis (Nutt.) Nutt. Stems clustered, loosely erect, 1–4 dm tall, simple or few-branched. Herbage shortly rough-hairy and often also glandular. Leaves tending to be basally disposed, but in larger plants sometimes more scattered on the lower 1/2 or 2/3 of the stem, mostly sinuate-pinnatifid to shallowly lyrate-pinnatifid, up to about 15 cm long and 3 cm wide. Heads solitary, the disk yellow, 1–2 cm wide; rays mostly 1–1.5 cm long. (n = 15) More or less all year, but mostly spring–midsummer. Pinelands and disturbed, sandy soil in peninsular Fla, w to Jefferson Co.

2. B. pumila (Michx.) Nutt. Stems clustered, 2–8 dm tall, conspicuously white-tomentose or -tomentose-puberulent, generally branched above. Leaves well scattered (most of the internodes 3 cm long or more), evidently petiolate (the upper less so), crenate, finely and closely tomentose or tomentose-puberulent (often whitened) beneath, less so (and often glandular) above, the middle ones with the blade narrowly ovate-oblong to rotund-ovate, 5–10 cm long and 2–6 cm wide, often truncate at the base. Heads several on rather short peduncles, the disk 1–2 cm wide, red to maroon; rays (1) 1.5–2 cm long. (n = 15) Spring–summer. Pinelands and other dry, sandy places; CP from n Fla to SC and sw Ala, s La and e Tex.

3. B. texana DC. Stems clustered, coarse, 4–12 dm tall, loosely spreading-hairy (often woolly-villous) branched above, densely leafy, most of the internodes generally less than 3 cm long. Leaves conspicuously crenate, rather loosely hairy at least beneath (seldom whitened), mostly triangular or triangular-ovate with truncate to subcordate base, 4–15 cm long and 2–6 cm wide, the middle and upper ones sessile or inconspicuously short-petiolate. Heads several on rather short peduncles, the disk 1–2 cm wide, red to maroon; rays 1–2 cm long. (n = 15) Summer. Dry woods and forest openings; Ozarkian sp., from s Mo to Okla, Ark, and Tex, and onto the GC in La.

30. PARTHENIUM L.

Bitter, aromatic herbs (ours) or shrubs. Leaves alternate, entire or toothed to dissected. Heads several or numerous in a terminal inflorescence, white, inconspicuously radiate, the rays 5, short, pistillate and fertile; involucral bracts relatively broad, dry and scarcely herbaceous, in 2 series of 5 each, the inner subtending the rays; receptacle small, conic or convex, chaffy throughout; disk flowers sterile, with undivided style. Ray achenes flattened parallel to the involucral bracts, ribbed on the margins; two adjacent receptacular bracts and disk flowers and the subtending bract adhering to each ripening achene and deciduous with it. Pappus of 2 or 3 short or elongate awns or scales, or obsolete. Rollins, Reed, 1950. The guayule rubber plant and its relatives. Contr. Gray Herb. 172: 1–72.

1 Perennial; leaves merely toothed, or some of them lyrate at the base.
  2 Plants with a tuberous-thickened, usually short root; heads relatively small, the disk mostly 4–7 mm wide.
    3 Herbage rather shortly and closely (generally inconspicuously) hairy, commonly strigose or puberulent to scabrous, or partly glabrous .... 1. *P. integrifolium.*
    3 Herbage loosely spreading-hairy, villous to scabrous-hispid, the stem often somewhat shaggy .... 2. *P. auriculatum.*
  2 Plants with a creeping rhizome, this sometimes coarse and woody, but not tuberous-thickened; heads relatively large, the disk mostly 7–10 mm wide; herbage usually loosely spreading-hairy as in no. 2 .... 3. *P. hispidum.*
1 Annual; leaves pinnatifid or bipinnatifid .... 4. *P. hysterophorus.*

1. P. integrifolium L. Perennial from a tuberous-thickened, usually short root. Herbage rather shortly and loosely (generally inconspicuously) hairy, commonly strigose or puberulent to scabrous, or partly glabrous. Stem simple or branched above, 3–10 dm tall. Leaves large and sometimes few, crenate-serrate, or sublyrate at the base, the basal ones long-petiolate, with lance-elliptic to broadly ovate blade commonly 7–20 cm long and 4–10 cm wide, the cauline with progressively shorter petiole and generally more or less reduced, the upper often sessile and clasping, but less conspicuously so than in the next 2 spp. Heads numerous in a flat-topped inflorescence, the disk about 4–7 mm wide; rays scarcely 2 mm long. Achenes obovate, black, about 3 mm long. (n = 36) Spring–summer. Prairies, fields, and dry woods; CP of Va and NC to PP of SC and Ga, w to Minn, ne Miss and nw Ark, avoiding GC and ME.

2. P. auriculatum Britton. Resembling *P. hispidum* in pubescence and approaching it in leaf form; otherwise as in *P. integrifolium*, from which it is sharply distinct at least at some points of contact. Spring. Dry woods and old fields; PP, BR, and RV, from n Va and adj. WVa to NC and Tenn. [*P. hispidum* var. *auriculatum* (Britt.) Rollins; *P. integrifolium* var. *auriculatum* (Britt.) Cornelius].

3. P. hispidum Raf. Perennial from a creeping rhizome, this sometimes coarse and woody, but not tuberous-thickened. Herbage usually loosely spreading-hairy, villous to scabrous-hispid, the stem often somewhat shaggy, or the pubescence seldom approaching that of no. 1. Middle and upper leaves tending to be broad-based and more or less strongly auriculate-clasping. Heads larger than in *P. integrifolium* (the disk mostly 7–10 mm wide) and averaging fewer. Achenes mostly 3–5 mm long. Otherwise as in *P. integrifolium*. (n = 36) Spring–summer. Prairies and dry woods; Ozarkian sp., from Mo and adj. Kans to Ark (including OU and ME) and adj. La, Okla, and Tex.

4. P. hysterophorus L. Annual, up to about 1 m tall, usually much branched but single-stemmed at the base, more or less hairy. Leaves pinnatifid or usually bipinnatifid, seldom more than 2 dm long and 1 dm wide. Heads small, numerous in an often leafy (not flat-topped) inflorescence, the disk only 3–5 mm wide. Achenes obovate, black, mostly 2–2.5 mm long. (n = 17) Spring–fall, or all year southward. A weed of waste places, native to tropical Am, widely intro. in our range, especially southward.

31. POLYMNIA L. LEAF CUP

Perennial herbs (all ours) or shrubs with large, opposite (or the upper alternate) leaves. Heads radiate or disciform; involucre a single series of green bracts; outer flowers pistillate and fertile, the corolla with or without an expanded, yellow to white ray; receptacle flat or nearly so, chaffy throughout, the bracts subtending the pistillate flowers larger and more herbaceous than those of the disk; disk flowers sterile, with undivided style. Ray achenes thick, not much compressed, glabrous or shortly hairy distally; pappus none.

1 Achenes prominently 3- to 6-ribbed and -angled; principal leaves distinctly pinnately lobed and veined; rays none, or white to ochroleucous and up to 1 (1.5) cm long.
  2 Achenes 3-ribbed and -angled; stem obviously hairy or glandular, at least above the middle; common species .... 1. *P. canadensis*.
  2 Achenes 4- to 6-ribbed and -angled; stem glabrous or nearly so, except sometimes for the fine branches of the inflorescence; rare species .... 2. *P. laevigata*.
1 Achenes impressed-striate, with many nerves; principal leaves palmately or pinnipalmately lobed and veined; rays yellow, mostly 1–2 (3) cm long, or rarely none (*Smallanthus*) .... 3. *P. uvedalia*.

1. P. canadensis L. Coarse perennial, 0.6–2 m tall, glabrate below, viscid-villous or stipitate-glandular above. Leaves large, thin, up to 3 dm long, broadly oblong to ovate, pinnately few-lobed, also toothed, the petiole wingless or winged only near the blade. Heads in congested cymes ending the slender branches, the disk 6–13 mm wide; involucral bracts lanceolate or lance-linear, narrower and sometimes shorter than the bracts that subtend the ray achenes; corolla of pistillate flowers minute and tubular, or expanded into a short, white or ochroleucous ray up to 10 (15) mm long. Achenes about 3–4 mm long, unequally 3-ribbed and -angled, not striate. (n = 15) Summer–fall. Moist woods, especially in calcareous regions; widespread in ne US, s in mt. provinces to NC, nw Ga, c Ala, and to OZ and OU of Ark, and in IP of Ky and Tenn. *P. radiata* (A. Gray) Small—S, the radiate form.

2. P. laevigata Beadle. Much like *P. canadensis*, but less hairy, the stem glabrous or nearly so, or shortly hairy only on the fine branches of the inflorescence. Heads averaging smaller, the disk mostly 4–10 mm wide. Achenes about 3 mm long, irregularly 4- to 6-ribbed and angled. (n = 15) Summer–fall. Rare but widespread, known from scattered localities in Ala, Fla, Ga, Mo, and Tenn.

3. P. uvedalia (L.) L. Coarse perennial, 1–3 m tall, the stem glandular or spreading-hairy beneath the heads, otherwise generally glabrous. Leaves large, sometimes more than 3 dm long, deltoid-ovate, subpalmately lobed and veined, scabrous-hispid to subglabrous above, more finely hairy and often glandular beneath, with broadly winged, sometimes runcinate petiole. Heads in moderately open, often leafy cymes, the disk ca. 1.5 cm wide; involucral bracts lance-ovate to ovate or elliptic, leafy, 1–2 cm long, much broader and generally longer than the outer receptacular bracts; rays yellow, 1–2 (3) cm long, or rarely reduced and inconspicuous. Achenes ca. 6 mm long, impressed-striate, with many nerves. (n = 16) Summer–fall. Woods and meadows; widespread in e US, from NY and Ill s to the Gulf and c Fla. *Smallanthus uvedalia* (L.) Mackenzie—S.

32. ACANTHOSPERMUM Schrank

Branching annuals with rather small, opposite, toothed or entire leaves. Heads solitary in the axils and forks of the stem, sessile or short-pedunculate, small, very shortly and inconspicuously radiate; involucre of 2 very dissimilar series, the outer set of 4–6 relatively broad, foliaceous, loose or spreading bracts, the inner set indurate, individually enclosing the achenes, each forming a prickly bur; receptacle small, convex, chaffy throughout, its bracts soft, loosely folded or convex and embracing the disk flowers; rays mostly 5–10, minute, yellow or ochroleucous, pistillate and fertile; disk flowers few, mostly 5–15, yellow or yellowish, sterile, with undivided style. Ray achenes thick, somewhat compressed parallel to radii of the head; pappus none.

1 Bur slightly compressed, strongly ribbed, invaginated at the tip so as to have a seemingly open orifice, without enlarged terminal prickles .... 1. *A. australe*.
1 Bur obviously compressed, obscurely ribbed, not invaginated; two terminal (apicolateral) prickles obviously enlarged and unlike the others.
  2 Leaf blades broad-based, up to about 3.5 cm long, abruptly contracted to a definite (commonly winged) petiole; prickles largely (but not wholly) confined to the margins and top of the bur .... 2. *A. humile*.
  2 Leaf blades more gradually narrowed to the sessile or shortly petiolar base, often more than 3.5 cm long; bur prickly all over .... 3. *A. hispidum*.

1. A. australe (Loefl.) Kuntze PARAGUAY BUR, SHEEP BUR. Stems prostrate and often rooting, up to 6 dm long or more, short-hairy, especially distally. Leaf blades rhombic-ovate or triangular, irregularly toothed above the cuneate, entire base, mostly 1.5–3.5 cm long and 1–3 cm wide, glandular-punctate and often inconspicuously short-hairy, on petioles 3–15 mm long. Heads 4–6 mm wide at anthesis; disk flowers commonly about (8) 13. Burs 7–9 mm long, ellipsoid-fusiform to nearly oblong, slightly compressed, densely glandular, strongly 5–7 ribbed, the ribs bearing one or two rows of hooked prickles 1–2 mm long, the top abruptly invaginated within an evident rim, producing a seeming orifice about 1 mm wide. (n = 11) Spring–fall, or all year southward. A weed in sandy soil and waste places; native to tropical Am, and widely intro. in our range, especially on CP.

2. A. humile (Swartz) DC. More or less erect, 2–10 dm tall, often diffusely branched, loosely spreading-hairy. Leaf blades ovate or deltoid, irregularly toothed (seldom entire), mostly 1–3.5 cm long and wide, gland-dotted beneath as well as hairy on both sides, abruptly contracted to a definite (commonly winged) petiole 4–18 mm long. Heads 3–4 mm wide at anthesis; disk flowers about 5. Burs cuneate, compressed-trigonous, glandular, uncinate-prickly on the angles and the apical margin, the sides unarmed or with only a few prickles, the body 3–4.5 mm long, the two large apicolateral prickles 2–4 mm long, one commonly straight, one hooked. All year. A weed in fields and waste places, native to the Greater Antilles; in our range known only from an old collection on ballast at Pensacola, Fla, but to be expected elsewhere as an occasional waif or adventive.

3. A. hispidum DC. Stem erect, 2–8 dm tall, often diffusely branched, long-spreading–hairy and also puberulent. Leaf blades commonly elliptic to ovate, toothed or entire, mostly (1) 2–10 cm long and 1–7 cm wide, gland-dotted beneath as well as pilose-hirsute on both sides, more or less gradually narrowed to a sessile or shortly petiolar base. Heads 4–5 mm wide at anthesis; disk flowers about 8. Burs cuneate, strongly compressed, glandular, prickly all over, the body 4–7 mm long, the two large apicolateral prickles 3–4 mm long, curved or nearly straight. (n = 11) Most or all year. A weed in fields and waste places; native to n S Am, intro. on CP in our area, especially Fla, Ga, and Ala.

33. IVA L. MARSH ELDER

Herbs or shrubs. Leaves opposite below, often alternate above, in our spp. entire or merely toothed. Heads more or less numerous, small, nodding, sessile or short-pedunculate in the axils of the upper leaves or bracts (inflorescence bractless in one sp.), disciform, the pistillate flowers 1–several, with short, tubular or obsolete corolla; involucre of a few equal or imbricate, more or less herbaceous bracts in 1–3 series; receptacle small, chaffy, its bracts commonly linear or linear-spatulate, often subtending the pistillate as well as the staminate flowers, these outer pales sometimes larger and more like the bracts of the involucre; disk flowers staminate, with undivided style, connate filaments, scarcely united anthers, and abortive or no ovary. Achenes obovate, thick but somewhat compressed parallel to the involucral bracts, glabrous to glandular or hispidulous. Pappus none. Jackson, R. C. 1960. A revision of the genus Iva L. Univ. Kans. Sci. Bull. 41: 793–875.

1 Involucral bracts distinct; pistillate flowers 2–6 in each head.
  2 Plants evidently perennial; maritime species; staminate flowers mostly (6) 8–20.
    3 Leaves mostly 4–10 cm long, only the upper (or bracteal) ones alternate; involucre mostly 2–4 mm high, with 4–5 (6) equal bracts .... 1. *I. frutescens.*
    3 Leaves mostly 1.5–4.5 (6) cm long, the middle as well as the upper ones alternate; involucre 4–7 mm high, with 6–9 imbricate bracts .... 2. *I. imbricata.*
  2 Plants annual (seldom short-lived perennial in no. 5, which has only 4–6 staminate flowers); nonmaritime.
    4 Leaves lanceolate to broadly ovate, mostly 2–15 cm wide; staminate flowers mostly 8–20.
      5 Heads axillary to evident bracts; involucral bracts 3–5, subtending the achenes, with or without a set of inconspicuous intervening receptacular bracts .... 3. *I. annua.*
      5 Heads not subtended by bracts; involucral bracts 5, separated from the achenes by a set of 5 well-developed, phyllarylike receptacular bracts .... 4. *I. xanthifolia.*
    4 Leaves linear, only 1–3 mm wide; staminate flowers mostly 4–6 .... 5. *I. microcephala.*

1 Involucral bracts united to form a toothed or lobed cup; usually 1 (2) pistillate and 1–4 staminate flowers in each head .... 6. *I. angustifolia.*

1. I. frutescens L. Perennial, somewhat fleshy shrub or coarse herb 0.5–3.5 m tall, puberulent or strigose at least above. Leaves generally all opposite except the reduced upper ones, the principal ones mostly 4–10 cm long, evidently short-petiolate. Involucre 2–4 mm high, of 4–5 (6) equal bracts; pistillate flowers 4–5 (6), with tubular corolla ca. 1 mm long tending to persist on the achene; staminate flowers mostly 6–20. Achenes copiously resin-dotted. (n = 17) Late summer–fall. Marshes and other moist places along the seashore; NS and Mass to s Fla and Tex. Most of our plants are var. *frutescens*, with the principal leaves lance-elliptic to rather narrowly lanceolate, mostly 4–7 cm long and 7–15 mm wide, 4–8 times as long as wide, with up to about 8 (seldom more) teeth on each side, or subentire. In Va this passes into the more n var. *oraria* (Bartlett) Fern. & Griscom, with larger, relatively broader leaves up to 10 cm long and 4 cm wide, less than 4 times as long as wide, and commonly with 8–17 teeth to a side.

2. I. imbricata Walter. Perennial, somewhat fleshy, glabrous herb or subshrub 3–6 (10) dm tall, commonly decumbent and branching at the base. Leaves oblanceolate to linear-oblong or elliptic, mostly 1–4.5 (6) cm × 4–10 mm, entire or with a few spreading teeth, scarcely or obscurely petiolate, only the lower ones opposite. Involucre 4–7 mm high, accrescent, its bracts 6–9, broad, rounded, imbricate, pistillate flowers 2–4, with tubular corolla 1–1.5 mm long tending to persist on the achene; staminate flowers mostly 8–15. Achenes copiously resin-dotted. (n = 17) Midsummer–fall. Dunes along the seashore; Va to s Fla, La Bahamas, and Cuba.

3. I. annua L. Taprooted annual 0.5–2 m tall; stem glabrous below, strigose near the middle, commonly spreading-hirsute above. Leaves chiefly opposite (only some of the reduced upper ones alternate), petiolate, the blade lanceolate to broadly ovate, trinerved, more or less serrate, acuminate, mostly 5–15 × 2–7 cm, scaberulous-strigose. Inflorescence of several or many spiciform branches, the heads sessile and solitary in the axils of herbaceous bracts, these lance-linear to ovate or elliptic, 5–20 mm long, often caudate-tipped, conspicuously ciliate-margined; involucre 2–3.5 mm high, its bracts 3–5, subtending the achenes, sparsely long-hirsute, broad and ciliate distally; pistillate flowers with tubular corolla ca. 1–1.5 mm long, tending to persist on the achene, sometimes subtended by very slender and inconspicuous receptacular bracts; staminate flowers mostly 8–17. Achenes sparsely to copiously resinous-dotted. (n = 17) Late summer–fall. A weed in waste places, especially in moist soil. Probably native mainly in the prairie and plains region of c US, s to the Gulf Coast, but now found nearly throughout our range. *I. ciliata* Willd.—F, G, S; *I. caudata* Small—S.

4. I. xanthifolia Nutt. Coarse, branching, taprooted annual 0.5–2 m tall, the stem glabrous below, viscid-villous in the inflorescence. Leaves chiefly opposite (only some of the upper ones generally alternate), long-petiolate, the blade ovate, often very broadly so, coarsely and often doubly serrate, trinerved or triplinerved, mostly 5–20 × 2.5–15 cm, strigose-scaberulous above, paler and often finely sericeous beneath. Inflorescence large, paniculiform, the numerous heads subsessile, not subtended by bracts; involucre 1.5–3 mm high, the 5 subherbaceous bracts somewhat larger than the 5 more membranous receptacular bracts (like an inner involucre) which partly enfold the achenes; corolla of the pistillate flowers less than 0.5 mm long, or obsolete; staminate flowers mostly 8–20. Achenes glabrous or distally hispidulous.

(n = 18) Late summer–fall. Weedy w Am sp., occasionally intro. eastward, in our range apparently only toward the n margin.

5. I. microcephala Nutt. Slender annual or occasionally short-lived perennial, simple to more often virgately branched, 4–10 dm tall, sometimes strigose to subglabrous. Leaves opposite below the middle or toward the base, otherwise alternate, linear, mostly 2.5–6.5 cm × 1–3 mm. Heads numerous, subtended by prominent, linear bracts; involucre of 4–5 distinct, more or less obovate bracts 1.5–2 mm long; pistillate flowers commonly 3, with tubular corolla ca. 1 mm long, generally subtended by slender receptacular bracts; staminate flowers mostly 4–6. Achenes distally hispidulous. (n = 16) Fall. Wet, low places, pine woods, and disturbed sites; CP from s SC to Fla, and w to Ala.

6. I. angustifolia Nutt. Annual or occasionally short-lived perennial, 4–10 dm tall, wholly erect or the stem(s) sometimes basally curved; herbage somewhat strigose. Leaves linear to narrowly elliptic, mostly 2–6 cm long and 1–8 mm wide, tapering to a sessile or shortly petiolar base. Heads axillary to linear bracts; involucre 2–3 mm high, spreading-hairy, its bracts connate to form a toothed or lobed cup; pistillate flowers 1 (2), not individually bracteate, with a tubular corolla ca. 1.5 mm long; staminate flowers 1–4. Achenes glabrous or distally hispidulous. (n = 16) Fall. Pinelands and moist low places, often in disturbed soil; GC of La to OU of Ark, w to Okla and Tex, and occasionally intro. elsewhere, as in Mo and at St. Marks, Wakulla Co., Fla, where reported by Jackson as *I. asperifolia* Less., a related, Mexican, maritime species, generally decumbent and rooting and only 2–4 dm tall.

34. AMBROSIA L. RAGWEED

Herbs (ours) or shrubs. Leaves opposite below, often alternate above, usually lobed or dissected (varying to entire in *A. bidentata*). Heads small, unisexual, rayless; staminate heads in a spiciform or racemiform, bractless inflorescence, with subherbaceous, 5–12-lobed involucre, flat receptacle with slender, filiform-setose bracts, monadelphous filaments, scarcely united anthers, and undivided style; pistillate heads borne below the staminate ones, in the axils of leaves or bracts, with closed, nutlike or burlike involucre bearing 1–several rows of tubercles or spines, the pistil 1 (–several), without corolla. Pappus none. *Franseria*—F, G, S. Our common spp. hybridize occasionally.

In addition to the following spp., two other perennials with creeping roots and dissected leaves have been reported from our range. *A. confertiflora* DC., with 2–3 rows of stout, hooked spines ca. 1 mm long on the pistillate involucres, is native to sw US and has been reported from Tenn—*Franseria confertiflora* (DC.) Rydb.—S. *A. tenuifolia* Spreng., much like *A. psilostachya* except for its more finely divided, mostly bipinnatisect leaves with linear segments, is a native of S Am that was once collected in 1900 at the mouth of the Mississippi.

1 Staminate heads short-pedunculate, and with the involucre not conspicuously oblique.
  2 Plants perennial.
    3 Maritime plant with creeping, rooting stems and finely divided leaves ..... 1. *A. hispida*.
    3 Nonmaritime plant with erect stems, creeping roots, and coarser, often only once pinnatifid leaves ..... 2. *A. psilostachya*.

2 Plants annual.
4 Leaves once or more commonly twice pinnatifid .... 3. *A. artemisiifolia.*
4 Leaves palmately 3–5 lobed, or undivided .... 4. *A. trifida.*
1 Staminate heads sessile, and with the involucre strongly oblique, produced above into a conspicuous lobe .... 5. *A. bidentata.*

1. A. hispida Pursh. Coarse, taprooted perennial with creeping stems up to several m long, the leafy flowering stems assurgent to 1–3 dm. Herbage (especially the stems) spreading-hairy, often coarsely and conspicuously so. Leaves mainly opposite, petiolate, the blade ovate in outline, mostly 2–7 cm long, 2–3 times pinnately dissected, with small segments, appearing lacy. Staminate involucres 2–3 mm high, more or less symmetrical. Fruiting involucres 3–4 mm long, with a single series of short, sharp, erect projections near the middle, or these obsolete. (n = 52) All year. Mostly on sea beaches and coastal dunes, occasionally a bit inland in hammocks; trop Fla and WI.

2. A. psilostachya DC. PERENNIAL RAGWEED. Similar to no. 3, but perennial from creeping roots, these often deep-seated; leaves thicker, short-petiolate or subsessile, usually only once pinnatifid, averaging narrower in outline; fruiting involucre merely tuberculate above, sometimes obscurely so. (n = 18, 36, ca. 54, 72) Late summer, fall. Waste places; species of the prairies and plains of c and sc US, extending into n La, and casually intro. here and there in other parts of our range. *A. rugelii* Rydb.—R, S, described as annual, but the type lacking a root.

3. A. artemisiifolia L. COMMON RAGWEED. Annual weed, 3–10 dm tall, branching at least above, variously hairy or in part subglabrous. Leaves opposite below, alternate above, once or more commonly twice pinnatifid, the blade narrowly to broadly ovate or elliptic in outline, 4–10 cm long, the middle and lower ones, at least, generally evidently petiolate. Staminate involucres 1.5–2 mm high, symmetrical or slightly oblique, inconspicuously nerved. Fruiting involucres 3–5 mm long, with a single series of short, sharp, erect spines near or above the middle. (n = 17, 18) Late summer–fall, or also spring–summer in s Fla. Waste places, throughout our range and most of the US and s Can. *A. elatior* L.—S; *A. glandulosa* Scheele—S; *A. monophylla* (Walt.) Rydb.—S.

4. A. trifida L. GIANT RAGWEED. Coarse annual weed up to 2 (5) m tall; stem spreading-hairy above, often glabrous below. Leaves opposite, petiolate, more or less scabrous, the blade broadly elliptic to more often ovate or suborbicular, often 2 dm long, serrate, palmately 3–5-lobed, or, especially in depauperate specimens, lobeless. Staminate involucres ca. 1.5 mm high, slightly oblique, unilaterally evidently 3-nerved. Fruiting involucres 5–10 mm long, several-ribbed, each rib ending in a tubercle or short spine. (n = 12) Late summer–fall. Moist soil and waste places throughout our range except penins. Fla, and widespread elsewhere in US. *A. aptera* DC—S.

5. A. bidentata Michx. Branching annual weed 3–10 dm tall, the stem conspicuously spreading-hirsute, especially above. Leaves numerous, sessile, opposite below, alternate above, lanceolate, acuminate, 2.5–7 cm × 4–10 mm, usually with a single pair of sharp large teeth below the middle, hirsute to scabrous, or the upper side glabrous. Staminate heads sessile, the involucre strongly oblique, its upper side produced into a conspicuous, retrorsely spreading, hispid-hirsute, lanceolate to triangular-ovate lobe, the bractless inflorescence thus appearing retrorse-bracteate. Fruiting involucre 5–8 mm long, villous-hirsute, several-ribbed, the ribs produced into short, stiff spines. (n = 17) Late summer–fall. Prairies, open woods, and disturbed soil; native to the prairies and Ozark region of c and sc US, found in all provinces of Ark and in n La, and occasionally intro. elsewhere in our range.

35. XANTHIUM L. COCKLEBUR

Coarse annual weeds. Leaves alternate, toothed or few-lobed to entire. Heads small, unisexual, solitary or clustered in the axils; staminate heads uppermost, many-flowered, their involucre of separate bracts in 1–3 series, the receptacle cylindric, chaffy, the filaments monadelphous, the pistil consisting mainly of the undivided style; involucre of the pistillate heads completely enclosing the 2 flowers, forming a conspicuous 2-chambered bur with hooked prickles; pistillate flowers lacking a corolla, and with the styles exsert from the involucre. Pappus none.

1 Leaves broad, generally cordate or deltoid at the base, spineless .... 1. *X. strumarium*.
1 Leaves lanceolate, tapering to the base, bearing a 3-forked axillary spine .... 2. *X. spinosum*.

1. X. strumarium L. COMMON COCKLEBUR. Plants 2–20 dm tall, appressed-hairy or subglabrous. Leaves long-petiolate, the blade broadly ovate to suborbicular or reniform, generally cordate or subcordate at the base, toothed and sometimes shallowly 3–5-lobed, often 15 cm long. Heads in several or many short, axillary inflorescences; bur broadly cylindric to ovoid, ellipsoid, or subglobose, (1) 1.5–3.5 cm long, covered with stout, hooked prickles, terminated by two straight (in var. *strumarium*) or in ours more or less incurved beaks. (n = 18) Summer–fall. Fields, waste places, floodplains, and lake and sea beaches; now a cosmopolitan weed, probably originally native to the New World. We have two vars., both essentially throughout our range. Var. *canadense* (Miller) T. & G., with the burs brownish or yellowish-brownish, 2–3.5 cm long, the lower part of the prickles conspicuously spreading-hairy as well as more or less stipitate-glandular, is more common on AC than inland. *X. echinatum* Murray—F, S; *X. italicum* Moretti—F; *X. pensylvanicum* Wallr.—F, S; *X. speciosum* Kearney—F, S. Var. *glabratum* (DC.) Cronq., with merely atomiferous-glandular or glandular-puberulent to subglabrous, commonly paler burs seldom more than ca. 2 cm long, is common in all provinces. *X. americanum* Walt.—S; *X. chinense* Mill.—F; *X. cylindraceum* Millsp. & Sherff—S; *X. echinellum* Greene—S; *X. globosum* Shull—F; *X. inflexum* Mackenzie & Bush—F.

2. X. spinosum L. Plants 3–12 dm tall, the stems strigose or puberulent. Leaf blades lanceolate, entire or with a few coarse teeth or pinnate lobes, tapering to each end, short-petiolate, 2.5–6 × 0.5–2.5 cm, sparsely strigose or glabrate above, except for the usually more hairy veins, densely silvery-sericeous beneath, and bearing a tripartite yellow spine 1–2 cm long in the axil. Burs mostly solitary or few in the axils, cylindric, ca. 1 cm long, beakless or with a single short beak, finely puberulent and provided with slender, hooked prickles. (n = 18) Summer. Waste places; now a cosmopolitan weed in the warmer parts of the world, and occasionally found in our range. *Acanthoxanthium spinosum* (L.) Fourr.—S.

36. MADIA Mol. TARWEED

1. **M. sativa** Molina. Tar-scented annual 2–10 dm tall, conspicuously spreading-hirsute and stipitate-glandular. Leaves opposite near the base, otherwise alternate, mostly linear or linear-oblong, commonly 3–18 cm × 3–12 mm, entire or nearly so, rather crowded, especially below. Heads radiate, the rays pistillate and fertile, yellow, commonly ca. (8) 13, 3–7 mm long, 3-lobed, relatively broad; involucral bracts subherbaceous, uniseriate, equal, enclosing the ray achenes, the involucre ovoid or broadly urn-shaped, 6–12 mm high and wide, appearing deeply sulcate; receptacle flat, with a single series of bracts between the ray and disk flowers, otherwise naked; disk flowers perfect and fertile; style branches flattened, externally hairy above, with introrsely marginal stigmatic lines extending nearly to the tip. Achenes black, radially somewhat compressed. Pappus none. (n = 16) Summer. Native to Chile and the Pacific states, and casually introduced in waste places here and there in our range. We have both of the 2 vars. Var. *sativa* has a relatively open, corymbiform to racemiform inflorescence. Var. *congesta* T. & G. (*M. capitata* Nutt.) has the heads tending to be aggregated into a number of glomerules.

37. JAMESIANTHUS Blake & Sherff

1. **J. alabamensis** Blake & Sherff. Fibrous-rooted perennial. Stem solitary, 6–15 dm tall, simple below the inflorescence, sparsely pubescent above with spreading, gland-tipped hairs. Leaves opposite, cauline, sessile or nearly so and somewhat auriculate-clasping, lance-elliptic, mostly 5–9 × 1–4 cm, remotely toothed, glabrous, or scabrous above and on the main veins beneath. Heads terminating the slender branches, yellow, radiate, the rays few (commonly ca. 8), pistillate and fertile, epappose, mostly 1–1.5 cm long; involucre glandular-hairy, ca. 3-seriate, imbricate, its bracts relatively broad and thin, the outer herbaceous; receptacle somewhat convex, naked; disk ca. 1 cm wide, its flowers perfect, the outer fertile, the inner often sterile; style branches flattened, with ventromarginal stigmatic lines and a short, externally hispidulous appendage. Achenes thick, several-nerved, hispidulous, with a pale cartilaginous collar. Pappus of the disk flowers of several slender, unequal, quickly deciduous capillary bristles. (n = 16) Late summer–fall. Streambanks in woods; Franklin and Colbert Cos., nw Ala.

38. BALDUINA Nutt. Nom. conserv. HONEYCOMB HEAD

Annual to perennial herbs. Leaves alternate, narrow. Heads 1–many on naked peduncles terminating the stem (and branches), radiate, the rays neutral, yellow; involucral bracts about 3-seriate, rather firm, greenish at least in part, the outer shorter; receptacle somewhat convex, chaffy, its bracts connate to form a toothed or cleft honeycomb in which the flowers are set; disk flowers numerous (30–200), perfect and fertile; style branches flattened, with inconspicuous ventromarginal stigmatic lines and an abrupt, slender appendage clothed at the base with a ring of short hairs. Achenes turbinate, villous. Pappus a ring of 7–12 small scales. (x = 18) *Actinospermum*

Ell.—S; *Endorima* Raf.—S. Parker, E. S., and S. B. Jones, 1975. A systematic study of the genus Balduina (Compositae, Heliantheae). Brittonia 27: 355–361.

1 Fibrous-rooted perennials; pappus mostly 1–2 mm long; heads relatively large, the disk mostly (1) 1.5–2.5 cm wide, the rays mostly 1.5–4 cm long.
 2 Disk yellow or somewhat reddish; basal leaves up to 10 cm long .... 1. *B. uniflora*.
 2 Disk dark purple; basal leaves usually longer, mostly 7–30 cm .... 2. *B. atropurpurea*.
1 Taprooted annual (possibly sometimes biennial); pappus less than 1 mm long; heads relatively small, the disk up to 1.5 cm wide, the rays mostly 1–2 cm long .... 3. *B. angustifolia*.

1. B. uniflora Nutt. Fibrous-rooted, mostly single-stemmed perennial 3–10 dm tall, simple or with a few long, suberect branches, puberulent (and often also atomiferous-glandular) at least on the long, stout, naked peduncle(s), often glabrous below. Basal leaves variously short and elliptic or spatulate to elongate and nearly linear, up to 10 cm long (petiole included) and 1 (2) cm wide; cauline leaves erect, mostly linear or oblanceolate, gradually reduced upward. Heads 1–4 (11), relatively large, the disk mostly (10) 15–25 mm wide, yellow to often somewhat reddish, the rays mostly ca. (5, 8) 13 (21), (1) 1.5–4 cm long, the disk corollas 5.5–8 mm long; honeycomb ridges of the receptacle toothed or cleft but not distinctly cuspidate. Pappus of several slender scales 1–2 mm long. (n = 36) Midsummer–fall. Moist to dry pine woods and savannas; CP from NC to n Fla and w to se La. *Endorima uniflora* (Nutt.) Barnhart—S.

2. B. atropurpurea Harper. Similar to *B. uniflora*, but with dark purple disk flowers, more often 2–4 heads, and typically longer lower leaves, these mostly 7–30 cm long and up to 1 cm wide. (n = 18) Late summer–fall. Pitcher plant bogs, commonly in wetter habitats than *B. uniflora*; CP from SC to Ga and n Fla.

3. B. angustifolia (Pursh) B. L. Robinson. Taprooted annual or perhaps sometimes biennial, 2–10 dm tall, with 1–several stems from the base, evidently branched above when well developed, atomiferous-glandular (especially on the slender peduncles) and sometimes also short-hairy, varying to glabrous. Leaves numerous, linear or nearly so, mostly 1–5 cm long and up to 1.5 (2.5) mm wide. Heads (1–) 5–many, relatively small, the disk mostly 6–15 mm wide, yellow, the rays ca. (5) 8 (13), 1–2 cm long, the disk corollas mostly 4–5 mm long; honeycomb ridges of the receptacle tending to be rather abruptly spinulose-cuspidate at the angles. Pappus of several broad, rounded scales less than 1 mm long. (n = 18) All year, but mostly summer–fall. Dry, sandy soil, especially in pinelands, on CP; throughout peninsular Fla, n and w to Ga and s Miss. *Actinospermum angustifolium* (Pursh) T. & G.—S.

39. GAILLARDIA Foug. BLANKET FLOWER

Taprooted herbs. Leaves alternate or all basal, entire or toothed to pinnatifid. Heads middle-sized to large, terminating the branches, often long-pedunculate, radiate or sometimes discoid, the rays yellow to partly or wholly red or purple, broad, 3-cleft, usually neutral; involucral bracts in 2–3 series, herbaceous above the chartaceous base, more or less spreading, becoming reflexed in fruit; receptacle convex to subglobose, provided with numerous setae that do not individually subtend the flowers, or the setae obsolete; disk corollas perfect and fertile, the corolla lobes woolly-villous; style branches flattened, with introrsely marginal stigmatic lines and a usually more or less elongate and externally hairy appendage. Achenes broadly obpyramidal, partly or wholly covered by a basal tuft of long, ascending hairs. Pappus of 6–10 awned scales.

1 Receptacle with well-developed setae about equaling the achenes .... 1. *G. pulchella.*
1 Receptacle essentially naked .... 2. *G. aestivalis.*

1. G. pulchella Foug. Glandular-villous annual or short-lived perennial, 1–6 dm tall, simple to more often freely branched, erect or often decumbent at the base, the littoral forms commonly somewhat succulent. Leaves variously entire or few-toothed to pinnately cleft or lobed, up to about 10 × 3 cm, sessile and often clasping, or the lower tapering to a petiole. Heads radiate, the rays 6–15, red-purple or the tip yellow, mostly 1–2 cm long, the disk brownish-purple, 1–2.5 cm wide; setae of the receptacle about equaling or a little longer than the achenes; lobes of the disk corollas narrowly triangular to acuminately subcaudate. (n = 17, 34) Summer–fall, or all year southward. In our range chiefly along sandy beaches, from s Va to Fla and La, and occasionally adventive in disturbed habitats inland; also extending inland from Tex to s Gr Pl. *G. drummondii* (Hook.) DC—S; *G. picta* Sweet—S, the littoral, somewhat succulent phase, perhaps properly to be treated as a distinct var.

2. G. aestivalis (Walter) H. Rock. Conspicuously to obscurely short-hairy annual or short-lived perennial, 1–6 dm tall, simple to more often freely branched. Leaves lance-linear to ovate-oblong, sessile, or the lower oblanceolate and petiolate, entire or slightly toothed to sometimes lyrate-pinnatifid, seldom more than 7 × 1 cm. Heads radiate or sometimes discoid, the rays 6–15, 1–2 cm long; disk 1–2 cm wide; receptacle essentially naked; lobes of the disk corollas caudate-tipped. (n = 17, 18) Spring–fall. Prairies and open woods, often in sandy soil; CP from SC to Fla and Tex, thence n to Kans. Var. *aestivalis,* with purple-brown disk and red or purple to partly or wholly yellow rays, or discoid, occurs nearly throughout the range of the species. *G. lanceolata* Michx.—G, S. Var. *flavovirens* (C. Mohr) Cronq., with yellow disk and rays, seldom discoid, in our range occurs mainly in La and sw Ark, but with scattered stations e to Fla. *G. chrysantha* Small—S; *G. lutea* Greene—F.

40. HELENIUM L. SNEEZEWEED, BITTERWEED

Herbs. Leaves alternate, very often decurrent on the stem. Heads 1–many, terminating the branches, mostly radiate, the rays pistillate or neutral, mostly yellow, cuneate, 3-lobed, not very numerous; involucral bracts in 2–3 series, subequal or the inner shorter, more or less herbaceous, spreading or often soon deflexed, the outer sometimes joined at the base; receptacle convex to ovoid or conic, naked; disk flowers very numerous, perfect, the corolla lobes glandular-hairy; style branches flattened, with ventromarginal stigmatic lines and dilated, subtruncate, penicillate tip. Achenes 4–5-angled, with as many intermediate ribs, pubescent on the angles and ribs, or glabrous. Pappus of 5–10 scarious or hyaline, often awn-tipped scales. Most or all of the species are poisonous (and unpalatable) to livestock. Bierner, M. W. Taxonomy of *Helenium* sect. Tetrodus and a conspectus of North American *Helenium* (Compositae).

Brittonia 24: 331–355. 1972. Rock, H. A. A revision of the vernal species of *Helenium* (Compositae). Rhodora 59: 101–116; 128–158; 168–178; 203–216. 1957.

1 Fibrous-rooted perennials.
  2 Rays pistillate; heads several to numerous; disk yellow .... 1. *H. autumnale.*
  2 Rays neutral (rarely wanting); heads and disk various.
    3 Disk red-brown or purple-brown; heads 1–many.
      4 Pappus scales (generally 5) shortly awn-tipped, ca. (0.5) 1 mm long overall; heads in well-developed plants generally several or many; disk corollas predominantly 4-merous .... 2. *H. flexuosum.*
      4 Pappus scales (5–10) awnless; heads 1–few; disk corollas 5-merous.
        5 Lower part of the stem spreading-villous; pappus ca. 0.5 mm long or less .... 3. *H. campestre.*
        5 Lower part of the stem glabrous or merely atomiferous-glandular; pappus 1–1.7 mm long .... 4. *H. brevifolium.*
    3 Disk yellow; heads 1 (–3).
      6 Pappus scales entire to somewhat lacerate, never fimbriate-fringed.
        7 Peduncle and achenes obviously pubescent, as well as often atomiferous-glandular.
          8 Middle and lower cauline leaves distinctly (though narrowly) decurrent, generally for well over 1 cm; peduncle loosely villous-puberulent or floccose-puberulent ....forms of 4. *H. brevifolium.*
          8 Middle and lower cauline leaves scarcely decurrent (less than 5 mm, or not at all); peduncle rough-puberulent .... 5. *H. pinnatifidum.*
        7 Peduncle and achenes glabrous, or merely atomiferous-glandular; leaves decurrent .... 6. *H. vernale.*
      6 Pappus scales strongly and irregularly fimbriate-fringed distally .... 7. *H. drummondii.*
1 Taprooted annuals or biennials.
  9 Leaves decurrent, the larger ones generally ca. 1 cm wide or more; disk corollas 4-merous; pappus scales awnless, 0.1–0.3 mm long .... 8. *H. quadridentatum.*
  9 Leaves not decurrent, very narrow, seldom more than 2 mm wide; disk corollas 5-merous, pappus scales awned, ca. 1.5 mm long overall .... 9. *H. amarum.*

1. H. autumnale L. Fibrous-rooted perennial, 5–15 dm tall, finely hairy or subglabrous. Leaves many, toothed, mostly elliptic to oblong or lanceolate, 4–15 × 1–4 cm, narrowed to a sessile or subpetiolar base, decurrent, the lower generally deciduous. Heads several or many in a leafy inflorescence, the disk yellow, hemispheric or subglobose, 1–2 cm wide; involucral bracts soon deflexed; rays ca. 13 to ca. 21, pistillate, mostly 1.5–2.5 cm long. Pappus of ovate or lanceolate scales tapering to a short awn, up to ca. 1 mm long overall. (n = 16, 17, 18) Summer–fall. Moist, low ground; throughout most of the US and adj. Can, in all provinces of our range. Our plants are var. *autumnale. H. latifolium* Mill.—S; *H. parviflorum* Nutt.—S.

H. virginicum Blake, of swampy places in Augusta Co., Va, is doubtfully distinct from *H. autumnale*. It differs in its not very leafy stem, with the persistent and elongate basal leaves larger than the progressively reduced and relatively few cauline ones, and in its white-hyaline pappus scales 1.5–2 mm long.

2. H. flexuosum Raf. Fibrous-rooted perennial, 2–10 dm tall, more or less puberulent or villous-puberulent, at least on the stem, the lower part of the stem generally spreading-villous. Leaves decurrent, but smaller, less numerous, and more erect than in no. 1, entire or subentire, the lowermost ones oblanceolate, commonly deciduous, the others oblong or lanceolate to lance-linear, sessile, generally not much reduced upward, except in the inflorescence, 3–12 × 0.5–2 cm. Heads in well-developed plants numerous in an open, corymbiform, leafy-bracteate inflorescence, the disk red-brown or purplish, subglobose or ovoid-globose, 6–15 mm wide; involucral bracts soon deflexed; rays ca. 8 to ca. 13, neutral, sometimes purplish at the base,

(0.5) 1–2 cm long, or rarely wanting; disk flowers predominantly 4-merous. Pappus scales 5 (–8), ovate or lanceolate, shortly awn-tipped, ca. (0.5) 1 mm long overall. (n = 14) Spring–fall. Moist ground and waste places; essentially throughout our range, n to NH and w to Kans and Tex. *H. nudiflorum* Nutt.—F, S; *H. polyphyllum* Small—S.

3. H. campestre Small. Like a sparingly branched, few-headed form of no. 2 (heads commonly 2–5), but the pappus of 5–10 short, blunt or rounded scales ca. 0.5 mm long or less. Heads averaging a bit larger, the rays 1.5–3 cm long, the disk 1–2 cm wide. (n = 14) Spring. Prairies and open woods; OU and ME of Ark.

4. H. brevifolium (Nutt.) A. Wood. Fibrous-rooted perennial, 3–8 dm tall, glabrous or merely atomiferous-glandular below, the peduncles and involucres loosely villous-puberulent or floccose-puberulent. Leaves relatively few, the basal tufted and generally persistent, oblanceolate or elliptic, obtuse or rounded, entire or slightly toothed, usually petioled, 2–17 cm long overall and 0.6–2 (3) cm wide; cauline leaves few, reduced, seldom exceeding the internodes, narrowly decurrent. Heads 1–3 (6), naked-pedunculate, the disk purple-brown or red-brown (rarely yellow), hemispheric to subglobose, 1–2 cm wide; involucral bracts generally soon deflexed; rays ca. 8 to ca. 13, neutral, 1–2.5 cm long, showy, often 1 cm wide. Pappus of 5–10 awnless, often distally rounded scales 1–1.7 mm long. Achenes appressed setose on the ribs. (n = 13, 14) Spring. Various mostly moist habitats, the inland localities chiefly in bogs; GC of w Fla, s Ala, s Miss, and se La, n irregularly to AC of se Va and mt. provinces of Ala, Ga, NC, and Tenn. *H. curtisii* A. Gray—F, S, the rare form with yellow disk.

5. H. pinnatifidum (Nutt.) Rydb. Fibrous-rooted perennial, 2–10 dm tall, rough-puberulent on the peduncle (at least distally) and often also on the involucre, less so or glabrous below. Basal leaves tufted and persistent, broadly linear to elliptic or oblanceolate, entire to coarsely toothed or somewhat pinnatifid, acute to obtuse or rounded, 3–25 × 0.5–2.5 cm overall; cauline leaves relatively few, reduced, the middle ones very shortly decurrent (less than 5 mm), the others not. Heads 1 (–3), naked-pedunculate, the disk yellow, hemispheric, (1) 1.5–2.5 cm wide; involucral bracts spreading or tardily deflexed; rays (13) 16–40, neutral, relatively narrow, ca. 5 mm wide. Achenes setose-hispidulous on the ribs. Pappus scales 5–10, 1–2 mm long, commonly erose or somewhat lacerate. (n = 16, 17) Spring. Wet pinelands, temporary pools, and pocosins; nearly throughout Fla, n irregularly (chiefly CP) to NC. *H. vernale* Walt.—S, misapplied.

6. H. vernale Walter. Much like no. 5, but the herbage glabrous or nearly so throughout except for some sessile glands. Basal leaves usually (not always) entire, more consistently acute; cauline leaves evidently decurrent (the middle ones for 2 cm or more). Achenes merely atomiferous-glandular. (n = 17) Spring. In wet habitats like those of no. 4, and sometimes occurring with it; CP of n Fla, n to se NC and w to se La. *H. helenium* (Nutt.) Small—S.

7. H. drummondii H. Rock. Fibrous-rooted perennial, 2–6 dm tall, puberulent toward the top of the peduncle, otherwise largely glabrous or merely atomiferous-glandular. Basal leaves tufted and persistent, broadly linear to linear-elliptic, entire to sometimes coarsely toothed or lobulate, gradually acute, 6–20 × 0.3–1.1 cm overall; cauline leaves progressively reduced, evidently decurrent. Heads 1 (–3), naked-pedunculate, the disk yellow, hemispheric, 1.5–2 cm wide; involucral bracts spreading or tardily reflexed; rays ca. 21, neutral, 1.5–2 cm long. Achenes pubescent to glabrous. Pappus scales 5–10, 2–3.5 mm long, deeply fringed-fimbriate distally. (n = 16) Spring.

Boggy places and pond margins on GC; c and sw La to Tex; n Fla? *H. fimbriatum*—S, the nomenclature recondite.

8. H. quadridentatum Labill. Taprooted annual or biennial, 3–10 dm tall, atomiferous-glandular, otherwise glabrous or nearly so. Lower leaves oblanceolate to obovate, entire to pinnatifid, up to 15 × 6 cm, the others more lanceolate to oblong or linear and gradually reduced upward, evidently decurrent. Heads several or many, terminating the branches, the disk yellow-brown or purplish, ovoid-conic to ellipsoid, 7–14 × 6–10 mm; involucral bracts soon deflexed; rays 10–15, pistillate, 5–12 mm long; disk flowers 4-merous. Achenes short-hairy. Pappus of ca. 6 minute, rounded scales ca. 0.1–0.3 mm long. (n = 13) All year, esp. spring. A weed in old fields and other disturbed sites on CP from s Ala to La, Tex, and Mex; Cuba; occasionally adventive elsewhere.

9. H. amarum (Raf.) H. Rock. Taprooted, glabrous or atomiferous-glandular annual, 2–5 dm tall. Leaves very numerous, not decurrent, mostly linear or linear-filiform, 1.5–8 cm long, seldom more than 2 mm wide, the basal and lower cauline often a bit wider, but soon deciduous. Heads on short, slender, naked peduncles extending above the leafy part of the plant; involucral bracts soon deflexed; rays 5–10, pistillate, 5–12 mm long: disk yellow, subglobose, 6–12 mm wide; disk corollas 5-merous. Pappus scales commonly 6–8, ca. 1.5 mm long overall, the awn tip about equaling the hyaline body. (n = 15) Summer–fall. Prairies, open woods, fields, and waste places, especially in sandy soil; widespread in our range, s to Mex and WI. *H. tenuifolium* Nutt.—F, S.

41. PALAFOXIA Lag.

Annual or perennial herbs or half-shrubs. Leaves alternate or the lower opposite, simple and mostly entire. Heads in an open, corymbiform or paniculiform inflorescence, discoid or in some extralimital species radiate, pink or purple to white; involucre campanulate to obconic or cylindric, its bracts imbricate or subequal in 1–3 series, narrow, more or less herbaceous throughout, or distally (or largely) scarious and whitish or anthocyanic; receptacle small, flat, naked; disk flowers relatively few, less than 40, perfect and fertile; style branches slender and elongate, slightly flattened, externally hispidulous throughout, with introrsely marginal stigmatic lines near the base or extending to the tip, without a clearly differentiated appendage. Achenes linear to clavate-obpyramidal, 4-angled. Pappus of 4–12 scales, generally with a strong midrib. Turner, B. L., and M. I. Morris. 1976. Systematics of *Palafoxia* (Asteraceae: Helenieae). Rhodora 78: 567–628.

The Texan species *P. hookeriana* T. & G. has been reported as a casual introduction in Lucedale, Miss. It is readily recognized by its well-developed ray flowers, which are 8–13 in number, pink, and deeply trifid. It is not provided for in the key.

1 Perennial herbs or half-shrubs; peduncles not glandular; flowers mostly (13) 15–30 per head.
  2 Pappus scales elongate, 4–6 mm long; corolla throat (as distinguished from the slender basal tube) very short, 1–2 mm long, obviously shorter than the tube and the lobes .... 1. *P. integrifolia*.
  2 Pappus scales short, 1–2 mm long; corolla throat cylindric, 3–6 mm long, obviously longer than the tube and the lobes .... 2. *P. feayi*.
1 Annual herbs; peduncles beset with conspicuous, tack-shaped glands; flowers 5–15 per head .... 3. *P. callosa*.

1. P. integrifolia (Nutt.) T. & G. Fibrous-rooted perennial herb or half-shrub mostly 0.5–1.5 m tall. Leaves numerous, rather small, with scabrous to merely strigillose or subglabrous, linear to lanceolate blade mostly 2.5–7 cm long and 2–13 mm wide, tapering to a short petiole. Involucre obconic, 8–11 mm high, at least the inner bracts distally whitish (pinkish)-scarious; disk flowers mostly (13) 15–26, the corolla lavender, its throat only 1–2 mm long, obviously shorter than the slender basal tube (3–6 mm) and the lobes (3–5 mm). Pappus 4–6 mm long. (n = 12) Spring–fall. Sandy pine and oak woods; Fla and adj. se and sc Ga. *Polypteris integrifolia* Nutt.—S.

2. P. feayi A. Gray. Fibrous-rooted perennial herb or half-shrub mostly 1–3 m tall. Leaves numerous, rather small, with scabrous to merely strigillose, elliptic or elliptic-ovate to broadly oblong blade mostly 2–6 cm long and 0.7–2.5 cm wide, on a definite short petiole mostly 2–6 mm long. Involucre campanulate or obconic, 6–9 mm high, its bracts commonly somewhat anthocyanic at least distally, but not scarious; disk flowers mostly 17–30, the corolla white to pinkish or pale lavender, its throat cylindric, 3–6 mm long, obviously longer than the slender basal tube (1.5–2.5 mm) and the lobes (2–2.5 mm). Pappus 1–2 mm long. (n = 12) Spring–fall. Sandy pine and oak woods; peninsular Fla.

3. P. callosa (Nutt.) T. & G. Slender, branching annual, mostly 1–6 dm tall, coarsely strigose, the inflorescence densely beset with conspicuous, dark, tack-shaped, stipitate glands. Leaves linear, up to ca. 6 cm × 3 mm, becoming reduced to mere filiform bracts in the much-branched, open inflorescence. Involucre obconic or campanulate, ca. 4–6 mm high, its bracts few, equal, strigose, herbaceous except for the pink-purple, more or less scarious tip; flowers mostly 5–15, the pink corollas surpassing the disk and loosely spreading, the throat very short, mostly less than 1 mm long, the lobes and tube more elongate. Achenes slender, 3–4 mm long. Pappus scales seldom more than 1 mm long, the midrib scarcely reaching the tip. (n = 10) Late summer–fall. Open, often dry and barren places; Mo and n Ark (OZ) to Tex and NM, and disjunct in the Black Belt of Miss.

42. FLAVERIA Juss.

Herbs or half-shrubs. Leaves strictly opposite. Heads yellow, small, 1–several-flowered, numerous, crowded into small glomerules, these scattered or aggregated into a compound terminal inflorescence; involucre of 1–several subequal, often yellowish bracts, sometimes with 1 or 2 smaller outer ones; receptacle small, naked; ray flower solitary, small, pistillate and fertile, or wanting; disk flowers 1–several, perfect and fertile, wanting from some heads of *F. trinervia*; style branches flattened, with introrsely marginal stigmatic lines, subtruncate and minutely papillate-penicillate at the tip. Achenes small, 10-ribbed. Pappus none, or of 2–4 small scales in some extralimital spp.

1 Principal involucral bracts (4) 5 (6); glomerules of heads forming a terminal, flat-topped inflorescence above the leafy part of the plant; fibrous-rooted perennial .... 1. *F. linearis.*
1 Principal involucral bracts 1–3; glomerules of heads scattered through the leafy part of the plant, not forming a terminal inflorescence; taprooted annuals.
  2 Principal involucral bracts 3; heads 2–several-flowered; individual glomerules of heads of compactly cymose structure, without a common receptacle ..... 2. *F. bidentis.*
  2 Principal involucral bracts 1–2; heads 1-flowered; heads of each glomerule grouped onto an evidently setose common receptacle .... 3. *F. trinervia.*

1. F. linearis Lag. Fibrous-rooted perennial herb or half-shrub, mostly 3–10 dm tall, essentially glabrous. Leaves entire, linear or nearly so, mostly 4–15 cm × 1–10 (15) mm. Glomerules of heads cymosely aggregated to form a terminal, flat-topped inflorescence above the leafy part of the plant; involucre 3–4 mm high, of (4) 5 (6) equal, several-nerved bracts and sometimes 1 or 2 minute outer ones; some heads with a single ligule 2–3 mm long, others discoid; disk flowers mostly 5–10 (13). (n = 18) All year. Pinelands, sand dunes, hammocks, shores, and disturbed sites, often maritime; c and s Fla to WI and Mex. *F. floridana* J. R. Johnston—S; *F. latifolia* (J. R. Johnston) Rydb.—S.

2. F. bidentis (L.) Kuntze. Taprooted annual 1–10 dm tall, essentially glabrous, divaricately branched. Leaves trinerved, narrowly elliptic or lance-elliptic, mostly 5–10 × 1–2 cm, acute, toothed, tapering to a narrow or subpetiolar base. Individual glomerules of heads 1–2 cm wide, compactly cymose in structure, without a common receptacle, terminal on foreshortened major and minor axes, some of these commonly surpassed by subtending branches; involucre 3–3.5 mm high, of 3 several-nerved principal bracts and 1 or 2 shorter outer ones; some heads with an erect, inconspicuous ligule scarcely 1 mm long; others discoid; disk flowers mostly 2–8. (n = 18) Summer–fall or all year. A weed of waste places; native to S Am, adventive chiefly about major ports in Fla, Ga, and Ala.

3. F. trinervia (Sprengel) C. T. Mohr. Taprooted, minutely puberulent or subglabrous annual 1–10 dm tall, divaricately branched. Leaf blades trinerved, elliptic to lanceolate or elliptic-oblanceolate, toothed, mostly 2–7 × 0.5–2.5 cm, tapering to a narrow or petiolar base up to 2 cm long. Heads individually 1-flowered, aggregated into secondary heads ca. 1 cm wide or less, the common receptacle with scattered setae half to fully as long as the primary heads; secondary heads commonly subtended by a secondary involucre of a few leafy bracts, and many of them surpassed by subtending branches; involucre of 1 or 2 bracts, 3–4 mm long; many of the outer heads bearing a ligule scarcely 1 mm long, the others bearing a disk flower. (n = 18) All year. Tropical Am weed of waste places; intro. into s Fla, and collected once on ballast at Mobile, Ala.

43. DYSSODIA Cav.

Aromatic herbs (ours) or subshrubs. Leaves opposite at least below, often alternate above, entire or more often pinnatifid or bipinnatifid, often with scattered embedded oil glands. Heads terminating the stem and branches, sessile or more often naked-pedunculate, radiate, the rays mostly yellow or orange, pistillate and fertile; involucral bracts beset with scattered embedded oil glands, 2–3-seriate, those of the outer series commonly looser and more herbaceous (and often smaller) than the somewhat chartaceous or coriaceous (and sometimes anthocyanic) inner ones, these sometimes united to form a cup; receptacle naked, flat or slightly convex (in our spp.) to conic; disk flowers perfect and fertile; style branches flattened, with introrsely marginal stigmatic lines, subtruncate and minutely penicillate (in our spp.) or with an elongate, slender appendage. Achenes slender, striate-angled or somewhat compressed. Pappus of 5–22 scales in 1–2 series, these sometimes (as in our spp.) cleft to below the middle into several bristles. Strother, John L. 1969. Systematics of *Dyssodia* Cavanilles (Compositae: Tageteae). Univ. Calif. Publ. Bot. 48: 1–88.

1 Involucral bracts all distinct, or united only at the base; stem leafy throughout; heads inconspicuously radiate, sessile or on peduncles mostly less than 3 cm long ..... 1. *D. papposa*.
1 Principal involucral bracts united most of their length to form a toothed cup; heads conspicuously radiate, arising above the leafy part of the stem on slender peduncles mostly 3–7 cm long ..... 2. *D. tenuiloba*.

1. D. papposa (Vent.) A. S. Hitchc. Malodorous, glabrous or somewhat hairy, taprooted annual 1–7 dm tall, much branched when well developed, leafy throughout, the leaves 1.5–5 cm long, pinnatisect to more often bipinnatisect, with slender segments. Heads generally numerous, sessile or on short peduncles up to ca. 3 cm long; involucre 5–8 mm high, the inner bracts distinct or united only at the base, with overlapping margins; rays commonly ca. 5 or 8 (13), inconspicuous, ca. 1–2 mm long; disk flowers mostly 12–50. Pappus of ca. 20 unequal scales in 2 series, each scale close-pinnately cleft into several slender bristles. (n = 12, 13) Late summer–fall. A weed of waste places, native to Mex and the Great Plains, now widely introduced elsewhere, in our range known from Ark and Tenn, reported from La, and to be expected occasionally elsewhere. *Boebera papposa* (Vent.) Rydb.—S.

2. D. tenuiloba (DC.) B. L. Robinson. Glabrous or somewhat hairy, branched, taprooted annual or short-lived perennial 1–3 dm tall. Leaves 1–3 cm long, pinnatisect, with slender segments. Heads arising above the leafy part of the stem on slender peduncles mostly 3–7 cm long; involucre 5–7 mm high, its inner bracts connate most of their length to form a toothed cup; rays 9–21 (often 13 or 16), spreading, 4–9 mm long; disk flowers numerous, commonly 75–100. Pappus of 10–12 similar scales, these palmately cleft into 3–5 slender bristles, the central one the longest. (n = 8, 13, 16, 26) Spring–fall. Disturbed sites; native to Tex and n Mex, adventive in Fla and s Miss. Our plants are var. *tenuiloba*. *Thymophylla tenuiloba* (DC.) Small—S.

44. PECTIS L.

Aromatic herbs, ours taprooted annuals (varying to short-lived perennial in no. 3), the stem in ours minutely pubescent in lines, or pubescent all around near the base. Leaves opposite, simple, sessile, usually entire but with several pairs of conspicuous, bristly cilia toward the base, commonly with some evident embedded oil glands, in our spp. otherwise glabrous except for the often scabrous margins. Heads terminal and axillary, individually pedunculate or often sessile in clusters, radiate, the rays pistillate and fertile, usually of the same number as the involucral bracts (5 in our spp.), yellow, sometimes also tinged with anthocyanin; involucre cylindric (in our spp.) its bracts uniseriate (5 in our spp.), distinct, beset with some scattered embedded oil glands, rounded-carinate at least below, and subtending the ray achenes; receptacle small, naked; disk flowers perfect and fertile; style minutely hispidulous, with very short, blunt branches. Achenes linear. Pappus various, in our spp. the ray flowers with 2 narrow, awn-tipped scales and sometimes some much reduced additional members, the disk flowers with mostly 5 narrow, awn-tipped scales. (x = 12)

1 Leaves broadly linear to linear-oblong or oblanceolate, commonly 2–5 mm wide, with more or less numerous, scattered, round oil glands beneath; heads sessile or subsessile as in no. 2 ..... 1. *P. prostrata*.
1 Leaves narrowly linear, 0.5–2 mm wide, with two rows of ellipsoid oil glands beneath, one row near each margin.

2 Heads sessile or subsessile, mostly in small clusters .... 2. *P. linearifolia*.
2 Heads borne singly on slender peduncles mostly 1–3 cm long .... 3. *P. leptocephala*.

1. P. prostrata Cav. Diffusely branched, forming mats or loose mounds up to 3 dm wide or 1.5 dm high. Leaves broadly linear to linear-oblong or oblanceolate, mostly (1) 1.5–3 cm long and 2–5 mm wide, with rather numerous scattered round oil glands beneath. Heads sessile or subsessile in small axillary clusters (or the lower solitary); involucre 6–7 mm high, its bracts lanceolate or oblong, keeled nearly or quite to the tip; rays ca. 1–2 mm long. Late summer–fall. Native in dry habitats from sw US to n S Am, and introduced into disturbed sites in s Fla.

P. humifusa Swartz, a related, West Indian species, has recently been found as a weed in Collier Co., Fla. It has oblanceolate to obovate leaves less than 1.5 cm long, and very broad, more or less elliptic involucral bracts with the keel tending to fade out 1 mm or more below the tip.

2. P. linearifolia Urban. Lemon-scented, erect or ascending, 1–4 dm tall, usually rather sparingly branched. Leaves as in no. 3. Heads mostly sessile or subsessile in small clusters borne in the axils or at the ends of short branches; involucre 5–6 mm high, its bracts keeled toward the base; rays ca. 2–3 mm long. Late summer–fall. Disturbed sites, sandy pinelands, and beaches; s Fla and reputedly Jamaica.

3. P. leptocephala (Cass.) Urban. Freely branched, erect or spreading-ascending, 1–4 dm tall. Leaves narrowly linear, 1–3 cm long and 0.5–2 mm wide, with a single row of ellipsoid oil glands near each margin beneath. Heads borne singly at the ends of the numerous branches, on slender peduncles mostly 1–3 cm long; involucre 4–6 mm high, its bracts keeled toward the base; rays ca. 2–3 mm long. Late summer–fall. Disturbed sites, sandy pinelands, and beaches; s Fla. and WI.

45. TAGETES L. MARIGOLD

Aromatic herbs, ours annual. Leaves opposite at least below, often (as in our spp.) alternate above, beset with scattered embedded oil glands, pinnately compound (simple in a few extralimital spp.). Heads terminating the stem and branches, often forming corymbiform clusters, radiate, the rays few (except in double forms), pistillate and fertile, yellow or orange to nearly red, commonly relatively broad; involucral bracts few, uniseriate, beset with scattered oil glands, connate most of their length to form a cylinder; receptacle flat, naked; disk flowers perfect and fertile; style branches more or less elongate, flattened, with introrsely marginal stigmatic lines and a short, often expanded, hirsutulous appendage. Achenes slender and elongate. Pappus of several very unequal, often more or less connate scales, generally 1 or 2 elongate and acute or awn-tipped, the others shorter and generally blunter.

1 Rays showy, mostly 1 cm long or more.
2 Heads borne on evidently fistulous peduncles, relatively large (measurements in descr.) .... 1. *T. erecta*.
2 Heads borne on scarcely fistulous peduncles, smaller (measurements in descr.) .... 2. *T. patula*.
1 Rays inconspicuous, ca. 1–2 mm long .... 3. *T. minuta*.

1. T. erecta L. "AFRICAN" OR COMMON MARIGOLD. Stout, glabrous annual 2–15 dm tall. Leaves pinnately compound; leaflets mostly 11–17, lanceolate, 1–5 cm long, sharply serrate, often with more than 7 teeth to a side. Heads on fistulous peduncles terminating the branches, relatively large, the involucre 1.5–2 cm high, with fairly numerous, well-scattered oil glands, the lower ones narrower and more elongate than the upper; rays mostly orange or yellow, 5–8 (or numerous in double forms), mostly 1–3 cm long, often as much as 1 cm wide. Achenes 7–10 mm long. Longer pappus scale(s) mostly 8–12 mm long, the shorter ones 2–7 mm long, often connate below. (n = 12) Late summer–fall. Native of Mex, widely cult. in double forms with yellow or orange heads 6–10 cm wide, the involucre ca. 2 cm high, and casually escaped into disturbed sites in our range.

2. T. patula L. "FRENCH" MARIGOLD. Much like *T. erecta*, but with smaller heads and scarcely fistulous peduncles. Plants seldom more than 6 dm tall. Leaflets more coarsely and remotely toothed, seldom with more than 7 teeth to a side. Involucre mostly 1–1.5 cm high, the lower half with relatively few oil glands; rays up to ca. 1.5 cm long. Achenes mostly 5–7 mm long. Longer pappus scale(s) mostly 5–8 mm long, the shorter ones 1.5–3 mm. (n = 24) Late summer–fall. Native of Mex, widely cult. in double forms with partly or wholly deep red-orange heads 3.5–5 cm wide, the involucre ca. 1.5 cm high, and rarely escaped into disturbed sites in our range.

3. T. minuta L. Glabrous annual 3–10 dm tall. Leaves pinnately compound; leaflets 9–17, lance-linear, 1.5–6 cm long, sharply serrate. Heads numerous, tending to form flat-topped clusters; involucre narrowly cylindric, 8–12 mm high, with 3–5 short teeth, few-flowered, the rays commonly 3, only 1–2 mm long. Achenes 5–6 mm long. Longer pappus scales 2–3 mm, the others less than 1 mm long. (n = 24) Fall. Native of S Am, now more or less established as a weed in disturbed habitats on AC and occasionally elsewhere in our range.

46. HYMENOPAPPUS L'Her.

Taprooted herbs, ours biennial. Leaves alternate, some or all of them once to thrice pinnatifid, the basal ones the largest. Heads mostly discoid; involucral bracts appressed, in 2–3 equal series, at least the inner with broad, obtuse, scarious, somewhat petaloid, whitish or yellowish tip, more or less herbaceous below; receptacle small, naked; disk flowers perfect; style branches flattened, with introrsely marginal stigmatic lines and very short, minutely hairy appendage. Achenes 4–5-angled, many-nerved. Pappus of 12–22 membranous or hyaline small scales, or rarely wanting. Turner, B. L. 1956. A cytotaxonomic study of the genus Hymenopappus (Compositae). Rhodora 58: 164–186; 208–242; 250–308.

1 Flowers white; leaves rather finely and openly once or twice pinnatifid .... 1. *H. scabiosaeus*.
1 Flowers anthocyanic, rarely white; leaves rather coarsely once or twice pinnatifid, or the basal ones sometimes simple .... 2. *H. artemisiaefolius*.

1. H. scabiosaeus L'Her. Plants 3–10 dm tall, floccose-tomentose and partly glabrate, becoming more villous-puberulent above. Leaves rather finely and openly once or twice pinnatifid, the lower ones 8–25 cm long including the short petiole, 3–12 cm wide, the cauline ones seldom with any of the segments more than 3 mm wide, nor with the undivided midstrip more than 5 mm wide, the radical ones sometimes less dissected. Heads in an open-corymbiform inflorescence, the disk 7–12 mm wide, white; petaloid tips of involucral bracts at least as long as the basal portion. Pappus scales 0.1–0.6 mm long. (n = 17) Spring. Prairies and open woods, often in sandy soil; CP and ME, from Fla irregularly to SC and Miss; nw Ark (OZ) and adj. Mo to Tex and Neb, and irregularly in Mo, Ill, and Ind. Ours is var. *scabiosaeus*.

2. H. artemisiaefolius DC. Much like no. 1; leaves rather coarsely once or twice pinnatifid, or the basal ones simple, some of the cauline ones generally with at least some of the ultimate segments more than 3 mm wide and often with the undivided midstrip more than 5 mm wide; flowers anthocyanic, rarely white; pappus 0.5–1 (1.5) mm long (n = 17) Spring–summer. Sandy pine and post-oak woods; CP of La, w to Tex. Ours is var. *artemisiaefolius*. Intergrades in La with no. 1, and perhaps better treated as a var. of that sp.

47. ANTHEMIS L. DOGFENNEL, CHAMOMILE

Aromatic or sometimes nearly inodorous herbs. Leaves alternate, incised-dentate to more often pinnately dissected. Heads terminating the branches, radiate or rarely discoid, the rays elongate, white or yellow, pistillate or neutral; involucral bracts subequal or more often imbricate in several series, dry, the margins more or less scarious or hyaline; receptacle convex to conic or hemispheric, chaffy at least toward the middle; disk flowers numerous, perfect, yellow; style branches flattened, with ventromarginal stigmatic lines, truncate, penicillate. Achenes terete or 4–5-angled, or sometimes more or less compressed but not callous-margined. Pappus a short crown, or more often none.

Several species, in addition to the two treated below, have been found as waifs or casual adventives in our range. Notable among these are *A. mixta* L. [annual; rays pistillate, white with yellow base; receptacular bracts pointed; tube of the disk corollas produced into a spurlike appendage at the base—*Ormenis mixta* (L.) Dum.—S] and *A. nobilis* L. (perennial; rays pistillate, white; receptacular bracts blunt).

1 Receptacle chaffy throughout; rays pistillate and fertile .... 1. *A. arvensis*.
1 Receptacle chaffy only toward the middle; rays sterile and usually neutral ...... 2. *A. cotula*.

1. A. arvensis L. Similar to no. 2, commonly a little more hairy, and not ill-scented. Leaves appearing a little less finely dissected, heads averaging a little larger; rays pistillate and fertile; receptacle chaffy throughout, its bracts softer, paleaceous, with short, cuspidate awn tip. Achenes not tuberculate. (n = 9) Spring–summer. Fields and waste places; native of Europe, naturalized over most of the US, and widespread in our range. The Mediterranean species *A. secundiramea* Biv. has been collected on railroad ballast in Va, but may not be fully established in our range. It has fertile rays, and a fully chaffy receptacle, as in *A. arvensis*, but the heads average smaller and the leaves are glandular-punctate beneath; it is sharply distinguished from *A. arvensis* by its tuberculate achenes.

2. A. cotula L. Taprooted, usually subglabrous, malodorous annual 1–6 dm tall. Leaves 2–6 cm long, 2–3 times pinnatifid, with narrow segments. Heads short-pedunculate, the disk 5–10 mm wide, becoming ovoid or short-cylindric at maturity; involucre sparsely villous; rays mostly 10–16, sterile and usually neutral, white, 5–11 mm long; receptacle chaffy only toward the middle, its firm, narrow, subulate bracts tapering to the apex. Achenes subterete, about 10-ribbed, strongly glandular-tuberculate. (n = 9) Spring–summer. Fields and waste places; native of Europe, naturalized over most of the US, and widespread in our range. *Maruta cotula* (L.) DC.—S.

48. ACHILLEA L. YARROW

1. A. millefolium L. Perennial, aromatic, often rhizomatous herbs, very sparsely to rather densely villous or woolly-villous. Leaves alternate, pinnately dissected with small ultimate segments, slender, the blade 3–15 × 0.5–2.5 cm, only the lower petioled. Heads numerous in a flat or round-topped, short and broad, paniculate-corymbiform inflorescence, radiate, the rays mostly 3–5, pistillate and fertile, white (pink), 2–3 mm long; involucre 3.5–5 mm high, its bracts imbricate in several series, dry, with scarious or hyaline margins; receptacle conic or convex, chaffy throughout; disk 2–4 mm wide, with 10–20 perfect and fertile flowers; style branches flattened, with ventromarginal stigmatic lines, truncate, penicillate. Achenes compressed parallel to the involucral bracts, callous-margined, glabrous. Pappus none. [n = 18 (native), 27 (intro.)] Spring–fall. Various habitats, especially in disturbed sites; circumboreal, s nearly throughout our range, except peninsular Fla. A highly variable polyploid complex, with both native and intro. forms in our range, not yet satisfactorily sorted out into infraspecific taxa. *A. asplenifolia* Vent.—S; *A. lanulosa* Nutt.—F; *A. occidentalis* Raf.—S.

49. CHRYSANTHEMUM L. CHRYSANTHEMUM

Herbs. Leaves alternate, entire or toothed to sometimes pinnatifid or subbipinnatifid. Heads generally radiate, the rays pistillate and fertile, white or sometimes yellow or anthocyanic; involucral bracts more or less imbricate in 2–4 series, dry, at least the margins and tip scarious or hyaline, the midrib sometimes greenish; receptacle flat or convex, naked; disk flowers perfect and fertile, yellow; style branches flattened, with ventromarginal stigmatic lines, truncate, penicillate. Achenes subterete or angular, 4–10-ribbed, or those of the rays with 2 or 3 angles or wings. Pappus a short crown, or none.

In addition to the 3 spp. treated below, 2 yellow-rayed annuals from the Old World have been found as waifs or adventives in our range. *C. coronarium* L., the garland chrysanthemum or crown daisy, has subbipinnatifid leaves with narrow rachis, and somewhat angular but not prominently ribbed and grooved disk achenes. *C. segetum* L., the corn chrysanthemum, has coarsely and sharply toothed or irregularly pinnatisect leaves with broad undivided midstrip, and subterete, ca. 10-ribbed and -grooved disk achenes.

1 Heads solitary or few, large, the rays 1–2 cm long, the disk 1–2.5 cm wide.
 2 Leaves somewhat basally disposed, the middle and upper ones progressively reduced; some of the leaves often somewhat lobed or cleft as well as toothed .... 1. *C. leucanthemum*.
 2 Leaves well distributed along the stem, the largest ones near or somewhat below the middle, all merely toothed .... 2. *C. lacustre*.

1 Heads several or numerous, small, the rays 4–8 mm long, the disk 5–9 mm wide .... 3. *C. parthenium*.

1. C. leucanthemum L. OX-EYE DAISY. Rhizomatous perennial, 2–8 dm tall, simple or nearly so, glabrous or inconspicuously hairy. Leaves somewhat basally disposed, the basal ones oblanceolate or spatulate, petiolate, 4–15 cm long overall, crenate and often also more or less lobed or cleft, the cauline reduced and becoming sessile, deeply and distantly blunt-toothed to sometimes more closely toothed or subentire. Heads solitary, terminating the stem (and long branches, if any), naked-pedunculate, relatively large, the disk 1–2 cm wide; rays 15–35, white, 1–2 cm long. Achenes terete, ca. 10-ribbed. (n = 9, 18, 27, 36) Spring–fall. Fields, roadsides, and waste places; native of Eurasia, naturalized throughout most of temperate N Am. *Leucanthemum leucanthemum* (L.) Rydb.—S.

2. C. lacustre Brotero. Somewhat similar to no. 1, but more robust and leafy-stemmed, with closely and sharply serrate, not at all pinnatifid leaves, the stem in well-developed plants few-branched above, the heads averaging a bit larger, the disk up to 2.5 cm wide. Basal leaves wanting or inconspicuous; lower cauline leaves narrowed to a petiolar base but mostly soon deciduous; middle cauline leaves well developed, elliptic or elliptic-oblanceolate, commonly 6–11 × 1.5–3 cm, tending to clasp at the base. (n = 9) Summer. Native of Europe, occasionally escaped from cult. into fields and moist, low ground in Va, NC, SC, and doubtless elsewhere in our range.

3. C. parthenium (L.) Bernh. FEVERFEW. Perennial from a taproot or stout caudex. Stem 3–8 dm tall, puberulent above. Leaves mainly cauline, finely hairy at least beneath, pinnatifid, with incised or again pinnate, rounded segments, evidently petiolate, the blade up to 8 × 6 cm. Heads several or many in a corymbiform inflorescence, relatively small, the disk 5–9 mm wide; rays 10–21, or more numerous in double forms, 4–8 mm long, white. Achenes subterete, ca. 10-ribbed. (n = 9) Summer. Waste places; native of Europe, occasionally intro. or escaped in our range and elsewhere in the US. *Matricaria parthenium* L.—S.

50. TANACETUM L. TANSY

1. T. vulgare L. COMMON TANSY. Coarse, aromatic perennial herb from a stout rhizome, glabrous or nearly so throughout, 4–15 dm tall. Leaves alternate, numerous, commonly 1–2 dm long and nearly half as wide, sessile or short-petiolate, punctate, pinnatifid, with evidently winged rachis, the pinnae again pinnatifid or deeply lobed, the pinnules often again toothed. Heads numerous (20–200) in a terminal, corymbiform inflorescence, yellow, disciform, the pistillate flowers with a short, tubular corolla; involucral bracts imbricate, dry, the margins and tips scarious; receptacle naked, broadly rounded-conic at maturity; disk 5–10 mm wide, its flowers perfect and fertile; style branches flattened, with ventromarginal stigmatic lines, truncate, minutely penicillate. Achenes 5-ribbed. Pappus a minute crown, almost obsolete. (n = 9) Summer–fall. Roadsides, fields, and waste places; native of the Old World, escaped from cult. and well established in much of the US, only at scattered stations in our range.

51. MATRICARIA L.

Herbs, often aromatic. Leaves alternate, pinnatifid or pinnately dissected. Heads small to middle-sized, terminating the branches or in a corymbiform inflorescence, radiate or discoid, the rays when present white, pistillate and usually fertile; involucral bracts dry, 2–3-seriate, not much imbricate, with more or less scarious or hyaline margins; receptacle naked, hemispheric to more often conic, or elongate; disk flowers perfect and fertile, yellow; style branches flattened, with ventromarginal stigmatic lines, truncate, penicillate. Achenes generally nerved on the margins and ventrally, nerveless dorsally, glabrous or roughened. Pappus a short crown or none.

1 Heads radiate; disk corollas 5-toothed .... 1. *M. chamomilla*.
1 Heads discoid; disk corollas 4-toothed .... 2. *M. matricarioides*.

1. M. chamomilla L. FALSE CHAMOMILLE. Aromatic annual from a quickly deliquescent taproot, glabrous, 2–8 dm tall. Leaves 2–6 cm long, bipinnatifid, the ultimate segments linear or filiform. Heads more or less numerous, the disk 6–10 mm wide; rays mostly 10–20, white, 4–10 mm long; disk corollas 5-toothed; receptacle conic, acute. Achenes with 2 nearly marginal and 3 ventral, raised but not at all winglike ribs, otherwise smooth. Pappus a short, toothed crown. (n = 9) Summer. Roadsides and waste places; native of Eurasia, now widely intro. across n US, and entering our range at least in Ark.

M. maritima L., Scentless False Chamomille, is another Eurasian weed that is widely intro. in n US and may be sought in our range. It is a nearly inodorous annual or short-lived perennial, with a hemispheric, rounded receptacle, much like *M. chamomilla* in aspect, but with distinctive achenes, these with 2 marginal and 1 ventral, strongly callous-thickened, almost winglike ribs, minutely roughened on the back and between the ribs. (n = 9, 18) *Chamomilla maritima* (L.) Rydb.—S.

2. M. matricarioides (Less.) T. C. Porter. PINEAPPLE WEED. Taprooted, branching, leafy, pineapple-scented, glabrous annual 0.5–4 dm tall. Leaves 1–5 cm long, 1–3 times pinnatifid, the ultimate segments short, linear or filiform. Heads several or many, discoid, the disk 5–9 mm wide; disk corollas 4-toothed; receptacle conic, pointed. Achenes with 2 marginal and 1 or sometimes more rather weak ventral nerves. Pappus a short crown. (n = 9) Summer–fall. Roadsides and waste places; native of w US, intro. eastward, and casual in our range, as in NC, Tenn, and Va.

52. ARTEMISIA L. WORMWOOD, SAGE

Herbs or shrubs, usually aromatic. Leaves alternate, entire to dissected. Inflorescence spiciform, racemiform, or paniculiform, the heads small, discoid or disciform, sometimes only with perfect flowers, sometimes the outer flowers pistillate, with tubular corolla, and the central flowers then sometimes functionally staminate; receptacle flat to hemispheric, naked or densely beset with long hairs; style branches flattened, with introrsely marginal stigmatic lines, truncate, penicillate. Achenes small, mostly glabrous. Pappus generally none.

1 Leaves evidently white-hairy at least on the lower surface; true perennials from a rhizome or woody caudex.

2 Receptacle not hairy.
3 Heads relatively large, the involucre 6–7.5 mm high .... 1. *A. stelleriana.*
3 Heads smaller, the involucre 2–5 mm high.
4 Principal leaves entire or once pinnatifid, or if bipinnatifid then with the ultimate segments well over 1.5 mm wide.
5 Divisions of the principal leaves again toothed, cleft, or lobed; leaves ordinarily with one or two pairs of stipulelike lobes at the base ...... 2. *A. vulgaris.*
5 Divisions of the leaves entire or nearly so, or the leaves entire; leaves usually without stipulelike lobes .... 3. *A. ludoviciana.*
4 Principal leaves 2–3 times pinnatifid, with narrow ultimate segments up to 1.5 mm wide .... 4. *A. abrotanum.*
2 Receptacle beset with numerous long hairs (unique among our spp.) .... 5. *A. absinthium.*
1 Leaves green, glabrous or inconspicuously hairy; plants annual, biennial, or short-lived perennial from a taproot.
6 Disk flowers fertile, with normal ovary .... 6. *A. annua.*
6 Disk flowers sterile, with abortive ovary (unique among our spp.) .... 7. *A. campestris.*

1. A. stelleriana Besser. BEACH WORMWOOD, DUSTY MILLER. Inodorous, rhizomatous perennial 3–7 dm tall, simple to the inflorescence, densely white-tomentose or floccose. Leaves 3–10 cm long (petiole included) and 1–5 cm wide, with a few relatively broad, rounded lobes, which may again be slightly lobed. Inflorescence elongate, narrow, and often dense; heads relatively large, the involucre 6–7.5 mm high, the disk corollas 3.2–4.0 mm long. (n = 9) Late spring–summer. Sandy beaches; native from Kamtchatka to Japan, escaped from cult. and established from Va to Que.

2. A. vulgaris L. MUGWORT. Rhizomatous perennial 5–15 dm tall, the stem glabrous or nearly so below the inflorescence. Leaves green and glabrous or nearly so above, densely white-tomentose beneath, chiefly obovate or ovate in outline, 5–10 × 3–7 cm, the principal ones cleft nearly to the midrib into ascending, acute, unequal segments that are again toothed or cleft, and ordinarily with 1 or 2 pairs of stipulelike lobes at the base. Inflorescence generally ample and leafy; involucre 3.5–4.5 mm high; disk corollas 2.0–2.8 mm long. (n = 8) Summer–fall. Fields, roadsides, and waste places; native to Eurasia, now established throughout most of e US, in our range s to Ga, Ala, and n La.

3. A. ludoviciana Nutt. Rhizomatous perennial 3–10 dm tall, simple up to the inflorescence, the stem more or less white-tomentose at least above. Leaves lanceolate or lance-elliptic, 3–10 cm long, entire or irregularly toothed to coarsely few-lobed or deeply parted, the undivided portion up to 1 (1.5) cm wide, persistently white-tomentose on both sides or becoming glabrous above. Involucre 2.5–3.5 mm high; disk corollas 1.9–2.8 mm long. (n = 9) Summer–fall. A widespread, variable sp. of dry lands in w US and n Mex, sporadic in disturbed sites in our range in 2 vars. Var. *ludoviciana* has entire to coarsely few-lobed leaves and a mostly compact and elongate inflorescence. Var. *mexicana* (Willd.) Fern., with many of the leaves deeply parted and with a strong tendency toward a more diffuse, often leafy inflorescence, may be native in its Ark (OU) occurrence.

4. A. abrotanum L. SOUTHERNWOOD. Perennial and more or less shrubby, 5–20 dm tall, much branched. Leaves 3–6 cm long, thinly tomentose beneath, green and glabrous or nearly so above, 2–3 times pinnatifid with elongate, linear or filiform, ascending segments 0.5–1.5 mm wide. Inflorescence ample; involucre 2–3.5 mm high, somewhat canescent. (n = 9) Late summer–fall. Roadsides and waste places; native to Eurasia, cult. and sparingly established throughout much of the US, and rarely in our range, as in NC. *A. pontica* L., a European species sparingly intro. in ne US as far s as Del, would key to *A. abrotanum*, from which it differs in being a rhizomatous perennial

4–10 dm tall, not much branched, with small, finely dissected leaves 1–3 cm long that are white-tomentose on both sides.

5. A. absinthium L. COMMON WORMWOOD, ABSINTHIUM. Perennial herb or near-shrub 4–10 dm tall, the stem finely sericeous or eventually glabrate. Leaves silvery-sericeous, sometimes eventually subglabrate above, the lower long-petiolate and 2–3 times pinnatifid, with mostly oblong, obtuse segments 1.5–4 mm wide, the blade rounded-ovate in outline, 3–8 cm long; upper leaves progressively less divided and shorter-petiolate, the segments often more acute. Inflorescence ample, leafy; involucre 2–3 mm high; receptacle beset with numerous long white hairs between the flowers. (n = 9) Summer–fall. Fields and waste places; native of Europe, now established across n US, s occasionally to NC and Tenn.

6. A. annua L. SWEET WORMWOOD. Sweet-scented glabrous annual, 3–30 dm tall, usually bush-branched. Leaves 2–10 cm long, mostly twice or thrice pinnatifid, the ultimate segments linear or lanceolate, sharply toothed. Inflorescence broad and open, paniclelike, the heads loose, often nodding, on evident short peduncles; involucre glabrous, 1–2 mm high. (n = 9) Late summer–fall. Fields and waste places; native of Eurasia, now naturalized in much of the US, in our range s to Va, Tenn, Miss, and Ark.

7. A. campestris L. Scarcely odorous, robust biennial or occasionally short-lived perennial from a taproot, mostly 5–12 dm tall, glabrous or subglabrous. Leaves mostly twice or thrice pinnatifid or ternate, with linear or linear-filiform segments seldom more than 2 mm wide, the basal rosette produced during the first (non-flowering) year and not persistent. Inflorescence more or less diffuse and paniclelike; involucre 2–4.5 mm high; disk flowers sterile, with abortive ovary. (n = 9) Summer–fall. Dunes and other very sandy places especially along the coast; circumboreal, taxonomically complex sp.; our plants as described here, irregularly distributed s to SC and even Fla, belong to ssp. *caudata* (Michx.) Hall & Clem. *A. caudata* Michx.—F, S.

53. SOLIVA R. & P. PIQUANTE, STICKERWEED

Low, subacaulescent or mat-forming annuals. Leaves small, alternate, commonly closely clustered beneath the heads, pinnatifid to tripinnatifid or with subpalmately cleft pinnae. Heads terminating the stems and commonly surpassed by the immediately subtending branches, either closely clustered at the base or scattered along the sympodial axes, disciform, the marginal flowers numerous, pistillate, with elongate, indurate-persistent style and without corolla, the disk flowers few, functionally staminate, with 4-lobed corolla; involucre of 1 or 2 series of dry and scarious to herbaceous bracts; receptacle naked, flat to conic-elevated. Achenes compressed at right angles to the radii of the head, with thin or corky, narrow or broad, often cross-ornamented wings. Pappus none. *Gymnostyles* Juss. Cabrera, A. L., Sinopsis del Género *Soliva* (Compositae), Notas Mus. La Plata (Bot). 14(70): 123–139. 1949.

1 Achenes (2.5) 3–4 mm wide, the body merely scabrous-hispidulous, the wings thin, unornamented; primary pinnae of the leaves almost palmately cleft ..... 1. *S. pterosperma*.
1 Achenes 1–2 mm wide, conspicuously long-hairy at the summit, the wings with conspicuous, raised transverse processes; leaves not with palmately cleft pinnae.
2 Leaves twice or thrice pinnatifid; distal part (ca. 0.7 mm) of the wings of the achene smooth, without raised processes .... 2. *S. anthemifolia*.

2 Leaves mainly only once pinnatifid, a few of the larger segments sometimes with 1 or 2 lateral teeth or lobes; wings of the achenes cross-ribbed to the summit .... 3. *S. stolonifera*.

1. S. pterosperma (Juss.) Less. Plants loosely hairy, diffusely branched, forming mats up to 2 dm wide. Leaves up to 3 × 1.2 cm overall, pinnatifid, the pinnae rather irregularly, almost palmately cleft into narrow segments, the basal leaves commonly soon deciduous. Heads mostly less than 1 cm wide at maturity, scattered along the sympodial axes; involucre of more or less herbaceous, ovate to elliptic bracts 3–4 mm long; receptacle small, conic-elevated; disk corollas not strongly narrowed at the base. Achenes (2.5) 3–4 mm wide, the body merely scabrous-hispidulous, the wings broad, thin, unornamented, broadest near or above the middle, constricted below but again flared toward the base, and prolonged distally into a pair of spinulose tips ca. 1 mm long flanking the straight, spinose-indurate style, this 1.5–2 mm long. Spring. A common weed in lawns, along roadsides, etc.; native of S Am, intro. on and near CP from NC to Fla, La, Ark and Tex. *S. sessilis* R. & P.—S, misapplied.

2. S. anthemifolia (Juss.) Less. Plants loosely hairy, tufted or forming mats up to 1.5 dm wide. Leaves up to 15 × 2.5 cm overall, twice or thrice pinnatifid, with narrow ultimate segments. Heads mostly less than 1 cm wide at maturity, all clustered at the base, or some scattered along the sympodial axes; involucral bracts dry, rather narrow, mostly 3–4 mm long; receptacle flat or nearly so; disk corollas very slender below, flaring above. Achenes cuneate-oblong, 1–2 mm wide, conspicuously long-hairy at the summit, corky-winged, the wings with evident transverse ridges, but the distal part (ca. 0.7 mm) smooth; style indurate-persistent but tending to be flexuous, 1–2 mm long. Spring. A weed in lawns, along roadsides, etc.; native to S Am, intro. on GC from Fla panhandle to Tex. *Gymnostyles anthemifolia* Juss.—S.

3. S. stolonifera (Brot.) Lour. Much like no. 2 but smaller and often more compact, the leaves up to 2.5 × 0.7 cm, merely pinnatifid, some of the pinnae sometimes with one or two lateral teeth or lobes. Involucral bracts mostly 2–3 mm long. Wings of the achenes prominently ornamented throughout, the transverse processes on the ventral side very strongly raised, those on the dorsal side lower, forming a continuous arching pattern around the top as well as along the sides, the wings sometimes distally prolonged into lateral spinules less than 0.5 mm long. Spring. A weed in lawns, along roadsides, etc.; native to S Am, intro. on and near CP from SC to Fla and Tex. *Gymnostyles nasturtiifolia* Juss.—S.

54. ARNICA L. ARNICA

1. A. acaulis (Walter) B.S.P. Fibrous-rooted perennial 2–8 dm tall from a short, simple, erect caudex. Herbage glandular and hirsute; basal leaves rosulate, sessile or nearly so, broadly elliptic to ovate or even rhombic, obscurely toothed or subentire, 4–15 × 1.5–8 cm; cauline leaves few and reduced, opposite or the uppermost ones alternate. Heads campanulate, 3–20 in a more or less corymbiform inflorescence, yellow, with mostly 10–15 relatively broad, pistillate rays 1.5–2.5 cm long; involucre ca. 10–12 mm high, its bracts herbaceous, subequal but more or less evidently biseriate; receptacle convex, naked; disk flowers perfect and fertile; style branches flattened, with ventromarginal stigmatic lines, truncate, penicillate. Achenes cylindric, 5–10-

nerved. Pappus of numerous white, barbellate, capillary bristles. (n = 19) Spring. Sandy pine woods and clearings, often in damp soil, chiefly on AC and PP; Del and s Pa to n Fla, w to Early Co., Ga, and Liberty Co., Fla.

55. SENECIO L. GROUNDSEL, RAGWORT, SQUAW-WEED, BUTTERWEED

Herbs (ours). Leaves alternate, entire to variously cleft or dissected. Heads radiate or discoid, the rays pistillate and fertile, yellow to orange, or wanting; involucral bracts more or less herbaceous, essentially equal, uniseriate or subbiseriate, often with some bracteoles at the base; receptacle flat or convex, naked; disk flowers perfect and fertile, yellow to orange; style branches flattened, with ventromarginal stigmatic lines, truncate, penicillate. Achenes subterete, 5–10-nerved. Pappus of numerous, usually white, entire or rarely barbellulate bristles. The reports of 22 and 23 pairs of chromosomes for some of our spp. may reflect facultative aneuploidy on a base of n = 20. Most of our radiate spp. have occasional discoid individuals. Interspecific hybridization is frequent, especially among spp. 3–9. Barkley, T. M. 1978. Senecio. Pp. 50–139 in North Amer. Flora, Ser. II. Part 10.

In addition to the following spp., three others may persist after cultivation or become locally established. *S. mikanioides* Otto, of S Afr, called German Ivy, is a vine with ivylike, shallowly lobed leaves and clusters of discoid, yellow heads. *S. confusus* Britten, of Mex and C Am, in our range probably viable only in s Fla, is a vine with lanceolate to ovate, often subcordate, merely toothed leaves and orange to flame-red heads with showy rays, *S. clivorum* Maxim. (*Ligularia clivorum* Maxim.), a Chinese sp. locally escaped at least in Md, has large, reniform-cordate, toothed basal leaves and a few large heads, the involucre glandular-hairy and somewhat floccose, 12–15 mm high, the orange-yellow rays 3–4 cm long.

1 Plants mostly perennial, or sometimes (as in *S. plattensis*) only biennial.
  2 Principal leaves 2–3 times pinnatifid into small and mostly narrow ultimate segments . . . . . 1. *S. millefolium*.
  2 Principal leaves subentire or toothed to once pinnatifid.
    3 Plants more or less persistently (sometimes thinly) tomentose, at least on the stem, lower surface of the leaves, and involucre.
      4 Tomentum fine and close; basal leaf blades up to 3.5 cm long . . . . . 2. *S. antennariifolius*.
      4 Tomentum looser; basal leaf blades often more than 3.5 cm long.
        5 Cauline leaves entire or more or less toothed, usually scarcely or not at all pinnatifid; plants perennial . . . . . 3. *S. tomentosus*.
        5 Cauline leaves evidently pinnatifid; plants biennial or sometimes short-lived perennial . . . . . 4. *S. plattensis*.
    3 Plants not tomentose, or not persistently so, except sometimes at the base and in the leaf axils.
      6 Plants generally with well-developed stolons (or very slender, superficial rhizomes); basal leaves narrowly obovate to orbicular . . . . . 5. *S. obovatus*.
      6 Plants not stoloniferous, or occasionally with very short stolons in no. 6; leaves various, but seldom obovate or orbicular.
        7 Basal leaves chiefly oblanceolate or elliptic and tapering to the petiole.
          8 Heads few, rarely more than 20; stem often glabrous at the base . . . . . 6. *S. pauperculus*.
          8 Heads many, mostly 20–100 or more; stem densely woolly at the base . . . . . 7. *S. anonymus*.
        7 Basal leaves broadly lanceolate to reniform, subtruncate to deeply cordate at the base.
          9 Basal leaf blades 1.75–3.5 times as long as wide . . . . . 8. *S. schweinitzianus*.
          9 Basal leaf blades 0.8–1.5 times as long as wide . . . . . 9. *S. aureus*.

1 Plants annual or winter-annual.
 10 Heads radiate; bracteoles, if present, not black-tipped.
  11 Plants distinctly fibrous-rooted .... 10. *S. glabellus.*
  11 Plants taprooted .... 11. *S. imparipinnatus.*
 10 Heads discoid; bracteoles present, evidently black-tipped .... 12. *S. vulgaris.*

1. S. millefolium T. & G. Perennial, 3–6 dm tall from a rhizome-caudex, glabrous or nearly so except usually for some persistent axillary tomentum. Leaves pinnately 2–3 times dissected into small and mostly narrow ultimate segments; basal or lower cauline leaves the largest, 10–30 cm (including the long petiole) × 3–10 cm, the others progressively smaller and becoming sessile, but still much dissected. Heads 10–many, the disk 5–8 mm wide, the involucre 4.5–6 mm high; rays 5–12 mm long. Achenes hairy on the angles. Spring–early summer. Dry, often rocky places in the mts.; BR of sw NC and adj. SC and Ga.

2. S. antennariifolius Britton. Perennial, 1.5–4 dm tall from a well-developed caudex. Lower leaf-surfaces very finely, densely, and persistently white-tomentose, the stem and involucres less so, the upper leaf surfaces commonly soon glabrate. Basal leaves tufted and conspicuous, the blade mostly elliptic or obovate, 1–3.5 cm × 6–20 mm, shallowly and rather remotely dentate to occasionally entire; cauline leaves few and much reduced, linear and entire or pinnatifid with linear segments. Heads 3–12, the disk 7–12 mm wide; involucre 5–8 mm high; rays 5–10 mm long. Achenes hispidulous on the angles. (n = 23) Spring–early summer. Shale barrens, RV of Va, WVa, and Md.

3. S. tomentosus Michx. Perennial, 2–7 dm tall from a short caudex, sometimes also stoloniferous, persistently floccose-tomentose until flowering time or later, generally very densely so at the base, the upper leaf surfaces sometimes soon glabrate. Basal leaf blades chiefly lance-ovate to elliptic or ovate and abruptly contracted to the petiole, up to 20 × 5 cm, crenate or subentire; cauline leaves conspicuously reduced upward, becoming sessile, entire or crenate, scarcely pinnatifid. Heads several or rather many, the disk 7–12 mm wide; involucre 4–6 mm high; rays 5–10 mm long. Achenes hispidulous on the angles. (n = 20, 23) Spring–early summer. Dry, open places and pine woods, especially in sandy soil; s NJ to Fla and Tex, chiefly on or near CP. *S. alabamensis* Britt.—S; *S. tomentosus* f. *alabamensis* (Britt.) Fern.—F, the least-hairy extreme.

4. S. plattensis Nutt. Biennial or sometimes short-lived perennial, 2–7 dm tall from a short caudex, sometimes also stoloniferous, more or less persistently floccose-tomentose until flowering time or later, at least as to the stem, lower leaf surfaces, and involucres. Basal leaves narrowly or broadly elliptic or ovate to suborbicular or broadly oblanceolate, crenate-serrate or some of them deeply lobed, the blade and petiole together up to 10 × 3 cm; cauline leaves conspicuously reduced upward, becoming sessile, more or less pinnatifid. Heads several or rather many, the disk 6–12 mm wide; involucre 4–6 mm high; rays 5–10 mm long. Achenes generally hispidulous on the angles. (n = 22, 23) Spring–early summer. Mostly in dry, open places; plains and prairies sp., extending e occasionally in suitable habitats to Va (RV), Tenn (RV), NC (BR) and n La, and possibly Ala.

5. S. obovatus Muhl. ex Willd. Perennial, 2–7 dm tall, with well-developed, slender stolons or superficial rhizomes and generally a short caudex, lightly floccose-tomentose when young, generally soon glabrate. Basal leaves narrowly obovate to orbicular, rounded at the tip, tapering or abruptly contracted to the petiolar base, crenate-serrate or sometimes (especially toward the base) more deeply cut, up to 20 cm (petiole included) × 6 cm; cauline leaves conspicuously reduced, becoming sessile and generally more or less pinnatifid. Heads several or rather many, the disk 7–14 mm wide; involucre 4–6 mm high, conspicuously shorter than the disk, the bracts often purple-tipped; rays 5–10 mm long. Achenes glabrous. (n = 22) Spring. Rich woods and rock outcrops, especially in calcareous situations; Vt to n Fla, w to Kans and Tex, all prov., but rare on CP and ME. *S. rotundus* (Britt.) Small—S; *S. obovatus* var. *elliottii* (T. & G.) Fern.—F, the more persistently hairy extreme.

6. S. pauperculus Michx. Perennial, 1–5 dm tall from a rather short caudex, occasionally also with very short, slender stolons or superficial rhizomes, lightly floccose-tomentose when young, generally soon glabrate, except often at the very base and in the axils. Basal leaves mostly oblanceolate to elliptic, generally tapering to the petiolar base, crenate or serrate to subentire, seldom more than 12 × 2 cm overall; cauline leaves more or less pinnatifid, the lower sometimes much larger than the basal, the others conspicuously reduced and becoming sessile. Heads seldom more than 20, the disk 5–12 mm wide; involucre 4–7 mm high, its bracts often purple-tipped; rays 5–10 mm long. Achenes glabrous or hispidulous on the angles. (n = 22, 23) Spring–early summer. Bogs and wet meadows; transcontinental sp. of Can and n US, extending s occasionally in our range, in various provinces, to NC, n Ga, and Tenn. *S. crawfordii* Britt.—F.

7. S. anonymus A. Wood. Perennial, 3–8 dm tall from a rather short caudex, densely and persistently woolly at the base, otherwise soon glabrate except in the axils. Basal leaves mostly elliptic-oblanceolate, tapering to the petiole, crenate or serrate, up to 30 cm (petiole included) × 3.5 cm; cauline leaves deeply pinnatifid, reduced and becoming sessile upward. Heads numerous, commonly 20–100 or more, small, the disk 5–9 mm wide; involucre 4–6 mm high; rays 4–8 mm long. Achenes hispidulous on the angles. (n = 22) Spring. Meadows, pastures, roadsides, and dry woods; s Pa to n Fla, w to Ky, Tenn, and c Miss, more common in the mt. provinces. *S. smallii* Britt.—F, G, R, S.

8. S. schweinitzianus Nutt. Blades of basal leaves lanceolate or lance-ovate, acute or acutish, subtruncate to shallowly cordate at base, sharply and generally rather finely toothed, up to 8 × 5 cm, commonly 1.75–3.5 times as long as wide. Otherwise much like *S. aureus*. (n = 23) Spring–summer. Sp. chiefly of moist meadows and swampy woods in NY, New England, and adj. Can, disjunct onto some high balds in BR in the vicinity of Roan Mt., NC and Tenn. *S. robbinsii* Oakes—F, G, R, S.

9. S. aureus L. Perennial, 3–8 dm tall from a branched, rhizomatous caudex or creeping rhizome, often with stoloniform but coarse, leafy basal offshoots, lightly floccose-tomentose when young, soon essentially glabrous. Basal leaves long-petiolate, some or all with more or less strongly cordate blade, these generally broadly rounded distally, crenate, up to 11 × 11 cm, mostly 0.75–1.5 (1.75) times as long as wide; cauline leaves conspicuously reduced and generally more or less pinnatifid, becoming sessile upward. Heads several or rather many, the disk 5–12 mm wide; involucre 5–8 mm high, its bracts often strongly purple-tipped; rays 6–13 mm long. Achenes glabrous. (n = 22) Spring–summer. Moist woods and swampy places; Lab to Minn, s to NC (PP and mts.), n Ga (mts.) n Ala, and c Ark (OU); isolated in Gadsden and Leon Cos., Fla. *S. gracilis* Pursh—S.

10. S. glabellus Poiret. YELLOWTOP. Fibrous-rooted annual or winter-annual, 1.5–8 dm tall, glabrous or obscurely floccose-tomentose in the axils. Stems mostly solitary and simple to the inflorescence. Leaves mostly once pinnatifid with generally rounded teeth and lobes, or the basal sometimes merely round-toothed, the largest ones at or near the base of the stem, up to 20 × 7 cm, progressively reduced upward. Heads more or less numerous, the disk 5–10 mm wide; involucre 4–6 mm high; rays 5–12 mm long. Achenes minutely hairy or glabrous. (n = 23) Spring–midsummer. Moist, open or shaded places, often a weed in low fields; NC to s Fla, w to SD and Tex.

11. S. imparipinnatus Klatt. Like a delicate form of no. 10 (seldom more than 4 dm tall), and sometimes with several stems from the base, differing consistently in having a definite slender taproot. Spring. Swamps and other moist, low places, often in disturbed sites; Tex and Okla, e to s La and s Miss.

12. S. vulgaris L. Taprooted annual, 1–4 dm tall, leafy throughout, sparsely crisp-hairy or subglabrous. Leaves coarsely and irregularly toothed to more often pinnatifid, 2–10 × 0.5–4.5 cm, the lower tapering to the petiole or petiolar base, the upper sessile and clasping. Heads several or many, discoid, 5–10 mm wide; involucre 5–8 mm high, with some short, black-tipped bracteoles at the base. Pappus very copious, equaling or generally surpassing the corolla. Achenes short-hairy chiefly along the angles. (n = 20) Spring–fall. A weed in disturbed garden soil and waste places, native to Eurasia, now widely distributed in n temperate regions, and s in our range to SC and Ga.

56. EMILIA Cass. TASSELFLOWER

Herbs, ours taprooted annuals. Leaves alternate, entire or toothed to pinnatifid or lyrate, from sessile and sagittate-clasping to evidently slender-petiolate. Heads terminating the branches or in a loosely corymbiform inflorescence, cylindric from a swollen base, strictly discoid, the flowers all tubular and perfect, variously pink or purple to crimson, scarlet, or bright to deep orange, or rarely white, but not yellow; involucre a single series of ca. 8 or ca. 13 equal, herbaceous bracts; receptacle naked; style branches very slender, somewhat flattened, with ventromarginal stigmatic lines and a short, slender, hispidulous appendage, or the appendage virtually obsolete (in *E. sonchifolia*) and the style branches then subtruncate and minutely penicillate. Pappus of slender capillary bristles. Achenes slender, beset with thick, short hairs on the 5 angles.

1 Involucre equaling or only very slightly shorter than the flowers, very slender, commonly 3–4 times as long as wide at anthesis; lower leaves often more or less lyrate-pinnatifid .... 3. *E. sonchifolia*.
1 Involucre evidently shorter than the flowers, coarser, up to about 3 times as long as wide; leaves entire or merely toothed, not pinnatifid.
  2 Involucre mostly 6–9 mm long, less than twice (often about 1.5 times) as long as wide .... 1. *E. coccinea*.
  2 Involucre mostly 9–14 mm long, mostly 2–3 times as long as wide .... 2. *E. fosbergii*.

1. E. coccinea (Sims) G. Don. Plants 2–10 dm tall, erect or lax, simple or with a few long branches, somewhat arachnoid-villous below, otherwise glabrous or nearly so. Leaves variable in form and size, the lower or lower and middle ones the largest, dentate or subentire, not lobed, often notably sagittate-clasping at the base. Involucre consisting of about 13 bracts, relatively short and coarse, mostly 6–9 mm long and less than twice (often ca. 1.5 times) as long as wide at anthesis; flowers numerous (commonly more than 50), conspicuously surpassing the involucre (commonly by 2–5 mm) and spreading out above it to form a disk mostly 1–1.5 (2) cm wide at anthesis, bright red to bright or dark orange; style appendages fairly well developed, slender, ca. 0.2–0.3 mm long. Open-pollinated. (n = 5) All year. A pantropical weed of Old World origin, occasionally found in disturbed habitats in s Fla.

2. E. fosbergii D. H. Nicolson. Vegetatively much like *E. coccinea,* but the leaves more consistently toothed. Involucre consisting of ca. 8 or ca. 13 bracts, thick-cylindric, mostly 9–14 mm long and 2–3 times as long as wide at anthesis (the width measured at midlength); flowers often fewer than in *E. coccinea,* conspicuously surpassing the involucre (commonly by 2–5 mm) and spreading out above it to form a disk mostly 7–15 mm wide at anthesis, variously pink or purple to bright or dark red, but not orange; style appendages fairly well developed, slender, ca. 0.2 mm long. Open-pollinated. (n = 10) Perhaps an allotetraploid of no. 1 and no. 3. All year. A neotropical or pantropical weed, established in disturbed habitats in peninsular Fla. Further study may turn up an older name for this species, based on Old World material. The name *E. javanica* (Burm.) C. B. Robinson, which has sometimes been used for it, seems not to apply.

3. E. sonchifolia (L.) DC. Vegetatively much like the other 2 spp., but averaging more slender and more branched, and the leaves often more or less lobed or lyrate-pinnatifid as well as toothed. Involucre consisting of about 8 bracts, slender, mostly 9–12 mm long and 3–4 times as long as wide at anthesis (the width measured at midlength); flowers relatively few, commonly 20–40, about equaling the involucre or surpassing it by up to 1 (2) mm, not spreading out much above it, the disk seldom more than 5 mm wide at anthesis, variously lavender or pink to purple (white), only rarely bright red; style appendages very short (ca. 0.1 mm) or obsolete. Self-pollinated. (n = 5) All year. A pantropical weed of Old World origin, established in disturbed habitats in c and s Fla.

57. GYNURA Cass., nom. conserv.

1. G. aurantiaca (Blume) DC. VELVET PLANT. Coarse, almost succulent, lax and branching, tender perennial herb, conspicuously purple-hairy (the cross-walls of the hairs purple), with alternate, toothed, ovate to elliptic, sessile or petioled leaves up to 10 × 5 cm, often auriculately expanded at the base of the petiole or sessile blade. Heads several in a loose, corymbiform inflorescence, 1–2 cm wide, orange, discoid, the flowers all tubular and perfect; involucre a single series of ca. 13 equal bracts 10–12 mm long, plus some loose, slender bracteoles at the base; receptacle naked; style branches flattened, with introrsely marginal stigmatic lines and an elongate (ca. 2 mm) slender, hispidulous appendage. Pappus of numerous capillary bristles. Achenes prismatic, ca. 5-angled and 10-ribbed. (n = 10) Spring. Native of the East Indies, cult. for its foliage, and occasionally escaped in s Fla.

58. CACALIA L. INDIAN PLANTAIN

Perennial herbs with fibrous or fleshy-fibrous roots, often more or less tuberous-thickened at the base of the stem. Leaves alternate, mainly petiolate, entire to toothed or lobed, or in some extralimital spp. dissected. Heads discoid, white to ochroleucous or chloroleucous or somewhat anthocyanic, but not yellow; involucre a single series of equal, more or less herbaceous, scarious-margined bracts, sometimes also with some smaller and looser basal bracteoles; receptacle naked, flat or convex, sometimes with a short, conic, central projection; flowers all tubular and perfect, the corolla more or less deeply lobed; style branches flattened, with broad ventromarginal stigmatic lines that may be almost confluent medially, truncate or with a very short, blunt appendage, penicillate. Achenes glabrous, cylindric to fusiform or somewhat flattened, several-nerved. Pappus of numerous capillary bristles. *Arnoglossum, Mesadenia, Odontotrichum, Synosma*. Kral, R., and R. K. Godfrey. Synopsis of the Florida species of *Cacalia*. Quart. J. Florida Acad. Sci. 21: 193–206. 1958.

1 Involucral bracts mostly 10–15; flowers numerous, commonly ca. 20–40.
  2 Leaves chiefly cauline, the larger ones hastate; plants mostly 1–2.5 m tall ..... 1. *C. suaveolens*.
  2 Leaves basally disposed, none of them hastate; plants well under 1 m tall .... 2. *C. rugelia*.
1 Involucral bracts 5; flowers 5.
  3 Involucral bracts with a median, scarious wing keel; stems strongly angulate or sulcate.
    4 Basal and lowest cauline leaves with truncate-ovate to cordate-ovate, mostly angulate-lobed to angulate-dentate blade; corollas usually pale lavender .... 5. *C. diversifolia*.
    4 Basal and lowest cauline leaves with elliptic to ovate, basally rounded or tapering, entire or merely toothed blade; corollas not anthocyanic.
      5 Basal and lowest cauline leaves mostly with the blade longer than the petiole; plants of dry, sandy habitats .... 6. *C. floridana*.
      5 Basal and lowest cauline leaves mostly with the blade shorter than the petiole; plants of wet, low habitats.
        6 Corollas mostly 6–7 mm long .... 7. *C. sulcata*.
        6 Corollas mostly 8–10 mm long .... 8. *C. plantaginea*.
  3 Involucral bracts not wing-keeled; stems (except in no. 3) terete or merely striate.
    7 Larger leaves palmately veined, the principal nerves divergent.
      8 Leaves green on both sides; stem conspicuously grooved, not glaucous .. 3. *C. muhlenbergii*.
      8 Leaves pale and glaucous beneath; stem glaucous, terete or slightly striate .... 4. *C. atriplicifolia*.
    7 Leaves strongly several-nerved, the nerves longitudinal, converging distally .... 9. *C. ovata*.

1. C. suaveolens L. Plants mostly 1–2.5 m tall, glabrous or nearly so. Stem striate or grooved, simple up to the inflorescence, leafy. Middle and lower leaves triangular-hastate, 5–20 cm long and nearly or fully as wide, sharply toothed, conspicuously petiolate; upper leaves progressively less hastate and with shorter, more winged petiole. Heads numerous in a flat-topped inflorescence, commonly 20- to 40-flowered, the disk 7–11 mm wide; principal involucral bracts mostly 10–15, ca. 1 cm long; some reduced but evident, loose, subulate outer bracts commonly also present: receptacle flat, deeply pitted. (n = 20) Midsummer–fall. Riverbanks and moist, low ground; RI and Conn to se Minn, s to Md, Ky, Tenn, and Ill, and in the mts. to Ga. *Synosma suaveolens* (L.) Britton—S.

2. C. rugelia (Shuttleworth ex Chapman) T. M. Barkley & Cronq. Plants 3–5 dm tall, floccose below, villous and glandular above, or eventually glabrate. Leaves basally disposed, the basal and lowermost cauline ones with long-petiolate, ovate, denticulate blade mostly 8–16 × 5–12 cm, distantly toothed, truncate or rounded to cordate at the base; cauline leaves few and much reduced, mostly sessile and bractlike. Heads few, relatively large, more or less nodding, the involucre of ca. 13 bracts 10–14 mm long; flowers numerous (ca. 30–40). (n = 28) Summer. Forest openings at high elev. in BR of NC and Tenn. *Senecio rugelia* (Shuttleworth ex Chapman) A. Gray—R, S.

3. C. muhlenbergii (Schultz-Bip.) Fern. Plants stout, 1–3 m tall, the stem conspicuously grooved. Leaves green on both sides, irregularly dentate and sometimes shallowly lobed, commonly more or less ciliate in the sinuses, palmately veined, the lower very large and long-petioled, up to 8 dm wide, sometimes reniform, the upper reduced and more flabellate or ovate. Heads numerous in a short and broad, flat-topped inflorescence, 5-flowered, narrowly cylindric; involucre 7–12 mm high, with 5 principal bracts. (n = 25) Summer. Open woods; BR and RV of Ga, Ala, NC, Tenn, and Va, n to NJ and Pa, and w to Minn and Mo and ne Miss. *Mesadenia reniformis* (Muhl.) Raf.—S.

4. C. atriplicifolia L. Much like no. 3. Stem glaucous, terete or slightly striate. Leaves pale and glaucous beneath, averaging smaller, proportionately longer, and more pointed, with fewer and larger teeth, or merely shallowly lobed, the sinuses smooth; receptacle with a short central beak. (n = 25, 26, 27, 28) Summer. Woods and moist or rather dry open places; NJ and Pa to s Minn, s to Ga (all), the panhandle of Fla, Miss, and Okla. *Mesadenia atriplicifolia* (L.) Raf.—S.

5. C. diversifolia T. & G. Plants up to 2 (3) m tall, glabrous or nearly so except for the brown, scalelike hairs toward the base of the basal leaves; stem evidently sulcate-angular. Leaves basally disposed, the basal and lowest cauline with ovate, basally truncate to subcordate, mostly 3–5-nerved blade up to 10 × 7 cm, generally more or less angulate-toothed or angulate lobed, shorter than the petiole, the others progressively reduced upward. Heads numerous, 5-flowered, narrowly cylindric; involucre 9–11 mm high, its principal bracts 5, whitish, wing-keeled; flowers pale lavender; receptacle with a short central beak. (n = 25) Spring–summer. River swamps, bottomland alluvium, and mesic woods along streams; nw peninsular Fla to the Fla panhandle, sw Ga, and se Ala. *Mesadenia diversifolia* (T. & G.) Greene—S.

6. C. floridana A. Gray. Plants 4–15 dm tall, largely glabrous or glabrate except that the petioles of the basal leaves are often pubescent with matted, tawny hairs, and the smaller branches of the inflorescence and base of the involucres are minutely fuscous-hairy; stem strongly angular-sulcate. Leaves basally disposed, the basal and lowest cauline ones with ovate or elliptic blade up to 18 × 13 cm, strongly several-nerved, the nerves converging distally, the margins toothed or undulate to entire, the petiole mostly shorter than the blade; cauline leaves progressively reduced and less petiolate upward. Heads numerous, 5-flowered, narrowly cylindric; involucre mostly 10–14 mm high, its principal bracts 5, pale, wing-keeled; flowers ochroleucous; receptacle with a short central beak. (n = 25, 26) Spring. Dry, sandy ridges with pine and scrub oak or pine and palmetto, or sometimes in old fields or along roadsides; c and ne Fla. *Mesadenia floridana* (A. Gray) Greene—S.

7. C. sulcata Fern. Much like no. 8; leaves with fewer (mostly 3–5) main veins; heads smaller, the involucre 7–10 (12) mm high; disk corollas ochroleucous, mostly 6–7 mm long, the slender basal tube mostly 3–3.5 mm long. (n = 25) Spring. Wet, low places; Fla panhandle, s Ala, and sw Ga. *Mesadenia sulcata* (Fern.) Small—S.

8. C. plantaginea (Raf.) Shinners. Plants 6–18 dm tall, stout, glabrous; stem striate-angled. Leaves basally disposed, thick and firm, entire or nearly so, with several (commonly 7–9) prominent longitudinal nerves converging toward the summit, basal and lowest cauline leaves conspicuously long-petioled, the blade 6–20 × 2–10 cm, commonly elliptic and tapering to the base, or deltoid-ovate and subtruncate in robust specimens; cauline leaves few, conspicuously reduced upward, becoming sessile or subsessile. Heads numerous, 5-flowered, narrowly cylindric; involucre mostly 12–15 mm high, its principal bracts 5, wing-keeled; receptacle with a short central beak; disk corollas ochroleucous, mostly 8–10 mm long, the slender basal tube mostly 5–6 mm. (n = 27) Spring. Wet prairies and marshy or boggy places, especially in calcareous regions; Mich and s Ont to Minn and Neb, s to Ala, La and Tex. *C. tuberosa* Nutt.—F, G, R. *Mesadenia tuberosa* (Nutt.) Britton—S.

9. C. ovata Walter. Plants 0.5–3 m tall, glabrous and often also glaucous, the stem terete, finely striate. Leaves basally disposed, the basal and lowest cauline ones with entire or toothed, lance-linear to ovate, basally tapering or rounded blade up to 30 × 12 cm, strongly several-nerved, the nerves longitudinal, converging distally, the petiole mostly shorter than the blade; cauline leaves progressively reduced upward, becoming sessile. Heads numerous in a flat-topped inflorescence, 5-flowered, narrowly cylindric; involucre 7–10 mm high, the principal bracts 5, pale, without any evident median wing or keel; receptacle with a poorly developed central beak. (n = 25) Late summer–fall. Wet, often sandy savannas and low woods; CP from SC to s Fla and w to Tex. *C. lanceolata* Nutt.—R; *Mesadenia lanceolata* (Nutt.) Raf.—S; *Mesadenia elliottii* Harper—S; *Mesadenia maxima* Harper—S.

59. ERECHTITES Raf. FIREWEED

1. E. hieracifolia (L.) Raf. Fibrous-rooted annual weed 0.1–2.5 m tall, glabrous or sometimes spreading-hairy throughout. Leaves numerous, well distributed along the stem, alternate, up to 20 × 8 cm, sharply serrate and sometimes also irregularly lobed, the lower oblanceolate to obovate and often petiolate, the others more elliptic, lanceolate, or oblong, and often some of them auriculate clasping. Heads turbinate-cylindric, with swollen base when fresh, disciform, whitish; involucre a single series of narrow, equal, more or less herbaceous bracts 10–17 mm long, often with a few loose bracteoles at the base; receptacle flat or nearly so, naked; pistillate flowers numerous, in several series, with elongate, filiform-tubular, eligulate corolla; central flowers perfect, with narrowly tubular, 5-toothed corolla; style branches flattened, with ventromarginal stigmatic lines and a set of short hairs at the base of the short, deltoid appendage. Achenes mostly 10–12-ribbed, 2–3 mm long. Pappus bright white, of numerous elongate, slender, eventually deciduous capillary bristles. (n = 20) Late summer and fall, or all year southward. Various wet or dry, often disturbed habitats; widespread in e US, s to tropical Am, and found throughout our range. Our plants, as here described, are var. *hieracifolia*.

60. TUSSILAGO L. COLTSFOOT

1. T. farfara L. Rhizomatous perennial 0.5–5 dm tall. Leaves basal, long-petioled, the blade cordate to suborbicular, with deep, narrow sinus, callous-denticulate and shallowly lobed, 5–20 cm long and wide, glabrous above, persistently white-tomentose beneath; stem merely scaly-bracted, the bracts alternate, ca. 1 cm long. Heads solitary and terminal, yellow, blooming before the basal leaves develop, cylindric at first, expanding with maturity and then up to 3 cm wide, radiate, the rays pistillate and fertile, very numerous in several series, narrow, not much exceeding the involucre and pappus; involucre a single series of equal, more or less herbaceous bracts, 8–15 mm high; receptacle naked; disk flowers sterile, with merely lobed style. Achenes linear, 5–10-ribbed. Pappus of numerous capillary bristles, that of the sterile flowers somewhat reduced. (n = 30) Early spring. Native of Europe, naturalized in disturbed and waste places in ne US, s to Va and WVa.

61. PETASITES Mill. SWEET COLTSFOOT

1. P. hybridus (L.) Gaertn., Mey. & Scherb. BUTTERFLY DOCK, BUTTERBUR. Perennial from a tuberous-thickened base, 1–4 dm tall, or up to 10 dm in fruit. Leaves basal, long-petiolate, the blade reniform or cordate, up to 3 dm wide or even more, callous-dentate or -denticulate, glabrous or nearly so above, arachnoid-tomentose and glabrate beneath; stem merely bracteate, the bracts alternate, 2–6 cm long. Heads numerous in a narrow, racemiform inflorescence, blooming before or as the basal leaves develop, purple, rayless; plants subdioecious, the flowers in the female heads all or nearly all pistillate and fertile, with tubular-filiform corolla, those in the male heads chiefly or entirely hermaphrodite but sterile, with puberulent, inconspicuously lobed style; involucre a single series of equal, more or less herbaceous bracts and commonly a few loose bracteoles, 3–4 mm high in the female heads, 6–8 mm in the male; receptacle flat, naked. Achenes linear, 5–10-ribbed. Pappus of numerous capillary bristles, that of the sterile flowers somewhat reduced. (n = 30) Spring. Native of Europe, escaped from cult. in moist places in ne US, s to Del.

62. HAPLOPAPPUS Cass., nom. conserv.

Taprooted herbs (all ours) or shrubs. Leaves alternate, simple and entire to pinnately dissected. Heads medium-sized to rather large, yellow, radiate (all ours) with pistillate and fertile rays, or sometimes discoid; involucral bracts generally greenish at least distally, in 2–several imbricate or subequal series; receptacle flat or somewhat convex, naked; disk flowers perfect and mostly fertile (the central ones sterile in *H. ciliatus*); style branches flattened, with ventromarginal stigmatic lines and short to elongate, externally hairy appendage. Achenes 4- to 5-angled or striate. Pappus of numerous generally sordid or brownish capillary bristles, these unequal but not divided into distinct long and short series. *Isopappus, Sideranthus*.

1 Heads relatively few and large, the involucre mostly 10–18 mm high, the disk commonly 1.5–3 cm wide.
  2 Achenes evidently hairy; herbage glandular and sometimes also hairy .... 1. *H. phyllocephalus*.
  2 Achenes and herbage glabrous .... 2. *H. ciliatus*.
1 Heads numerous and relatively small, the involucre mostly 5–7 mm high, the disk less than 1 cm wide .... 3. *H. divaricatus*.

1. H. phyllocephalus DC. Aromatic, glandular and sometimes also hairy annual or casually biennial or short-lived perennial, 2–10 dm tall, freely branched (often even from the base), leafy throughout. Leaves oblanceolate to narrowly obovate, firm and often fleshy, mostly 2–7 cm long, the middle ones commonly (5) 8–12 mm wide (including the usually prominent, salient teeth or short lobes), the upper only slightly reduced, subtending and often surpassing the heads. Heads terminating the branches, relatively large, hemispheric, the disk commonly 1.5–2 cm wide; involucre 10–14 mm high, its bracts not much imbricate, commonly loose or recurved at the slender tip, the larger ones well over 1 mm wide; rays ca. 25–40, 9–14 mm long; style appendages shorter than the stigmatic portion. Achenes evidently hairy. Pappus copious, persistent. (n = 6) Summer–fall. Dunes and waste places near the coast; Gulf Coast from s Fla to Tex and Mex. *Sideranthus megacephalus* (Nash) Small—S.

2. H. ciliatus (Nutt.) DC. Glabrous annual or biennial, mostly 4–15 dm tall, commonly simple to near the top, leafy to the summit. Leaves sessile, clasping, oblong to ovate or elliptic-obovate, rounded or obtuse at the tip, mostly 3–8 × 1–4 cm, spinose-dentate. Heads few, large, the disk mostly 1.5–3 cm wide; involucral bracts slightly or moderately imbricate, the outer ones (at least) with long, loose or spreading green tip, at least the larger bracts generally well over 1 mm wide; rays numerous, ca. 25–50, 1–2 cm long; style appendages generally shorter than the stigmatic portion. Achenes glabrous, the central ones sterile. Pappus rather scanty, eventually deciduous. (n = 6) Late summer–fall. Open or waste places; Tex and NM to n La and w Mo.

3. H. divaricatus (Nutt.) A. Gray. Annual weed 2–10 dm tall, simple below, generally much branched above, subglabrous or more often coarsely scabrous or spreading-hairy, sometimes also glandular. Leaves oblanceolate or linear-oblanceolate, with a few spiny teeth or occasionally entire, 3–7 cm × 3–10 mm, or the lower sometimes larger. Heads numerous in a diffuse inflorescence, small, the disk less than 1 cm wide; involucre 5–7 mm high, its bracts well imbricate, not over 1 mm wide; rays 7–11 (in ours), 4–6 mm long; style appendages longer than the stigmatic portion. (n = 4, 5, 6, 7). Midsummer–fall. Waste places, especially in sandy soil; CP and PP from Va to n Fla, w to Tex, and inland to OU of Ark and to Kans and Okla. *Isopappus divaricatus* (Nutt.) T. & G.—S.

63. CHRYSOPSIS Elliott, nom. conserv. GOLDEN ASTER*

Herbs. Leaves alternate, entire or merely toothed. Heads 1–many, yellow, radiate (in all ours) or seldom discoid, with pistillate and fertile rays; involucral bracts imbricate, greenish at least in part; receptacle flat or a little convex, naked; disk flowers numerous, perfect and fertile; style branches flattened, with ventromarginal stigmatic lines and slender, externally short-hairy, more or less elongate appendage. Achenes more or less compressed, several- to many-nerved. Pappus double, the inner of elongate capillary bristles, the outer of short, coarse bristles or scales. *Pityopsis* Nutt.—S. Included in *Heterotheca* Cass. by R. Species 4–9 are intergradient and in need of critical study. Work currently in progress by John T. Semple may necessitate the revival of some names here reduced to synonymy.

*I gratefully acknowledge my use of an unpublished thesis on the section *Pityopsis* by Frank Bowers. The interpretation here presented is my own, and is not to be attributed to Dr. Bowers.

1 Achenes relatively short and broad, narrowly to broadly obovate or oblong-ovate, or turbinate; leaves either pinnately veined or (when very narrow) with only the midrib evident, neither grasslike nor parallel-veined (Section *Chrysopsis*).
  2 Herbage rather coarsely (often sparsely) hairy (as well as finely glandular in part), hirsute or hispid-hirsute or hirsute-strigose; plants spreading by slender creeping rhizomes .... 1. *C. camporum*.
  2 Herbage very finely and softly hairy or more or less glandular, or both, but without coarse, eglandular hairs; plants annual to perennial, but without slender, creeping rhizomes.
    3 Plants taprooted annuals; herbage softly and finely spreading-hairy and finely glandular, not at all floccose .... 2. *C. pilosa*.
    3 Plants biennial or perennial, either fibrous-rooted or with a mostly short and quickly deliquescent taproot; herbage variously glandular or hairy (often in considerable part floccose), or largely glabrous, but not with the pubescence of *C. pilosa*.
      4 Involucre evidently glandular or hairy or both; achenes, except in *C. latisquamea*, without any raised, reddish-translucent ribs.
        5 Plants floccose-tomentose or woolly at least up to the tops of the peduncles, the wool commonly also extending onto the involucre; rays mostly ca. 34, or fewer on some of the later heads .... 4. *C. gossypina*.
        5 Plants floccose-tomentose or woolly or loosely long-hairy only below the inflorescence (if at all), the wool (when present) not extending onto the glandular peduncles and involucres.
          6 Involucral bracts relatively narrow, the green tip not expanded, less than 1.3 mm wide; achenes without any raised, reddish-translucent ribs.
            7 True perennial, fibrous-rooted from a caudex or short, woody rhizome; herbage below the inflorescence at first loosely long-villous with soft and flexuous but not notably tangled hairs, later tending to be more or less glabrate, not glandular .... 3. *C. mariana*.
            7 Biennial or short-lived perennial from a more or less well-developed taproot; herbage below the inflorescence persistently stipitate-glandular (especially above) or loosely floccose-tomentose with intricately tangled hairs (especially below), not becoming glabrate .... 5. *C. scabrella*.
          6 Involucral bracts notably broad, the green tips of at least some of them somewhat expanded, 1.3–2.5 mm wide; achenes with 2–several raised, reddish-translucent ribs in addition to the more obscure ordinary veins .... 6. *C. latisquamea*.
      4 Involucre glabrous or nearly so, often resinous or minutely puncticulate-pustulate, but not glandular or hairy; achenes, except sometimes in *C. trichophylla*, generally with 2–several raised, reddish-translucent ribs in addition to the more obscure ordinary veins.
        8 Involucral bracts narrowly acute to shortly caudate-acuminate, appressed or somewhat loose, but not conspicuously prolonged.
          9 Leaves narrow, the lower ones mostly 3–7 mm wide, the others narrower .... 8. *C. hyssopifolia*.
          9 Leaves wider, the lower ones mostly 8–18 mm wide .... 7. *C. trichophylla*.
        8 Involucral bracts with very loose or reflexed, conspicuously elongate, caudate-subulate tip .... 9. *C. subulata*.

1 Achenes slender and elongate, linear or fusiform; leaves tending to be grasslike and parallel-veined, or (when very narrow) with only the midrib evident (section *Pityopsis*).
  10 Leaves basally disposed, the lowest ones much the largest, the others progressively reduced.
    11 Cauline leaves relatively numerous, generally more than 10; heads in well-developed plants generally more than 6 .... 10. *C. graminifolia*.
    11 Cauline leaves few, mostly 2–7; heads few, mostly 1–6 .... 11. *C. oligantha*.
  10 Leaves well distributed along the stem, the basal ones wanting or not much enlarged.
    12 Peduncles and involucres conspicuously glandular; leaves sericeous; plants only 1–3 dm tall .... 12. *C. ruthii*.
    12 Peduncles and involucres nearly or quite without glands; leaves and habit various.

13 Leaves very numerous, the internodes mostly well under 1 cm long, the stem not flexuous or angled from one node to the next; plants essentially glabrous .... 13. *C. pinifolia*.
13 Leaves not very numerous, the principal internodes mostly ca. (0.5) 1–3 cm long, the stem more or less flexuous or zigzag from one node to the next; plants thinly silky .... 14. *C. flexuosa*.

1. C. camporum Greene. Plants perennial, taprooted and spreading by slender, creeping rhizomes, robust, mostly 4–10 dm tall, rather coarsely hairy (as well as finely glandular in part), hirsute or hispid-hirsute or hirsute-strigose, often only thinly so. Leaves numerous, chiefly cauline, elliptic or oblong to ovate and sessile, or the lower more oblanceolate or obovate and subpetiolate, mostly 3–7 cm × 8–20 mm, generally with a few sharp, small teeth. Heads terminating the branches or forming a corymbiform inflorescence, hemispheric, the disk mostly 1.2–2.5 cm wide; involucre 8–11 mm high, its bracts imbricate, gradually tapering to an often minutely purple tip; rays from ca. 21 to ca. 34, ca. 1 cm long or a bit more. Achenes rather narrowly obovate, 6- to 12-nerved. Summer–fall. Typically a prairie sp. of Mo to Ind and Ill, but recently intro. and becoming very abundant along roadsides and in fields in parts of our area, as in Tenn, NC, Ga, and Ala. *C. villosa* (Pursh) Nutt. var. *camporum* (Greene) Cronq.—G.

2. C. pilosa Nutt. Taprooted annual, mostly 2–9 dm tall, minutely glandular-puberulent and sparsely to densely pilose throughout with long, soft, slender, spreading hairs. Leaves numerous, chiefly cauline, narrowly oblong to broadly oblanceolate, seldom more than 8 cm × 13 mm, entire or toothed, sessile, or the lower shortly subpetiolate. Heads hemispheric, terminating the branches, the central one characteristically overtopped by the lateral ones, the disk 1–2 cm wide; involucre 7–9 mm high; rays from ca. 13 to ca. 21, ca. 1 cm long. Achenes oblong-obovate, 10- to 15-nerved. (n = 4, 5) Midsummer–fall. Dry, often sandy places; Mo to La, w to Kans, Okla, and Tex, and occasionally intro. eastward, as far as NC and Fla. *C. nuttallii* Britton—S; *Heterotheca pilosa* (Nutt.) Shinners—R.

3. C. mariana (L.) Elliott. Fibrous-rooted perennial from a short, woody rhizome or branching caudex, 2–8 dm tall, loosely shaggy-villous with elongate, flexuous but not notably contorted and tangled hairs when young, later tending to be more or less glabrate. Leaves basally disposed, the lower oblanceolate to obovate, petiolate, up to 18 × 3.5 cm (petiole included) generally denticulate; middle and upper leaves smaller, lanceolate to oblong or elliptic, sessile. Heads few to rather numerous in a corymbiform, commonly somewhat congested inflorescence, nearly hemispheric, the disk 1–2 cm wide; involucre and peduncles stipitate-glandular, the involucre 7–10 mm high, rays from ca. 13 to ca. 21, ca. 1 cm long. Achenes obovate, 3- to 5-nerved. (n = 12) Late summer–fall, or all year southward. Woods and sandy places; CP, PP, and mt. provinces, from s NY to s Ohio, s to c Fla and s Ala, and thence w on GC to La, and irregularly in Tenn to ME. *Heterotheca mariana* (L.) Shinners—R.

4. C. gossypina Nutt. Biennial from a short, quickly deliquescent taproot, or more often a rather short-lived perennial from a fibrous-rooted caudex, 3–8 dm tall, commonly with several basally decumbent stems, or the stem with several long branches from near the base, seldom with a single erect, strict stem. Herbage more or less densely floccose-woolly throughout, at least up to the tops of the peduncles, the wool commonly also extending onto the otherwise glabrous or glandular involucres. Leaves numerous, oblanceolate to oblong-elliptic, obtuse or rounded, entire or nearly so, sessile or, especially the lower, narrowed to a subpetiolar base, mostly less than

6 cm × 18 mm, gradually reduced upward. Heads few in an open, corymbiform inflorescence, hemispheric, the disk mostly 1.2–2.5 cm wide; involucre 7–11 mm high, only slightly or moderately imbricate, the loose or spreading green tips of the phyllaries slender and often rather elongate; rays mostly ca. 34, or fewer on some of the later heads, ca. 1 cm long. Achenes narrowly obovate, obscurely several-nerved. (n = 9) Fall. Sandy places, often with pine or scrub oak, sometimes on beaches; CP from s Va to n Fla and w to sw Ala. *C. arenicola* Alexander—S; *C. decumbens* Chapm.—S; *C. longii* Fern.—F; *C. pilosa* (Walt.) Britton—S, misapplied; *Heterotheca gossypina* (Michx.) Shinners—R.

5. C. scabrella T. & G. Biennial or short-lived perennial from a short, quickly deliquescent taproot, 2–8 dm tall, with one or several erect or decumbent stems, the herbage coarsely stipitate-glandular or with spreading, gland-tipped hairs at least in the inflorescence, sometimes floccose-woolly (instead of glandular) near the base or up to the bottom of the inflorescence. Leaves numerous, the lowermost ones oblanceolate or spatulate, 3–8 cm × 6–20 mm, often deciduous, the others more oblong to lance-ovate or nearly linear and mostly sessile. Heads several or many, terminating the branches of a corymbiform inflorescence, the disk 1–2 cm wide; involucre densely stipitate-glandular, 7–11 mm high, its bracts well imbricate, slender, the green tip ca. 1 mm wide or less, acute to subcaudately short-acuminate; rays mostly ca. 21, ca. 1 cm long. Achenes narrowly obovate, obscurely several-nerved. Fall. Dry, sandy soil with pine and scrub oak; throughout peninsular Fla, and w through the panhandle to s Ala; apparently also in Moore Co., NC. *C. floridana* Small—S; *C. lanuginosa* Small—S.

6. C. latisquamea Pollard. Biennial or short-lived perennial from a rather short, stout taproot, 2–8 dm tall, with one or several erect stems, the herbage conspicuously floccose-tomentose up to the stipitate-glandular inflorescence. Leaves numerous, the lowest oblanceolate to oblong-obovate, up to 6 × 2 cm, but often deciduous, the others more oblong, mostly broad-based and sessile, 2–4 cm × 5–12 mm. Heads several or many in a corymbiform inflorescence, the disk 1–2 cm wide; involucre 8–12 mm high, with notably wide bracts, at least some of them with somewhat expanded green tip 1.3–2.5 mm wide; rays mostly ca. 13, ca. 1 cm long. Achenes obovate, obscurely several-nerved and with several prominent, raised, reddish-translucent ribs. Late summer–fall. Sandy places with oak and pine; n and c peninsular Fla, from Columbia and Madison Cos. to Hillsboro and Orange Cos.

7. C. trichophylla Nutt. Short-lived perennial from a short, fibrous-rooted caudex or a short, quickly deliquescent taproot, 3–8 dm tall, floccose below, usually glabrous or glabrate above, not or scarcely glandular. Leaves numerous, the lower oblanceolate, often broadly so, tapering to a shortly petiolar base, mostly 4–8 cm × 8–18 mm, the others more oblong or elliptic-oblong and sessile, gradually reduced upward. Heads few to many, terminating the branches of a corymbiform inflorescence, the disk 1–2 cm wide; involucre essentially glabrous, 8–11 mm high, its bracts well imbricate, slender, narrowly acute to shortly caudate-acuminate, but not prolonged as in *C. subulata*, often a little loose; rays mostly ca. 21 (to 34), ca. 1 cm long. Achenes oblong-obovate, several-nerved, with or without one or more raised, reddish-translucent ribs. (n = 5) Late summer. Sandhills on CP, locally from NC to Fla. *Heterotheca trichophylla* (Nutt.) Shinners—R.

8. C. hyssopifolia Nutt. Biennial or short-lived perennial from a short, quickly deliquescent taproot, sometimes further perennating by slender, creeping roots, 2–10 dm tall, glabrous or sparsely to moderately long-hairy (often only the basal leaves evidently long-hairy), the leaves and involucral bracts often finely whitish-granular-punctate, but not glandular. Leaves very numerous and small, entire or with a few sharp teeth, the lower oblanceolate or linear-oblanceolate, 3–5 (7) cm × 3–7 mm, often deciduous, the others more linear or lance-oblong and sessile, gradually reduced upward. Heads several or many, terminating the branches of a loosely corymbiform inflorescence, or the inflorescence more compact and crowded, somewhat umbelliform; disk 8–15 mm wide; involucre glabrous, 6–9 mm high, its bracts narrow, strongly imbricate, acuminate or narrowly acute, but not caudate prolonged, the inner sometimes a little loose at the tip; rays mostly ca. 13 to ca. 21, ca. 1 cm long. Achenes turbinate, obscurely several-nerved and with 2–several conspicuous, raised, reddish-translucent, distally swollen ribs. Summer. Sandy soil, with pine and scrub oak; peninsular Fla, and w through the panhandle to s Ala and s Miss. *C. gigantea* Small—S.

9. C. subulata Small. Biennial or short-lived perennial from a usually quickly deliquescent taproot, 3–7 dm tall, branched above, loosely long-hairy, more densely and conspicuously so below, not evidently glandular. Basal and lower cauline leaves oblanceolate, mostly 5–7 cm (including the petiolar base) × 5–11 mm, the others numerous, more oblong or linear and sessile, often somewhat auriculate-clasping. Heads several or many in a corymbiform inflorescence, terminating the branches, the disk 1–2 cm wide; involucre glabrous, sometimes resinous, 7–10 mm high, its bracts narrow, with very loose or reflexed, conspicuously elongate, caudate-subulate tip; rays mostly ca. 21, ca. 1 cm long. Achenes turbinate, obscurely several-nerved and with 2–several conspicuous, reddish-translucent, distally swollen ribs. Spring–summer. Sandy soil, especially in pinelands; peninsular Fla, and w to Liberty Co.

10. C. graminifolia (Michx.) Elliott. Fibrous-rooted perennial, mostly 3–9 dm tall, often with slender, stoloniform rhizomes, more or less silvery-silky with appressed hairs, at least below, sometimes more glandular upward. Leaves basally disposed, the larger ones elongate, parallel-veined, grasslike, up to 35 × 2 (3.5) cm, the others numerous (generally more than 10) and progressively reduced upward. Heads (1–) several or many, terminating the branches, turbinate or campanulate. Achenes linear. (n = 9, 18) Sandy, usually dry places; Del to s Fla and Bahama Isl, w to s Ohio, e Ky, sw Tenn, c Ark, se Okla, and e Tex, and thence s to Guatemala. Late summer–fall. A highly complex sp., here considered to consist of three vars.:

Peduncles silky, not glandular; involucre more or less silky, at least below, sometimes glandular above.
- Heads larger, the involucre mostly 8–12 mm high, the rays mostly ca. 13, 7–12 mm long, the disk flowers mostly 30–50, with corolla 6.5–9 mm long (dry); tetraploid; nearly the range of the sp., but rare on PP, uncommon in Ark and La, and absent from Okla. *C. nervosa* (Willd.) Fern.—F; *Heterotheca correllii* (Fern.) Ahles—R; *H. graminifolia* (Michx.) Shinners—R; *H. nervosa* (Willd.) Shinners—R; *Pityopsis graminifolia* (Michx.) Nutt.—S; *P. tracyi* Small—S. . . . var. *graminifolia*.
- Heads smaller, the involucre mostly 5–8 (10) mm high, the rays mostly ca. 8, less often ca. 13, mostly 4–8 mm long, the disk flowers mostly 15–30, with corolla 4–6.5 mm long, diploid or seldom tetraploid; CP from NC to s Fla, w to Tex, and n into OU and OZ of Ark and se Okla. *Pityopsis microcephala* Small—S. . . . . var. *microcephala* (Small) Cronq.

Peduncles, as well as the involucre, evidently stipitate-glandular, scarcely silky; heads mostly of a size with var. *microcephala*; diploid; Va to n Fla and s Miss, commonest on PP. *C. graminifolia* (Michx.) Ell.—F, misapplied; *Heterotheca adenolepis* (Fern.) Ahles—R; *Pityopsis aspera* (Shuttlew.) Small—S. . . . . var. *aspera* (Shuttlew.) A. Gray.

11. C. oligantha Chapman. Fibrous-rooted perennial, mostly 1.5–6 dm tall, commonly with clustered stems, silvery-silky below, stipitate-glandular above. Leaves basally disposed, as in *C. graminifolia*, the largest ones up to 30 × 1.5 cm; cauline leaves few, generally 2–7, progressively reduced upward. Heads 1–6, conspicuously pedunculate, campanulate; involucre 7–11 mm high, stipitate-glandular; rays ca. (8) 13 (21), 9–13 mm long. Achenes linear. (n = 18) Spring. Sandy, wet or dry pinelands; w Fla and adj Ga and Ala. *Pityopsis oligantha* (Chapm.) Small—S.

12. C. ruthii Small. Fibrous-rooted perennial with clustered stems 1–3 dm tall, silvery-sericeous (or partly glabrate) throughout except for the stipitate-glandular peduncles and involucre. Leaves numerous, chiefly cauline, linear or lance-linear, mostly 2–6 cm × 2–4 mm, the lower cauline ones generally soon deciduous, the basal ones, when present, often tufted but not enlarged. Heads 1–several; involucre 6–10 mm high; rays ca. 8 (13), 6–10 mm long. Achenes linear-fusiform. (n = 9) Fall. Rocks along the Hiwassee R Gorge in Polk Co., Tenn. *Pityopsis ruthii* (Small) Small—S.

13. C. pinifolia Elliott. Fibrous-rooted, more or less rhizomatous perennial, 2–6 dm tall, glabrous or nearly so throughout. Leaves mostly or all cauline, numerous and crowded, narrowly linear, straight or nearly so, flat or folded, the larger ones mostly 4–8 cm × 1–2 mm, gradually reduced upward. Heads several, terminating the subnaked or linear-bracteate branches; involucre 6–9 mm high, the inner bracts often distally ciliolate, otherwise glabrous; rays ca. 13, 6–8 mm long. Achenes linear-fusiform. (n = 9) Late summer–fall. Fall-line sandhills of NC, SC, and Ga. *Heterotheca pinifolia* (Ell.) Ahles—R; *Pityopsis pinifolia* (Ell.) Nutt.—S.

14. C. flexuosa Nash. Fibrous-rooted, shortly rhizomatous perennial, 2–6 dm tall, thinly silky, the stem more or less flexuous or zigzag from one node to the next. Leaves mainly cauline, linear or lance-linear or the lower oblanceolate, 3–7 cm × 3–7 mm, not very numerous, the principal internodes mostly (0.5) 1–3 cm long. Heads several, terminating the mostly leafy-bracteate branches or peduncles; involucre 8–11 mm high, thinly silky-puberulent or subglabrous; rays ca. 13, 5–9 mm long. Achenes linear-fusiform. (n = 9) Late summer–fall. Sandy oak and pine woods; Gadsden, Leon, Liberty and Wakulla Cos., Fla. *Pityopsis flexuosa* (Nash) Small—S.

64. HETEROTHECA Cass. GOLDEN ASTER, CAMPHOR WEED

1. H. subaxillaris (Lam.) Britt. & Rusby. Taprooted, weedy annual or biennial, 2–25 dm tall, erect or in some habitats low and spreading, glandular, especially above, and spreading-hairy. Leaves simple, alternate, ovate or oblong, dentate or subentire, the lower ones petiolate, mostly deciduous, the middle and upper (or at least the upper) sessile and clasping. Heads yellow, radiate, the rays mostly 20–45, 5–10 mm long, pistillate and fertile; involucre mostly 6–8 mm high, its bracts narrow, imbricate; receptacle flat or nearly so, naked; disk 0.8–1.5 (2) cm wide, its flowers perfect and fertile; style branches flattened, with ventromarginal stigmatic lines and a slender, externally papillate-hairy appendage. Ray achenes thick, 3-angled, essentially glabrous, without pappus; disk achenes somewhat compressed or compressed-quadrangular, with prominent marginal and more obscure lateral nerves, evidently villous-hirsute, with reddish-tawny, double pappus like that of *Chrysopsis*. (n = 9) Midsummer–fall, or all year southward. Dry, often sandy places, especially in disturbed sites; widespread in sw US and adj. Mex, e to all provinces in Ark, and e on CP to Fla and thence n mainly on CP to Del and LI, perhaps not native northeastward. Harms, Vernon L. 1965. Biosystematic studies in the *Heterotheca subaxillaris* complex (Compositae: Astereae). Trans. Kansas Acad. Sci. 68: 244–257.

65. GRINDELIA Willd. GUM PLANT, TARWEED, ROSIN WEED

Taprooted herbs. Leaves alternate, strongly to obscurely resinous-punctate. Heads several or many, hemispheric, medium-sized to rather large (disk 1–3 cm wide), terminating the branches, yellow, radiate (ours) or seldom discoid, with mostly 12–45 pistillate and fertile rays; involucre more or less resinous or gummy, its bracts firm, herbaceous-tipped, imbricate or subequal; receptacle flat or convex, naked; disk flowers numerous, yellow, the inner and often also the outer sterile; style branches flattened, with ventromarginal stigmatic lines and externally hairy, lance-linear or occasionally very short appendage. Achenes compressed to subquadrangular, scarcely nerved. Pappus of 2–several firm, deciduous, often minutely serrulate awns. (x = 6)

1 Involucral bracts loose but not squarrose, not markedly imbricate; teeth of the leaves bristle-tipped, or the leaves sometimes entire .... 1. *G. lanceolata*.
1 Involucral bracts distinctly squarrose-reflexed, especially the outer, markedly imbricate; teeth of the leaves blunt or rounded .... 2. *G. squarrosa*.

1. G. lanceolata Nutt. Biennial, glabrous or sometimes sparsely hairy below, branched above, 3–15 dm tall. Leaves scarcely punctate, sharply serrate or serrulate with bristle-tipped teeth, or sometimes entire, acute to acuminate, the middle ones linear or lance-oblong, occasionally broadly so and amplexicaul, 4–11 cm × 4–28 mm, those of the floriferous branchlets often shorter and broader. Heads several, the disk mostly 1.5–2 cm wide; involucral bracts only slightly resinous, loose but not squarrose-reflexed, subequal or at least not markedly imbricate; rays ca. 15–30, 10–16 mm long. Achenes 4–6 mm long. Pappus awns mostly 2, entire, generally about equaling the disk floret. (n = 6) Midsummer–fall. Dry, open places, often on limestone; c Tenn (IP) and nc Ala (CU) to Ark (OU, OZ), Mo, Kans, Okla, and Tex.

2. G. squarrosa (Pursh) Dunal. Biennial, glabrous, branched at least above, 1–10 dm tall. Leaves abundantly punctate, closely and evenly callous-serrulate; main middle and upper leaves in our phase ovate or oblong, up to 7 × 2 cm, 2–4 times as long as wide, mostly obtuse or acutish. Heads several or numerous, the disk mostly 1–2 cm wide; involucral bracts strongly resinous, imbricate in several series, the tips (especially of the outer) squarrose-reflexed; rays mostly 25–40, 7–15 mm long. Achenes 2–3 mm long. Pappus awns 2–8, finely serrulate or subentire, 1/2 to 7/8 as long as the disk florets. (n = 6, 12) Midsummer–fall. Open or waste places; plains and cordilleran sp., widely but spottily introduced or adventive in our range and elsewhere in e US. Our phase, as described here, is var. *squarrosa*. A more narrow-leaved phase, with the middle and upper leaves mostly 5–8 times as long as wide, may also be expected occasionally in our range; this is var. *serrulata* (Rydb.) Steyerm.

66. SOLIDAGO L. GOLDENROD*

Fibrous-rooted perennial herbs from a rhizome or caudex. Leaves alternate, not glandular-punctate (except in *S. odora*, in which the punctae are anatomically unlike those of *Euthamia*). Heads several to more often numerous, relatively small, in axillary clusters or more often forming a thyrsoid to paniculiform, less often racemiform or corymbiform inflorescence, radiate, with 1–18 pistillate and fertile yellow rays (white in *S. bicolor* and *S. ptarmicoides*), or discoid in *S. brachyphylla*; involucral bracts more or less imbricate in several series, or rarely subequal, more or less chartaceous at the base, commonly with more herbaceous green tip, in a few species longitudinally evidently striate-nerved; receptacle small, flat or a little convex, naked or occasionally with a few phyllarylike bracts near the margin (internal to the rays), commonly more or less alveolate; disk flowers seldom more than 25 (–60), perfect and fertile, yellow (white in *S. ptarmicoides*); style branches flattened, with ventromarginal stigmatic lines and lanceolate, externally hairy appendage. Achenes several-nerved, subterete or angled, glabrous or hairy. Pappus of numerous capillary, usually white bristles, these much reduced in *S. sphacelata*. (x = 9)

Goldenrods can be divided into several groups on the basis of three sets of characters that are independently distributed with respect to each other: the nature of the underground parts, the nature and distribution of the leaves, and the nature of the inflorescence. Several of the terms explained below are used without further comment in the specific descriptions.

In many species the rhizome is very short, stout, and densely rooting, sometimes being more nearly a caudex than a rhizome proper. Such species often have several stems clustered together, although the stems may also be solitary. In other species the rhizome is elongate and generally more slender, and the stems are usually scattered. Some few species have both a short, stout rhizome or caudex and more elongate, slender, sometimes stoloniform rhizomes.

In many species the leaves are *basally disposed*. The radical and lowermost cauline leaves are relatively large and usually more or less persistent, with the blade either gradually or abruptly contracted to a definite petiole. The cauline leaves (either numerous or often few) are progressively reduced and less petiolate upward, those near and above the middle of the stem usually being sessile or nearly so and of different shape from those below, often relatively as well as actually narrower. In other species the leaves are *chiefly cauline*. The radical leaves are small and relatively inconspicuous, or more often wanting. The lowermost cauline leaves are reduced and generally soon deciduous, so that the stem appears to be naked toward the base at flowering time. The largest leaves are somewhat above the base but evidently below the middle of the stem. The middle and upper leaves are gradually reduced but essentially similar in shape to the largest ones. Most species of this group have numerous, sessile or subsessile leaves, but in some the leaves are fewer, or obviously petiolate, or both. A few species fall between these two habit types, or range from one to the other. Measurements of leaf length in the descriptions include the blade and petiole, unless only the blade is specified.

*I gratefully acknowledge the advice and counsel of Dr. Gary Morton in the preparation of the treatment of *Solidago*. The conclusions presented here are my own; they do not in all cases fully reflect his views.

Inflorescences are mostly of three general types. In one group the heads are in *axillary* clusters or in a terminal, more or less elongate *thyrse* that is straight, cylindrical, and not at all secund, or the inflorescence consists of several such thyrsoid branches. In another group the inflorescence is *paniculiform*, with at least the lower branches recurved-secund, or is slender and elongate (sometimes *racemiform*) and more or less one-sided, or at least nodding at the tip. In a third group the inflorescence is short and broad, flat- or round-topped, but not at all secund, and is said to be *corymbiform*.

Hybrids are common, often between species that do not seem closely allied. In the field these are generally less abundant than the parents and often can be readily recognized; their recognition in the herbarium is much more difficult.

Key to the Groups of Species

1 Inflorescence corymbiform (see commentary under generic description) .... Group I.
1 Inflorescence otherwise.
  2 Inflorescence either of small, axillary clusters or a terminal, simple or branched thyrse, neither nodding at the summit nor with recurved-secund branches.
    3 Leaves basally disposed (see commentary after generic description) .... Group II.
    3 Leaves chiefly cauline .... Group III.
  2 Inflorescence terminal and paniculiform (or unilaterally racemiform), either nodding at the summit, or with at least the lower branches more or less strongly recurved-secund, or both.
    4 Pappus bristles very short and firm, much shorter than the achene; basal and lowermost cauline leaves evidently cordate and slender-petiolate .... Group IV.
    4 Pappus of well-developed capillary bristles, usually as long as or longer than the achene; leaves not at once distinctly cordate (at most subcordate) and slender-petiolate.
      5 Leaves, or many of them, with conspicuously cordate-clasping sessile base, or with broadly winged, basally flared and cordate-clasping petiole Group V.
      5 Leaves not cordate-clasping.
        6 Plants maritime, with more or less fleshy leaves, and without long, stoloniform rhizomes .... Group VI.
        6 Plants neither maritime nor with markedly fleshy leaves, except often in *S. stricta*, which has elongate, slender stoloniform rhizomes.
          7 Leaves basally disposed .... Group VII.
          7 Leaves chiefly cauline .... Group VIII.

Group I

1 Rays yellow; pappus bristles not clavellate-thickened.
  2 Involucral bracts not striate-nerved; plants mostly 1–4 dm tall, at high altitudes in the mts. .... 1. *S. spithamea*.
  2 Involucral bracts evidently striate-nerved; plants mostly (3) 5–20 dm tall, not at high altitudes.
    3 Leaves relatively narrow, the largest ones commonly 0.7–1.5 cm wide, the middle cauline ones mostly 10–30 times as long as wide; rays 1–4; disk flowers 7–13 .... 51. *S. nitida*.
    3 Leaves relatively broad, the largest ones commonly 2–10 cm wide, the middle cauline ones mostly 2–6 times as long as wide; rays 7–14; disk flowers 17–35 .... 52. *S. rigida*.
1 Rays white; pappus bristles, or many of them, clavellate-thickened toward the tip .... 53. *S. ptarmicoides*.

Group II

1 Involucral bracts, or at least the outer ones, with conspicuously squarrose tip ... 2. *S. squarrosa*.
1 Involucral bracts appressed or a little loose, but not squarrose.
  2 Involucral bracts relatively broad, the middle ones mostly 1.5–2.5 mm wide at midlength; style appendages 0.9–1.2 mm long .... 3. *S. glomerata*.
  2 Involucral bracts narrower, the middle ones not more than ca. 1.5 mm wide at midlength; style appendages up to 0.7 mm long.

3 Involucral bracts very narrow, ca. 0.5 (–0.75) mm wide at midlength, tapering very gradually to a slender, pointed or minutely rounded tip.
4 Herbage finely pubescent throughout with minute, stiffly spreading hairs, or glabrate toward the base; rays mostly 9–16 .... 4. *S. puberula*.
4 Herbage glabrous below the inflorescence, or the leaves scaberulous above and the stem irregularly or decurrently short-hairy; rays mostly 6–9 ... 5. *S. roanensis*.
3 Involucral bracts broader, the middle ones commonly (0.75) 1–1.5 mm wide at midlength, often blunter.
5 Plants with elongate, stoloniform rhizomes; herbage glabrous throughout 15. *S. stricta*.
5 Plants without elongate rhizomes; herbage variously hairy or glabrous.
6 Achenes persistently short-hairy.
7 Plants essentially glabrous; flowering in late summer and fall.
8 Heads relatively few-flowered, with mostly 2–10 rays and 6–16 disk flowers; heads often but not always more than 25.
9 Plants relatively small, mostly 2–6 (7) dm tall; rays mostly 7–10 6. *S. spathulata*.
9 Plants larger, mostly 7–15 dm tall; rays mostly 2–7 .... 13. *S. gracillima*.
8 Heads relatively many-flowered, with mostly 8–13 rays and 20–25 disk flowers; heads few, mostly 5–25 .... 14. *S. pulchra*.
7 Plants softly and rather shortly spreading-hairy throughout; flowering in the spring .... 36. *S. verna*.
6 Achenes essentially glabrous, at least at maturity.
10 Leaves evidently pubescent, at least beneath; stem usually evidently spreading-hairy.
11 Rays yellow .... 8. *S. hispida*.
11 Rays silvery-white .... 9. *S. bicolor*.
10 Leaves below the inflorescence (and commonly also the stem) essentially glabrous, except for the sometimes scabrous margins and ciliate petioles.
12 Leaves very narrow, the largest ones not more than ca. 1.5 cm wide; plants essentially glabrous even in the inflorescence ...... 7. *S. plumosa*.
12 Leaves wider, the largest ones generally more than 1.5 cm wide; plants evidently puberulent in the inflorescence.
13 Lowest leaves (petiole included) seldom more than 7 times as long as wide; petioles of the lower leaves scarcely or not at all sheathing; upland species.
14 Inflorescence very narrow, spiciform-thyrsoid, and often also interrupted, or with well-spaced, scarcely crowded branches of similar nature; middle cauline leaves mostly 0.5–2 cm wide 10. *S. erecta*.
14 Inflorescence denser and broader, with more crowded, densely flowering branches, middle cauline leaves often more than 2 cm wide .... 11. *S. speciosa*.
13 Lowest leaves mostly 7–15 times as long as wide; petioles of the lowest leaves with more or less sheathing base; bog species ... 12. *S. uliginosa*.

Group III

1 Achenes glabrous, at least at maturity; outer involucral bracts tending to be squarrose-tipped; leaves entire or few-toothed .... 17. *S. petiolaris*.
1 Achenes persistently short-hairy: involucral bracts not squarrose-tipped; leaves generally more or less toothed.
2 Leaves relatively broad, the blade mostly 1–2.2 (2.5) times as long as wide and abruptly contracted to the definite though usually winged petiole.
3 Stem glabrous below the inflorescence .... 18. *S. flexicaulis*.
3 Stem conspicuously spreading-hairy .... 19. *S. albopilosa*.
2 Leaves narrower, mostly (2.2) 2.5–10 times as long as wide and tapering to the sessile or only obscurely short-petiolate base.
4 Leaves mostly (2.2) 2.5–3 (3.5) times as long as wide; stem striate-angled and grooved, as in the next two spp. .... 20. *S. flaccidifolia*.
4 Leaves mostly 3–10 times as long as wide.
5 Stem striate-angled, grooved, not glaucous.
6 Heads relatively small and delicate (see description); inflorescence generally appearing more axillary than terminal .... 21. *S. curtisii*.
6 Heads relatively large and coarse (see description); inflorescence generally appearing more terminal than axillary .... 22. *S. lancifolia*.
5 Stem terete, glaucous .... 23. *S. caesia*.

Group IV

A single species .... 38. *S. sphacelata*.

Group V

A single species .... 37. *S. auriculata*.

Group VI

A single species .... 16. *S. sempervirens*.

Group VII

1 Stem and often also the leaves evidently (though sometimes only very shortly) pubescent with loose or spreading hairs.
2 Disk flowers few, mostly 3–8; plants flowering in late summer and fall.
3 Rays mostly 3–9.
4 Herbage densely and finely puberulent with minute, loosely spreading hairs .... 24. *S. nemoralis*.
4 Herbage evidently hirsute with elongate hairs (or in part glabrous) .... 35. *S. ulmifolia*.
3 Rays none, or seldom 1–2 .... 33. *S. brachyphylla*.
2 Disk flowers numerous, mostly 14–27; plants flowering in spring .... 36. *S. verna*.
1 Stem (and often also the leaves) glabrous or nearly so below the inflorescence.
5 Leaves strongly scabrous above; stem more or less strongly angled, at least below .... 25. *S. patula*.
5 Leaves glabrous, or faintly scabrous, or occasionally somewhat strigose above; stem terete or striate, not angled.
6 Basal and lowermost cauline leaves gradually tapering to the petiole (or sometimes more abruptly so in *S. juncea*, which has some chaff toward the margin of the receptacle).
7 Inflorescence mostly much longer than broad, or of a few elongate, slender, racemiform branches; plants of wet or poorly drained sites.
8 Plants puberulent in the inflorescence; achenes glabrous .... 12. *S. uliginosa*.
8 Plants glabrous even in the inflorescence; achenes hairy, though sometimes sparsely so.
9 Plants with long slender, stoloniform rhizomes; rays mostly 3–7 and disk flowers mostly 8–12 .... 15. *S. stricta*.
9 Plants without long, slender rhizomes.
10 Heads few, ca. 5–25, many-flowered, with ca. 8–13 rays and 20–25 disk flowers .... 14. *S. pulchra*.
10 Heads numerous, generally more than (20) 30, fewer-flowered, with ca. 2–7 rays and 6–16 disk flowers .... 13. *S. gracillima*.
7 Inflorescence paniculiform, about as broad as long; upland plants.
11 Heads (many or all of them) with some slender, phyllarylike receptacular bracts internal to the outer (ray) flowers; rays mostly 7–13; disk flowers mostly 8–14.
12 Leaves scarcely or not at all triple-nerved; basal leaves 2–7.5 cm wide; achenes persistently short-hairy .... 26. *S. juncea*.
12 Leaves more or less strongly triple-nerved; basal leaves often less than 2 cm wide (or wanting); achenes glabrous or sparsely hairy .. 27. *S. missouriensis*.
11 Heads generally without receptacular bracts; rays mostly 3–8; disk flowers mostly 3–9.
13 Upper leaves more or less spreading or reflexed, and bearing evidently axillary fascicles of reduced leaves .... 28. *S. pinetorum*.
13 Upper leaves closely ascending or appressed, and without axillary fascicles .... 29. *S. gattingeri*.
6 Basal and lowermost cauline leaves rather abruptly contracted to the petiole; heads without chaff.
14 Leaves glabrous or occasionally strigose; disk flowers mostly 8–14.
15 Plants without slender, stoloniform rhizomes .... 30. *S. arguta*.
15 Plants with slender, stoloniform rhizomes in addition to the more deep-seated main rhizome or caudex.
16 Lower leaves elliptic to obovate (ovate), acute or obtuse to attenuate base; upper cauline leaves sharply reduced and usually appressed .... 31. *S. ludoviciana*.

16 Lower leaves mostly ovate, acute or acuminate, with truncate to obtuse base; upper cauline leaves gradually reduced; rarely appressed .... 32. *S. tarda*.
14 Leaves loosely hirsute at last on the midrib and main veins beneath; disk flowers mostly 4–7 .... 35. *S. ulmifolia*.

Group VIII

1 Leaves chiefly petiolate, with broadly ovate or elliptic blade 1.3–2 times as long as wide .... 39. *S. drummondii*.
1 Leaves sessile or nearly so, generally more than twice as long as wide.
2 Leaves not triple-nerved (or sometimes obscurely so in *S. tortifolia*).
3 Plants perennial from a branched caudex or short rhizome, without elongate rhizomes.
4 Leaves more or less evidently toothed, neither translucent-punctate nor anise-scented.
5 Leaves essentially glabrous .... 34. *S. delicatula*.
5 Leaves loosely hirsute at least on the midrib and main veins beneath .. 35. *S. ulmifolia*.
4 Leaves entire, minutely translucent-punctate, at least ordinarily anise-scented when bruised (unique in the genus) .... 40. *S. odora*.
3 Plants perennial from elongate creeping rhizomes.
6 Stem more or less hairy, at least above the middle; plants mostly not of swamps.
7 Leaves narrow, linear or lance-linear to narrowly oblong, the larger ones 2–7 (10) mm wide; stem uniformly puberulent at least above the middle; rays mostly 2–6 .... 41. *S. tortifolia*.
7 Leaves wider, lance-ovate to lance-elliptic or elliptic-oblong, the larger ones 1–4 cm wide; stem spreading-hirsute, at least above the middle; rays mostly 6–12.
8 Leaves sessile and more or less clasping, obscurely serrulate or subentire, not at all rugose .... 42. *S. fistulosa*.
8 Leaves merely subsessile, not clasping, more or less strongly toothed and rugose-veiny .... 43. *S. rugosa*.
6 Stem glabrous below the inflorescence; swamp plants .... 44. *S. elliottii*.
2 Leaves more or less strongly triple-nerved (least so in *S. radula*).
9 Plants glabrous even in the inflorescence; heads, or many of them, with some phyllarylike receptacular bracts near the margin (internal to the rays) .... 27. *S. missouriensis*.
9 Plants puberulent at least in the inflorescence; receptacular bracts generally absent.
10 Rays relatively few, mostly 4–8.
11 Leaves scabrous-hirsute to sometimes subglabrous, thick and firm, relatively broad, commonly 2–5 times as long as wide, often more than 1.5 cm wide .... 45. *S. radula*.
11 Leaves essentially glabrous, narrow, commonly 5–10 times as long as wide, not more than ca. 1.5 cm wide.
12 Involucre mostly 4–5 mm high; rays 2–3 mm long .... 46. *S. shortii*.
12 Involucre 2–3 mm high; rays 1–2 mm long .... 47. *S. rupestris*.
10 Rays more numerous, commonly (8) 10–17.
13 Leaves evidently puberulent across the surface beneath, more scabrous above .... 48. *S. canadensis*.
13 Leaves glabrous, or merely with a line of short hairs along the midrib and main veins beneath.
14 Stem not glaucous, and commonly puberulent for some distance below the inflorescence.
15 Heads tiny, the involucre mostly 2–3 mm high; rays mostly 7–10 (11); disk flowers (2) 3–6 (7) .... 47. *S. rupestris*.
15 Heads larger, the involucre mostly 3–5 mm high; rays mostly 10–15 and disk flowers 6–10 .... 49. *S. leavenworthii*.
14 Stem glaucous, glabrous below the inflorescence .... 50. *S. gigantea*.

1. S. spithamea M. A. Curtis. SKUNK GOLDENROD. Plants somewhat mephitic. Stems 1–4 dm tall from a short, stout rhizome or branched caudex, rough-puberulent or shortly spreading hirsute, or glabrate below. Leaves basally disposed, glabrous or nearly so, sharply serrate, the largest ones with elliptic to ovate or subrhombic blade mostly 5–10 × 1.5–4 cm. Inflorescence densely corymbiform, up to 10 (15) cm wide; involucre 5–6 mm high, with firm, green-tipped, rather narrow bracts; rays ca. 8 (13), 2–3.5 mm long; disk flowers mostly 20–60; style appendages 0.5–0.7 mm long. Achenes short-hairy or eventually subglabrate. (n = 27) Late summer. Rock crevices at upper altitudes in BR of NC and Tenn, notably on Grandfather Mt. and Roan Mt.

2. S. squarrosa Muhl. Stems 3–15 dm tall from a branched caudex, glabrous or nearly so below the rough-puberulent inflorescence. Leaves basally disposed, glabrous, or somewhat scabrous above, the larger ones with broadly oblanceolate to obovate, elliptic, or elliptic-ovate, sharply serrate blade mostly 5–20 × 2–10 cm, tapering or sometimes abruptly contracted to the long petiole. Inflorescence narrow and elongate, generally leafy-bracteate at least below; involucre 5–9 mm high, its bracts firm, at least the outer with squarrose, commonly herbaceous tip, the middle ones mostly 1–1.5 mm wide at midlength; rays 10–17, 3.5–5 mm long; disk flowers mostly 13–24. Achenes glabrous. (n = 9) Late summer. Rocky woods; NB to s Ont, Ind, and Ohio, s to Del, the mt. provinces of Va (and adj. Md and WVa), and BR of NC.

3. S. glomerata Michx. Plants somewhat mephitic. Stems 4–12 dm tall from a short, stout rhizome or branched caudex, glabrous below the puberulent inflorescence. Leaves basally disposed, glabrous or nearly so, serrate to subentire, the larger ones with elliptic to elliptic-obovate blade 7–20 × 4–9 cm. Inflorescence leafy-bracteate, at least below, elongate and thyrsoid, or of elongate, thyrsoid branches; heads relatively large and coarse; involucre 5–8.5 mm high, with firm, broad bracts, the middle ones mostly 1.5–2.5 mm wide at midlength; rays ca. 8, 4–5 mm long; disk flowers ca. 13 to ca. 34; style appendages 0.9–1.2 mm long. Achenes glabrous or nearly so. (n = 54) Late summer. Forest openings and rocky places at high altitudes in BR of NC and Tenn.

4. S. puberula Nutt. Plants 2–10 dm tall from a branched caudex, covered with minute, stiffly spreading viscidulous hairs, or glabrate toward the base. Leaves basally disposed, the larger ones broadly oblanceolate to elliptic or obovate, serrate, obtuse or acute, mostly 5–15 × 1–3.5 cm, the others more lance-elliptic to lance-linear and entire. Inflorescence thyrsoid, dense, often leafy-bracteate, with stiffly ascending, not at all secund branches, or unbranched in small plants; involucre 3–5 mm high, its bracts narrow, ca. 0.5 mm wide or less at midlength, with slender, acuminate, more or less subulate tip; rays 9–16; disk flowers mostly (8) 10–18. Achenes glabrous or occasionally sparsely hairy. (n = 9) Late summer–fall. Mostly in open places, generally on sandy or acid soil or rocks; NS and s Que through the coastal and Appalachian states to n Fla, s Ala, and reputedly La. N plants, s to Va and the mts. of NC and Tenn, are var. *puberula*, as principally described above. The well-marked var. *pulverulenta* (Nutt.) Chapm., on CP from s Va to n Fla, s Ala, and reputedly La, has more numerous and smaller leaves (the middle cauline ones commonly 1–4 cm long) that are often less hairy on the upper surface than on the lower, and has more evenly tapering, scarcely subulate involucral bracts that average a mite wider (to 0.75 mm wide at midlength). *S. pulverulenta* Nutt.—S.

5. S. roanensis T. C. Porter. Stems 2–10 dm tall from a branched, sometimes elongate caudex, hirsute-puberulent in the inflorescence and sometimes irregularly or decurrently so below. Leaves basally disposed, thin, glabrous, or scaberulous above, tending to be acuminate, the larger ones mostly 6–15 × 2–5 cm (or the very basal sometimes smaller and deciduous), with elliptic to elliptic-obovate or subrhombic, serrate blade, the others mostly rhombic to lance-elliptic. Inflorescence terminal, elongate and narrow, leafy-bracteate below, not secund; involucre 4–5 mm high, its bracts thin and slender (less than 0.75 mm wide at midlength), tapering to a narrowly acute or minutely obtuse tip; rays 6–9, 2–3 mm long; disk flowers mostly 8–12. Achenes glabrous to sometimes sparsely hairy. (n = 9) Late summer, fall. Woods and clearings; mt. provinces from Va, Md, and WVa to Tenn, NC, SC, and Ga. *S. maxoni* Pollard—F.

6. S. spathulata DC. Plants 2–7 dm tall from a branched caudex, essentially glabrous. Leaves basally disposed, the larger ones oblanceolate, often narrowly so, 4–15 × 0.5–1.5 (2.5) cm. Inflorescence narrowly thyrsoid to almost racemiform, or with a few ascending branches, not at all secund, inconspicuously bracteate; involucre 4.5–6.5 mm high, its bracts acutish or obtuse, often glutinous, the middle and upper ones tending to be scarious-margined, ca. (0.75–) 1 mm wide at midlength; rays 7–10; disk flowers mostly 9–16. Achenes persistently short-hairy. (n = 9, 18, 27) Late summer. Transcontinental, taxonomically complex sp., in our range represented only by ssp. *randii* (Porter) Cronq. var. *racemosa* (Greene) Cronq., as described above, on rocks along the Potomac R in Va, Md, and DC, and reputedly along the Cumberland R in Ky. *S. racemosa* Greene—F.

7. S. plumosa Small. Like a robust form of *S. spathulata* var. *racemosa* (plants 4–13 dm tall; basal leaves up to 20 or reputedly 30 cm long and 1.5 cm wide), but with broader, more branched, loosely thyrsoid-paniculate inflorescence and glabrous achenes. Late summer. Rocks along the Yadkin R, NC; not recently collected.

8. S. hispida Muhl. Plants 3–10 dm tall from a branched caudex, generally spreading-hirsute throughout, or the hairs occasionally mostly appressed. Leaves basally disposed, the larger ones broadly oblanceolate to obovate or elliptic, toothed or entire, 8–20 × 1.5–6 cm. Inflorescence elongate and narrow, generally more or less leafy-bracteate below, not at all secund, some of the lower branches sometimes elongate and stiffly ascending; involucre 4–6 mm high, its bracts obtuse or rounded, commonly ca. 1 mm wide or less (averaging a bit narrower than in *S. erecta*), tending to be rather yellowish, the green tip often ill-defined; rays 7–14, usually deep yellow; disk flowers 7–16. Achenes glabrous. (n = 9) Late summer–fall. Dry woods and open, often rocky places; widespread in ne US and adj. Can, s in our range rather uncommonly to Ga (mt. provinces), Ark (OU), and possibly n La.

S. porteri Small, known only from the type collection, near Monticello, Ga (PP), in 1846, suggests a robust form of *S. hispida*, but is less hairy (stem glabrous below, rather sparsely spreading-hirsute above) and has a somewhat wider and looser inflorescence. The proper disposition of this name is uncertain.

9. S. bicolor L. SILVERROD. Much like *S. hispida*, but with white or whitish rays; involucre 3–5 mm high, its bracts whitish or light stramineous except for the generally well-defined light green tip. (n = 9) Late summer–fall. Dry woods and open, often rocky places; widespread in ne US and adj. Can, s generally to NC, Tenn, and Mo, and to OU of Ark and mt. provinces of Ga and Ala, and occasionally to s Miss and n La. Hybridizes extensively with *S. hispida* and *S. erecta*, but retains its populational identity over large areas.

10. S. erecta Pursh. Plants 3–12 dm tall from a branched caudex, essentially glabrous below the puberulent inflorescence. Leaves basally disposed, the larger ones broadly oblanceolate to obovate or elliptic, mostly 7–30 × 1.5–5 cm, the middle cauline ones commonly 0.5–2 cm wide. Inflorescence elongate and narrow, often interrupted below, not at all secund, sometimes with a few long, straight or arching, cylindrical branches like the main axis; involucre 3.5–6.5 mm high; rays 5–9, averaging less deeply yellow than in *S. hispida;* disk flowers commonly 6–10. Achenes glabrous, seldom less than 2.5 mm long. (n = 9) Late summer–fall. Dry woods; coastal Mass; NJ to Ind, s to c Ga, c Ala, and ne Miss, mostly avoiding CP southward.

11. S. speciosa Nutt. Plants mostly 5–15 dm tall from a stout, woody caudex, coarsely puberulent in the inflorescence, otherwise glabrous or slightly scabrous. Leaves thick and firm, numerous, entire or the lower slightly toothed, sometimes large and basally disposed, the persistent lower ones then often broad and abruptly petiolate, as much as 30 × 10 cm, sometimes all smaller and nearly uniform in size, the lower then generally deciduous. Inflorescence dense and simple or more commonly with rather crowded, stiffly ascending branches, not at all secund; involucre mostly 3–5 mm high, its bracts obtuse or rounded, glutinous, yellowish; rays 6–8; disk flowers mostly 7–9. Achenes glabrous, seldom more than 2 mm long. (n = 9, 18) Late summer–fall. Open woods, fields, prairies, and plains; New England to Minn and Wyo, s to SC, n Ga, Tenn, Miss, Ark (OZ), and Tex, and occasionally intro. elsewhere. Most of our plants are the eastern var. *speciosa,* which is robust and, when well-developed, very broad-leaved, the middle cauline leaves commonly 2 cm wide or more, the lower leaves larger and generally persistent. *S. conferta* Mill.—S, misapplied; *S. harperi* Mackenzie—S. The more western var. *rigidiuscula* T. & G., typically of the plains states, is smaller, more rigid, and often more scabrous, with smaller leaves seldom more than 2 cm wide, the lower generally deciduous; it extends e occasionally to Tenn. *S. rigidiuscula* (T. & G.) Porter—S; *S. speciosa* var. *angustata* T. & G., the more slender extreme of var. *speciosa,* with persistent lower leaves, misapplied here by F.

12. S. uliginosa Nutt. Similar to *S. gracillima;* leaves sometimes wider, up to 7 cm; cauline leaves not especially numerous; inflorescence puberulent; involucre 3–5 mm high; rays 1–8; achenes generally glabrous. (n = 18) Fall. Bogs; se Can and ne US, s to Md, Ohio, Ind, and in the mts. to NC and Tenn. *S. uniligulata* (DC.) Porter—S. Some of the NC material is transitional to *S. gracillima.*

13. S. gracillima T. & G. Plants 7–15 dm tall from a stout, fairly short rhizome, essentially glabrous throughout. Leaves basally disposed, the larger ones oblanceolate to narrowly elliptic, entire or serrate, tapering to the elongate, somewhat clasping petiole, 10–30 × 1–4.5 cm, 6–15 times as long as wide; cauline leaves notably numerous. Inflorescence elongate and narrow, varying from straight and nonsecund to more often secund or with short, secund branches and recurved tip, or with several long, slender, ascending, evidently secund branches; involucre (3.5) 4–7 mm high; rays mostly (2) 3–7; disk flowers 6–16. Achenes rather thinly strigose. (n = 9) Late summer–fall. Swamps and other moist places; CP to mt. provinces, from Va and WVa to Ga, Ala, and n Fla. *S austrina* Small—F, G, S; *S. perlonga* Fern.—F. *S. flavovirens* Chapm., from brackish marshes near Apalachicola, may be this species, or perhaps a hybrid with *S. sempervirens. S. simulans* Fern., a broad-leaved extreme, approaching *S. uliginosa.*

14. S. pulchra Small. Plants slender, single-stemmed from a short caudex or crown, 3–10 dm tall, glabrous throughout. Basal leaves tufted, oblanceolate or elliptic, petiolate, 3–12 × 0.7–1.5 cm; cauline leaves abruptly reduced and bractlike. Inflorescence slender, sometimes unilaterally racemiform and apically recurved, sometimes more erect and nonsecund; involucre 3.5–5 mm high; rays well developed, 8–13; disk flowers mostly 20–25. Achenes puberulent. Late summer–fall. Moist, sandy savannas in Brunswick Co., NC. Obviously allied to *S. gracillima,* but seemingly distinct.

15. S. stricta Aiton. Glabrous perennials 3–20 dm tall with a short, simple caudex and long stoloniform rhizomes. Leaves basally disposed, thick and firm, the lowest ones oblanceolate or elliptic-oblanceolate, sometimes very narrowly so, 6–30 × 0.3–2 (5) cm, entire or obscurely serrate; cauline leaves abruptly reduced and sessile, entire, the middle and upper ones numerous, erect, often scarcely more than bracts. Inflorescence narrow, elongate, naked, sometimes nodding at the tip, the short branches occasionally recurved-secund; heads on slender, flexuous, minutely bracteate peduncles; involucre 4–6 mm high; rays mostly 3–7; disk flowers commonly 8–12. Achenes hairy, sometimes sparsely so. (n = 27) Late summer and fall, or all year southward. Sandy, usually moist places, especially among pines, or sometimes in coastal salt marshes, where it hybridizes with *S. sempervirens*; CP from NJ to Fla, Tex, WI, and s Mex. *S. petiolata* Mill.—S, misapplied.

16. S. sempervirens L. SEASIDE GOLDENROD. Plants somewhat succulent, 4–20 dm tall, usually with a very short and compact caudex, essentially glabrous, or scabrous-puberulent in the inflorescence. Leaves basally disposed, entire, the largest ones oblanceolate, 10–40 × 1–6 cm, the cauline ones generally rather numerous. Inflorescence dense, paniculiform, sometimes leafy at the base, at least the lower branches more or less recurved-secund; involucre 3–7 mm high, its bracts acute or acuminate; rays 3–5 mm long. Achenes hairy. (n = 9, 18) Late summer and fall, or all year southward. Saline places along the coast from se Can to tropical Am. Var. *sempervirens*, with relatively large heads (involucre 4–7 mm high, rays 12–17, disk flowers 17–22) is northern, occurring s to NJ and locally to Va. Var. *mexicana* (L.) Fern., with smaller heads (involucre 3–4 mm or seldom 5 mm high, rays 7–11, disk flowers 10–16), and commonly also with narrower leaves, is southern, seldom occurring much n of our range. *S. mexicana* L.—S.

17. S. petiolaris Aiton. Plants 4–15 dm tall from a stout caudex, sometimes with long slender rhizomes as well, the stem finely puberulent or scabrous-puberulent at least above. Leaves chiefly cauline, thick and firm, entire or few-toothed, glabrous or scabrous above, glabrous or short-hairy beneath (hairs mostly 0.1–1.4 mm long), numerous, sessile or nearly so, lance-linear to more commonly lance-elliptic, elliptic, or ovate, mostly 3–15 × 0.5–3 cm. Inflorescence narrow and generally elongate, usually more or less leafy-bracteate, the lower clusters sometimes elongate and stiffly ascending but not secund; peduncles mostly 3–15 mm long; involucre 4.5–7.5 mm high, the outer bracts acute (often very strongly so) and tending to be squarrose-tipped; rays (5) 7–9, 3–7 mm long; disk flowers (8) 10–16. Achenes glabrous or nearly so. (n = 9) Late summer–fall. Woods and open places, especially in sandy soil; NC to Mo and Neb, s to n Fla, La, and NM. Two well-marked but wholly intergradient vars.: Var. *angusta* (T. & G.) A. Gray, essentially Ozarkian (and s to La), has the leaves strongly glutinous (as if varnished), the lower surface glabrous or merely scabrous-hispidulous along the midrib and main veins, the involucre atomiferous-glandular to occasionally glabrous. *S. angusta* T. & G.—G. Var. *petiolaris*, widespread, but rare or wanting in the range of var. *angusta*, has the leaves scarcely glutinous, the lower surface softly puberulent, the involucre puberulent (often viscidulous) to occasionally glabrous. *S. milleriana* Mackenzie—S.

The name *S. buckleyi* T. & G. has been applied to a heterogeneous group of plants from the Ozarkian and s Appalachian regions, differing from *S. petiolaris* most notably in their larger, thinner, more toothed, unvarnished leaves, with slightly longer hairs. The proper taxonomic position of these plants is uncertain.

18. S. flexicaulis L. Plants 3–12 dm tall from elongate rhizomes, the stem striate-angled and grooved, glabrous below the inflorescence. Leaves chiefly cauline, sharply toothed, acuminate, usually hirsute beneath at least on the midrib and main veins, mostly with ovate or elliptic blade 7–15 × 3–10 cm, 1–2.2 (2.5) times as long as wide, abruptly contracted to the broadly winged petiole. Inflorescence a series of mostly short clusters, the lower axillary to foliage leaves, but these progressively reduced upward, the terminal part of the inflorescence often appearing as a naked thyrse; involucre 4–6 mm high, the outer bracts obtuse, the inner broadly rounded; rays mostly 3–4; disk flowers 5–9. Achenes short-hairy. (n = 9, 18) Late summer–fall. Woods; NS and NB to ND, s to Va, Ky, n Tenn, and Mo, and in the mt. provinces to NC, Ga, se Tenn, and Ark.

19. S. albopilosa L. Braun. Much like *S. flexicaulis*, but shorter and weaker (mostly 3–5 dm tall), the stem conspicuously spreading-lanate, the leaves more hairy, inclined to be subcordate at the base, and averaging smaller (blade 4–9 × 2.5–5 cm); rays 4–5. (n = 18) Late summer. Under overhanging cliffs in Powell and Menifee Cos., Ky.

20. S. flaccidifolia Small. Much like *S. curtisii*, but the principal leaves wider, mostly (2.2) 2.5–3 (3.5) times as long as wide, and the inflorescence usually in large part terminal, slender, and thyrsoid (sometimes branched), with only the lower clusters of heads evidently surpassed by their reduced subtending leaves, but varying to fully axillary and leafy; plants often with elongate rhizomes. Intermediate in leaf form between *S. flexicaulis* and *S. curtisii*, but forming homogeneous local populations in either the presence or the absence of these species, and extending well s of the range of both in Ga; perhaps an allopolyploid. (n = 18) Late summer–fall. Moist woods and edges of clearings; mt. provinces from sw Va and se Ky to n Ga and Ala, and s in Ga onto PP and even CP, w to n Miss (GC). *S. latissimifolia* Mill.—misapplied by S; included in *S. caesia* by F, R.

21. S. curtisii T. & G. Plants 3–15 dm tall from a short and caudexlike to occasionally more elongate rhizome, the stem striate-angled and grooved, glabrous or hairy. Leaves chiefly cauline, more or less serrate, acuminate, glabrous to evidently hairy, numerous, lanceolate to narrowly elliptic, mostly 10–18 × 1–4 (4.5) cm, 3–10 times as long as wide, tapering to the sessile or obscurely short-petiolate base. Inflorescence largely or wholly of axillary clusters shorter than their subtending leaves (or in robust forms many of the clusters themselves elongate and leafy-bracteate), the terminal segment sometimes thyrsoid and nearly naked; involucre mostly (2.5) 3–5 (6) mm high, glabrous or short-hairy but not glandular, its bracts 1-nerved or obscurely nerved to more or less evidently 3-nerved, seldom any of them as much as 1 mm wide at or above midlength; rays mostly (2) 3–5 (6); disk flowers mostly 4–9. Achenes hairy. (n = 9) Late summer–fall. In woods at relatively moderate elevations, up to ca. 5200 ft, in the mt. provinces of Va, WVa, Tenn, and Ga. *S. pubens* M. A. Curtis (ex T. & G,)—S, the phase with hairy stem and involucre.

22. S. lancifolia (T. & G.) Chapman. Much like *S. curtisii*, and perhaps not sharply distinct, but differing in a series of well-correlated characters. Rhizome commonly elongate. Inflorescence in considerable or large part terminal and thyrsoid, with much reduced leaves shorter than the clusters of heads, only the lower part consisting of axillary clusters; heads relatively large and coarse, the involucre mostly 4.5–7 mm high, tending to be finely granular-glandular or glandular-puberulent; involucral bracts strongly 3-nerved, the larger ones ca. 1 mm wide or more at or above midlength; rays mostly 5–8; disk flowers mostly 6–12. Late summer. At upper elevations, mostly 5000 ft and above, in BR and less commonly RV, from sw Va to w NC and e Tenn. Included in *S. curtisii* by R.

23. S. caesia L. Plants 3–10 dm tall from a short, stout, caudexlike rhizome, sometimes with long, creeping rhizomes as well; stem terete, glaucous. Leaves chiefly cauline, more or less serrate, acuminate, glabrous or slightly hairy above and along the midrib beneath, lanceolate or lance-elliptic, 6–12 × 1–3 cm, 3–10 times as long as wide, tapering to the sessile or obscurely short-petiolate base. Inflorescence chiefly axillary (sometimes branched) as in *S. curtisii*; involucre glabrous, 3–4.5 mm high, its bracts narrow, obtuse or rounded, tending to be obscurely several-nerved; rays (1–) 3–4 (5); disk flowers mostly 5–7. Achenes hairy. (n = 9) Late summer–fall. Woods; widespread in e US and adj. Can, and found throughout most or all of our range except s Fla.

24. S. nemoralis Aiton. Plants 1–10 dm tall from a branched caudex; herbage densely and finely puberulent with minute, loosely spreading hairs. Leaves basally disposed, weakly or scarcely triple-nerved, the larger ones oblanceolate or a bit broader, 5–25 × 0.8–4 cm, more or less toothed. Inflorescence paniculiform, sometimes long, narrow, and merely nodding at the tip, varying to more ample and with long, divergent, recurved-secund branches; involucre 3–6 mm high, its bracts imbricate, glabrous except for the ciliolate margins; rays short, 5–9; disk flowers mostly 3–6. Achenes short-hairy. (n = 9, 18) Late summer–fall. Dry woods and open places, especially in sandy soil; NS to Fla, w to Alta and Tex. Three geographic vars.:

Heads smaller, the involucre mostly 3–4.5 mm high; pubescence faintly viscidulous; achenes hirtellous or strigose.
- Inflorescence generally rather compact, or narrow and elongate; upper leaves usually rather gradually reduced; plants not very robust; n phase, passing into var. *haleana* in the n part of our range; scattered similar specimens from farther s may be depauperate plants of var. *haleana* .... var. *nemoralis*.
- Inflorescence ample and open, with long, divergent branches; upper leaves numerous and conspicuously reduced; plants relatively robust; the common phase in our region. .... var. *haleana* Fern.

Heads larger, the involucre mostly 4.5–6 mm high; pubescence not at all viscidulous; achenes subsericeous; w phase, in our range only in Ark [var. *decemflora* (DC.) Fern.—F.] .... var. *longipetiolata* (Mackenzie & Bush) Palmer & Steyerm.

25. S. patula Muhl. Stems 5–20 dm tall from a short caudex, glabrous below the rough-puberulent inflorescence, angular at least below. Leaves basally disposed, glabrous beneath, strongly scabrous on the upper surface, the lower with somewhat sheathing petiole. Inflorescence paniculiform, generally with widely spreading, recurved-secund branches, in smaller plants sometimes narrower, denser, and elongate, but still secund; involucre 3–4.5 mm high, its bracts acute (especially the outer) to obtuse (especially the inner); rays 5–12; disk flowers 8–23. Achenes sparsely hairy. (n = 9) Late summer–fall. Swamps and wet meadows; Vt to Wis, s to Ga, Miss, and La. Two vars.:

Plants relatively robust and large-leaved, the lower leaves with elliptic, elliptic-ovate, or elliptic-obovate, sharply toothed blade mostly 8–30 × 4–10 cm, the middle and upper leaves not especially numerous, gradually reduced but still generally toothed; n. var., extending s to Del, Md, and in the mts. to n Ga (*S. rigida* L., misapplied by S) .... var. *patula*.

Plants more slender and smaller-leaved, the lower leaves both relatively and actually narrower, up to 5 or 6 cm wide, less strongly toothed, the upper ones notably numerous, much reduced, and commonly entire; mostly on CP and adj. PP, from Va to Ga, w to Ala, Miss, La, and se Tex. (*S. salicina* Ell.—S) .... var. *strictula* T. & G.

26. S. juncea Aiton. Plants 3–12 dm tall from a stout, branched caudex or short rhizome, commonly with more or less deep-seated creeping rhizomes as well, essentially glabrous, or sometimes more or less short-hirsute on the leaves or in the inflorescence. Leaves basally disposed, the larger ones 15–40 × 2–7.5 cm, with rather narrowly elliptic, acuminate, more or less serrate blade tapering to the long petiole. Inflorescence dense, mostly about as broad as long, with recurved-secund branches; involucre 3–5 mm high; rays minute, mostly 7–12; disk flowers mostly 9–14; receptacle with some chaffy, slender, phyllarylike bracts internal to the disk flowers. Achenes short-hairy. (n = 9) Summer, one of the first spp. to bloom. Dry, open places and open woods, especially in sandy soil; NS and NB to Minn, s to Va, Tenn, and ne Miss, and in the mts. to n Ga and n Ala.

27. S. missouriensis Nutt. Plants (3) 5–10 dm tall from creeping rhizomes, sometimes with a caudex as well, glabrous throughout. Leaves firm, strongly triple-nerved (at least the middle and lower), entire or some (especially the lower) serrate, the lowest ones oblanceolate and conspicuously serrate but soon deciduous, the others slightly to strongly reduced upward, lance-elliptic to broadly linear, tapering to a sessile or obscurely petiolar base, often with axillary fascicles of much reduced leaves. Inflorescence paniculiform, with more or less strongly recurved-secund branches, mostly short and broad; involucre 3–5 mm high, its bracts firm, broadly rounded to occasionally acutish; rays 7–13; disk flowers mostly 8–13; receptacle commonly with some chaffy bracts as in *S. juncea*. Achenes glabrous or sparsely hairy. (n = 9) Summer. Prairies and other dry, open or sparsely wooded places; widespread in w US, entering our range in the prairies of Ark (OZ, ME), and occasionally intro. farther e. Our plants, as described here, are var. *fasciculata* Holz.—F. *S. glaberrima* Martens—S.

28. S. pinetorum Small. Plant 4–11 dm tall from a branched caudex, slender, glabrous throughout, or some of the leaves more or less ciliate-margined. Leaves basally disposed, the lower linear-oblanceolate, serrate or subentire, more or less strongly triple-nerved; upper leaves spreading or reflexed, and bearing axillary fascicles of much reduced leaves. Inflorescence paniculiform, with recurved-secund branches, commonly as broad as long; involucre 3–4 mm high, its bracts obtuse or rounded, with evident midrib; rays 3–7; disk flowers 5–9. Achenes glabrous, or slightly hairy distally. (n = 9) Late summer–fall. Open places and dry woods, especially in sandy soil; AC and PP from se Va through NC to adj. SC.

29. S. gattingeri Chapman. Plants 5–10 dm tall from a branched caudex, slender, essentially glabrous throughout. Leaves basally disposed, entire or slightly toothed, the lower oblanceolate, tapering to the petiolar base, and more or less strongly triple-nerved, but often deciduous, commonly 8–17 × 1–2 cm; middle and upper leaves reduced and less prominently or scarcely triple-nerved, erect or closely ascending, becoming minute and bractlike, without axillary fascicles. Inflorescence paniculiform, with recurved-secund branches, commonly about as broad as long; involucre 3–5 mm high, yellowish, its bracts broad, obtuse to broadly rounded distally, the midrib obscure or wanting; rays 5–8; disk flowers 3–9. Achenes glabrous or nearly so. (n = 9) Late summer–fall. Cedar barrens and limestone ledges and glades; c Tenn (IP) to Mo and possibly Tex.

30. S. arguta Aiton. Stems 5–15 dm tall from a stout, branched caudex, glabrous except for the somewhat puberulent inflorescence. Leaves basally disposed, glabrous, or slightly scabrous above, or sometimes strigose or strigillose, toothed to subentire, the larger ones 10–30 × 3–12 cm, the broadly elliptic or ovate blade rather abruptly contracted to the long petiole. Inflorescence paniculiform, with recurved-secund branches, sometimes elongate and narrow, more often broad and open, with long, divergent branches; involucre 3–7 mm high, its bracts acute or obtuse; rays 2–8; disk flowers 8–20. Achenes glabrous or hairy. (n = 9, 18) Late summer–fall. Open woods and dry meadows; Me to Fla, w to Ky, Mo, and La. Four vars.:

Achenes glabrous; northern, from s Me to Va and less commonly NC, w occasionally to Ky, Tenn, and s Mo; n = 9 . . . . var. *arguta*.
Achenes hairy, at least distally; more southern, from Md, Va, WVa, and Ky s.
  Shale-barren ecotype of w Md, w Va, e WVa, and e Ky, with relatively very firm leaves, the basal ones commonly more or less truncate at the base; n = 9; *S. harrisii* Steele—F, S . . . . var. *harrisii* (E. S. Steele) Cronq.
  Series of non-shale-barren ecotypes with less firm, basally somewhat tapering leaves.
    Leaves glabrous; n = 9, 18, the tetraploids with somewhat larger, more numerously flowered heads than the diploids; Va to n Fla, w to WVa, Ky, and occasionally to s Mo, Ark, and La; *S. boottii* Hook—F, G, S, misapplied; *S. yadkinensis* (Porter) Small—F, S (name misapplied; the type appears to be part of a hybrid swarm involving *S. gracillima* and *S. arguta* var. *caroliniana*) . . . . var. *caroliniana* A. Gray.
    Leaves, especially the lower cauline ones, strigose or strigillose; drier places; s Mo, Ark, La, and Miss, and sporadically to Tenn, Ala, Ga, and SC; n = 9; plants perhaps reflecting hybridization with some hairy species; *S. dispersa* Small—S . . . . var. *boottii* (Hook.) Palmer & Steyerm.

31. S. ludoviciana (A. Gray) Small. Much like *S. arguta* var. *caroliniana*, but with well-developed, slender, stoloniform rhizomes in addition to the more deep-seated main short rhizome or caudex. Leaves strigose or more often glabrous, the lower ones elliptic to obovate (ovate), acute or obtuse, with obtuse attenuate base, the upper sharply reduced and usually appressed. (n = 9, 18) Late summer–fall. Mostly in dry, open woods; w La and w Ark to Tex. *S. strigosa* Small, mainly, of F, G, S.

32. S. tarda Mackenzie. Much like *S. ludoviciana*, differing as indicated in the key. (n = 27) Late summer–fall. In sandy soil in more xeric places than *S. arguta*; s NJ and se Pa to n Fla and Ala, mainly CP. *S. ludoviciana* (A. Gray) Small, in part—F, G.

Dr. Gary Morton has called my attention to some specimens from n Fla and s Ga that differ from typical *S. tarda* in their narrower, basally more tapering lower leaves. At least some of these plants are tetraploid, rather than hexaploid as in *S. tarda*. The proper taxonomic status of these plants remains to be determined.

33. S. brachyphylla Chapman. Stems mostly solitary or paired from a short, caudexlike rhizome, 5–12 dm tall, loosely hirsute-puberulent. Leaves numerous and small, glabrous to somewhat hairy, entire or shallowly toothed, the basal ones wanting or with oblanceolate or spatulate to ovate or rotund blade 2–4 cm long on a petiole 3–5 cm; cauline leaves mainly sessile or subpetiolate and elliptic or lance-elliptic to ovate, the middle ones mostly 2.5–5 (6.5) × 1–2 (2.5) cm, the upper numerous and often bractlike. Inflorescence open-paniculiform, with several elongate, divaricate, recurved-secund branches; involucre 3–5 mm high; rays none, or seldom 1–2 and short; disk flowers 4–8. Achenes short-hairy. (n = 9) Late summer–fall. Open woods; SC to n Fla, w to Ala and probably Miss, on PP and CP.

34. S. delicatula Small. Resembling *S. ulmifolia*, but essentially glabrous, and with thicker, firmer, less conspicuously veiny leaves. (n = 9) Late summer–fall. Open woods; Tex to Okla and Kans, entering our range in sw Ark (OU).

35. S. ulmifolia Muhl. Stems mostly solitary or paired from a caudex or short rhizome, 4–12 dm tall. Leaves numerous, thin, sharply serrate, loosely hirsute at least on the midrib and main veins beneath; basal leaves wanting or with well-developed, elliptic to elliptic-obovate blade abruptly contracted to the petiole, but generally soon deciduous; lowermost cauline leaves tending to be soon deciduous and smaller than the persistent ones just above, which are ovate or rhombic-ovate to elliptic or lance elliptic, acute or acuminate, broadly short-petiolate or tapering and subsessile, mostly 6–12 × 1.2–5.5 cm, the leaves thence more or less reduced upward, those at the base of the inflorescence generally small, relatively broad, and often numerous. Inflorescence paniculiform, with recurved-secund branches, these generally few, elongate, and divergent, the heads crowded; involucre 2.5–4.5 mm high; rays 3–5, minute; disk flowers 4–7. Achenes short-hairy. (n = 9) Late summer–fall. Woods; NS to Ga and n Fla, w to Minn, Kan, and Tex. The widespread var. *ulmifolia* has the stem glabrous or nearly so below the inflorescence; the var. *palmeri* Cronq., with the stem evidently spreading-hirsute, largely replaces var. *ulmifolia* in OZ and OU of Ark, and extends also into Miss.

36. S. verna M. A. Curtis ex T. & G. Plants 5–12 dm tall from a short, stout rhizome, softly and rather shortly spreading-hairy throughout. Leaves basally disposed, the lower with elliptic to more often broadly ovate (or even subcordate), toothed blade mostly 3–7 × 1.5–4 cm, on a petiole of equal or lesser length. Inflorescence paniculiform, with obscurely to evidently recurved-secund lower branches; heads evidently slender-pedunculate; involucre 4–5 mm high; rays mostly 7–12, 3–6 mm long; disk flowers mostly 14–27. Achenes evidently strigose-puberulent. (n = 9) Spring (–early summer). In woods and open places on AC of NC and adj. SC.

37. S. auriculata Shuttlew. ex Blake. Plants 4–15 dm tall from a short, stout rhizome or caudex, velutinous or loosely hirsutulous throughout, or the upper surface of the leaves more scabrous. Leaves basally disposed, the lower with broadly ovate or subcordate, sharply toothed blade 3.5–12 × 3–7 cm on a petiole of equal or greater length, but often deciduous, the others gradually reduced upward, and with shorter, more winged, basally flaring and amplexicaul petiole, those farther up the stem sessile, cordate-clasping, and often entire. Inflorescence paniculiform, with crowded heads on several long, divaricate, recurved-secund branches; involucre 3–5 mm high, its bracts evidently striate-nerved; rays 1–3, short; disk flowers 4–8. Achenes short-hairy. (n = 9) Late summer–fall. Woods; Ala (all) w to La and s Ark (ME), n to Tenn (IP), and irregularly e and s to n Fla, Ga, and w SC (PP). *S. amplexicaulis* Martens, misapplied by Torrey & Gray; *S. notabilis* Mackenzie—R, S.

38. S. sphacelata Raf. Stems 5–12 dm tall from a short and caudexlike to somewhat elongate rhizome, densely spreading-puberulent to occasionally subglabrous. Leaves basally disposed, more or less densely spreading-puberulent beneath, sparsely so or more often glabrous above, the basal ones tufted and persistent, conspicuously petiolate, with serrate, cordate blade 4–12 × 4–11 cm, the cauline ones progressively reduced, less petiolate, and less cordate upward. Inflorescence paniculiform, with a few widely spreading, secund branches, the narrow heads densely crowded and often subglomerate; involucre 3–4.5 mm high, its bracts firm, somewhat keeled, shortly green-tipped; rays 3–6, short; disk flowers 3–6. Pappus bristles firm, reduced, much shorter than the hairy achenes. (n = 9) Late summer, fall. Open woods and rocky places, especially in calcareous soil; mt. provinces and adj. PP of Va, NC, Ga, and Ala, w to Ill and w Ky and Tenn (IP). *Brachychaeta sphacelata* (Raf.) Britton—S.

39. S. drummondii T. & G. Stems 3–10 dm tall from a stout, branched caudex, uniformly pubescent with short, spreading hairs, occasionally glabrate near the base. Leaves chiefly cauline, broadly ovate or elliptic-ovate, all except sometimes the uppermost evidently short-petiolate, more or less triple-nerved, but also pinnately veined, finely and usually densely spreading-hairy at least on the lower side, generally only those near the inflorescence reduced, the others 1.3–2 times as long as wide, the larger ones 3.5–9 × 2.5–7 cm. Inflorescence paniculiform, with recurved-secund branches, or the heads apparently sometimes drooping; involucre 3–4.5 mm high, its bracts obtuse or rounded; rays well developed, 3–7; disk flowers 4–7. Achenes short-hairy. (n = 9) Late summer–fall. Cliff crevices and rocky woods, especially in calcareous soil; Mo, Ill, Ark (OZ), and reputedly La.

40. S. odora Aiton. Stems 6–16 dm tall from a short, stout caudex, rough-puberulent in the inflorescence and in lines decurrent from at least the upper leaf bases, or all the way around. Leaves chiefly cauline, sessile, entire, glabrous except for the scabrous margins, finely translucent-punctate, anise-scented when bruised, or seldom inodorous, not prominently veined. Inflorescence paniculiform, with recurved-secund branches; involucre 3.5–5 mm high, its bracts slender, acute, yellowish; rays 3–5 (6), fairly showy; disk flowers 3–5. Achenes short-hairy or subglabrous. (n = 9) Late summer–fall. Dry, open woods, especially in sandy soil; widespread in e US (n to NH and Vt), and found essentially throughout our range. Two vars.:

Stem pubescent in lines or strips decurrent from the margins of (at least the upper) leaf bases; principal leaves mostly 4–11 × 0.5–1.5 (2) cm, (4) 5–15 times as long as wide; widespread, but hardly entering pen. Fla. .... var. *odora*.

Stem pubescent all the way around, or sometimes with a short glabrous or subglabrous strip beneath each leaf base; principal leaves mostly (1.5) 3–7 × 0.8–2 cm, 2–5 (6) times as long as wide, perhaps not so consistently anise-scented as in var. *odora*;—largely or wholly confined to pen Fla. *S. chapmannii* A. Gray—S. .... var. *chapmannii* (A. Gray) Cronq.

41. S. tortifolia Elliott. Stems 3–13 dm tall from elongate creeping rhizomes, uniformly puberulent above the middle. Leaves chiefly cauline, very numerous, sessile, linear to lance-linear or narrowly oblong, the larger ones 2.5–7 cm × 2–7 (10) mm, glabrous or more or less scabrous on one or both surfaces, not prominently veined (but sometimes obscurely triple-nerved), usually at least some of the lower ones remotely serrulate. Inflorescence paniculiform, with recurved-secund branches; involucre 2.5–3.5 mm high, yellowish; rays 2–6, small; disk flowers 2–6. Achenes short-hairy. (n = 9) Late summer–fall, or all year southward. Dry, usually sandy soil, often in pinelands; CP from Va to s Fla, w to Tex.

42. S. fistulosa Miller. Stems 7–15 dm tall from elongate creeping rhizomes, stout, conspicuously spreading-hirsute, at least above the middle. Leaves chiefly cauline, numerous, crowded, strongly hirsute on the midrib beneath, and often less densely so across the surface, less hairy or more often glabrous above, sessile, broad-based and somewhat clasping, lance-ovate to elliptic-oblong, obscurely serrulate or subentire, the larger ones 3.5–12 × 1–3.5 cm. Inflorescence paniculiform, usually dense, with recurved-secund branches; involucre glabrous, 3.5–5 mm high, its bracts thin and slender; rays 7–12, small; disk flowers 4–7. Achenes short-hairy. (n = 9) Late summer and fall, or all year southward. Wet or dry places, often in pinelands; NF to s Fla, w to La, mainly on CP, but also on Stone Mt., Ga (PP).

43. S. rugosa Miller Stems 3–15 (25) dm tall from elongate creeping rhizomes, evidently spreading-hirsute. Leaves chiefly cauline, numerous, crowded, slightly to very strongly rugose-veiny, not triple-nerved, glabrous or scabrous above, hirsute at least on the midrib and main veins beneath, lance-elliptic to lance-ovate or rhombic-elliptic, serrate, subsessile, not clasping, the larger ones 3.5–13 × 1.3–4 cm. Inflorescence paniculiform, with recurved-secund branches; involucre 2.5–4 (5) mm high, its bracts slender, not more than ca. 0.6 mm wide: rays 6–11, small; disk flowers (3) 4–8. Achenes short-hairy. (n = 9, 18) Late summer–fall. Various habitats; NF to n Fla, w to Mich, Mo, and Tex. *S. altissima* L.—S, misapplied. Two sspp.:

Leaves relatively thin, not very strongly rugose, tending to be sharply toothed and acuminate; pubescence tending to be long and relatively soft; involucral bracts mostly acute or acutish; rays 8–11. Mostly in relatively moist, often wooded places; northern ssp., passing into ssp. *aspera* along the n border of our range; ours is var. *villosa*. . . . . ssp. *villosa*.

Leaves relatively thick and firm, strongly rugose-veiny, tending to be blunt-toothed or even subentire, and often merely acutish at the tip, the pubescence tending to be relatively short and harsh; involucral bracts generally obtuse or rounded; rays 6–8. Mostly in rather dry places; the common phase in our range, n occ. to Mass and Mich. *S. celtidifolia* Small—S. Subspecies perhaps to be divided into several vars. . . . . ssp. *aspera* (Aiton) Cronq.

44. S. elliottii T. & G. Plants stout, (4) 10–30 (40) dm tall from elongate creeping rhizomes, wholly glabrous, or the branches of the inflorescence puberulent. Leaves chiefly cauline, numerous, sessile or nearly so (the base of the blade often subauriculately rounded to a very short and inconspicuous petiole), elliptic or lance-elliptic, evidently to obscurely serrate or the upper entire, not triple-nerved, evidently to obscurely veined but not much rugose, the larger ones mostly 6–15 × 1.5–3.5 cm. Inflorescence paniculiform, sometimes conspicuously leafy-bracteate, with short or elongate, slightly to strongly recurved-secund branches; involucre 4–6 mm high, its bracts obtuse or rounded, relatively broad, the larger ones 0.7–1.2 mm wide; rays 6–10 (12); disk flowers 4–7. Achenes short-hairy. (n = 9, 27) Late summer–fall, or all year southward. Fresh or brackish swamps on CP from Fla to Mass; NS. *S. edisoniana* Mackenzie—S; *S. mirabilis* Small—S.

45. S. radula Nutt. Plants 4–12 dm tall from a caudex, at least sometimes with creeping rhizomes as well. Stem scabrous to shortly and loosely hirsute. Leaves chiefly cauline, numerous, firm, elliptic or lance-elliptic to rather narrowly elliptic-obovate, subsessile, obscurely to evidently toothed or the upper entire, more or less evidently triple-nerved, subglabrous, or more commonly scabrous-hirsute (seldom more softly spreading-hairy), mostly 2–5 times as long as wide, the larger ones 3–8 × 1–3 cm. Inflorescence paniculiform, with densely flowered, more or less recurved-secund branches, or occasionally simple and nodding; involucre glabrous, 3.5–5.5 mm high, its bracts relatively broad and firm, acutish to more often obtuse or broadly rounded; rays 4–7, 2–3.5 mm long; disk flowers 4–6. Achenes short-hairy. (n = 9) Late summer–fall. Open, rocky places and dry woods, especially in calcareous soil; Mo and s Ill to Okla, La, and Tex, and disjunct on PP of NC and in BR and PP of Ga.

46. S. shortii T. & G. Stems 6–13 dm tall from a short, stout rhizome, scabrous-puberulent at least above the middle. Leaves chiefly cauline, numerous and rather crowded, firm, glabrous, triple-nerved, remotely serrulate, narrowly elliptic or lanceolate, acuminate or sharply acute, tapering to the subsessile or obscurely short-petiolate base, the larger ones 7–10 × 1–1.5 cm. Inflorescence paniculiform, with recurved-secund branches; involucre 4–5 mm high, its bracts firm, acute or obtusish; rays 5–8, 2–3 mm long; disk flowers 5–9. Achenes short-hairy. (n = 18) Late summer–fall. Dry, open places; IP of Ky at Blue Licks, in Fleming, Nicholas, and Robertson Cos., and on Rock Island, at the Falls of the Ohio, near Louisville, Ky.

47. S. rupestris Raf. Stems 5–15 dm tall from elongate creeping rhizomes, glabrous below, slightly puberulent above the middle, more definitely so in the inflorescence. Leaves chiefly cauline, numerous, crowded, thin, triple-nerved, glabrous, or occasionally puberulent on the midrib and main veins beneath, remotely serrulate or subentire, sessile or subsessile, slender, tapering to both ends, mostly 5–12 × 0.6–1.2 cm. Inflorescence paniculiform, with recurved-secund branches; involucre 2–3 mm high; rays mostly 7–10 (11), 1–2 mm long; disk flowers 2–7. Achenes short-hairy. (n = 9) Midsummer–early fall. River banks from Pa, Md, and DC to IP of Tenn, Ky, and s Ind.

48. S. canadensis L. Perennial from elongate creeping rhizomes, mostly 8–20 dm tall, the stem densely spreading-puberulent at least above the middle. Leaves chiefly cauline, numerous and crowded, triple-nerved, lance-linear to lance-elliptic or narrowly elliptic, sessile, tapering to both ends, the larger ones 5–15 × 0.7–2.2 cm, densely spreading-puberulent beneath, merely scabrous (or even subglabrous) above. Inflorescence paniculiform, with strongly recurved-secund branches; involucre (2) 2.5–4.5 mm high, its bracts thin and slender, acute or acuminate, yellowish, without well-defined green tip; rays short and slender, mostly (8) 10–17; disk flowers 2–8. Achenes short-hairy. (n = 9, 18, 27) Late summer–fall. Moist or dry, open places and thin woods; throughout our range, except s Fla. A transcontinental, geographically diversified species. Typical *S. canadensis* is northeastern, and is not known in our range. Most of our plants have the involucre 3–4.5 mm high, with ca. 13 (10–16) rays, and are hexaploid; these are var. *scabra* T. & G., as principally described above. *S. altissima* L.—F, R; *S. hirsutissima* Mill.—S. Some very small-headed plants (involucre only 2–3 mm high; rays often fewer, sometimes only 8) are diploid; such plants have been called var. *hargeri* Fern., but the correlation between ploidy level and head size is not yet firmly established.

49. S. leavenworthii T. & G. Perennial from elongate creeping rhizomes, mostly (5) 10–20 dm tall, the stem commonly rough-puberulent for some distance below the inflorescence, especially in strips beneath the leaves. Leaves chiefly cauline, numerous and crowded, triple-nerved, lance-linear to lance-elliptic, tapering to a sessile or subsessile base, the larger ones mostly 6–15 × 0.5–1.8 cm, glabrous except for the scabrociliate margins, or seldom with a line of short hairs along the three main veins beneath. Inflorescence paniculiform, with recurved-secund branches; involucre 3–5 mm high, its bracts mostly firmer, blunter, and greener than those of *S. canadensis*; rays mostly 10–15; disk flowers 6–10. Achenes short-hairy. (n = 18) Late summer and fall, or all year southward. Various habitats on CP from s NC to s Fla.

50. S. gigantea Aiton. Perennial from elongate creeping rhizomes, (5) 10–20 dm tall, the stem glabrous and glaucous beneath the puberulent inflorescence. Leaves chiefly cauline, numerous and crowded, strongly triple-nerved, lance-elliptic or narrowly elliptic, acuminate, tapering to the sessile or obscurely petiolate base, glabrous or with a line of hairs along the 3 main veins beneath, the larger ones 6–17 × 1–4.5 cm. Inflorescence paniculiform, with recurved-secund branches; involucre 2.5–4 mm high, its bracts mostly firmer, blunter, and greener than those of *S. canadensis*; rays mostly (8) 10–17; disk flowers 6–10. Achenes short-hairy. (n = 9, 18, 27) Late summer–fall. Moist, open places; transcontinental, occurring nearly or quite throughout our range. *S. serotina* Retz.—S. The var. *gigantea*, with the leaves shortly hairy on the three main veins beneath, and the var. *serotina* (Kuntze) Cronq. (*S. gigantea* var. *leiophylla* Fern.—F) with the leaves essentially glabrous, both occur throughout most of our range. The populational structure of the species is complex and needs reconsideration.

51. S. nitida T. & G. Plants 3–10 dm tall, essentially glabrous below the rough-puberulent inflorescence, the stems clustered. Leaves basally disposed, elongate and narrow, the lower ones tapering to a long-petiolar base, mostly 10–25 × 0.5–1.5 cm, sometimes deciduous, the others numerous but evidently reduced upward, the middle cauline ones mostly 10–30 times as long as wide. Inflorescence compactly corymbiform, mostly flat-topped, 4–10 cm wide; involucre 4.5–6 mm high, its bracts firm, striate-nerved, the inner broadly rounded distally; rays 1–4, 3–5 mm long; disk flowers 7–13. Achenes glabrous, plump, several-nerved. (n = 9) Mid- and late summer. Prairies and open woods; GC of La and adj. Miss and s Ark to se Okla and Tex.

52. S. rigida L. Plants 5–15 dm tall from a branching caudex or short stout rhizome. Leaves basally disposed, the larger ones with elliptic, elliptic-oblong, or broadly lanceolate to broadly ovate, rounded to acutish blade 6–25 × 2–10 cm, often exceeded by the conspicuous long petiole; middle cauline leaves sessile or nearly so, 2–6 times as long as wide. Inflorescence dense, corymbiform, 5–25 cm wide; involucre 5–9 mm high, its bracts firm, broadly rounded, strongly striate; rays 7–14, 3–5 mm long, disk flowers 17–35. Achenes glabrous, turgid or angular, 10–20-nerved. (n = 9) Late summer–fall. Prairies and other dry, open places, especially in sandy soil; Conn and NY to SC (PP), Ga (RV, GC), and Ala (GC), w to Alta and NM. We have two of the three vars.:

Herbage more or less densely short-hairy throughout. Nearly throughout the more eastern part of the range of the species, but irregular and uncommon toward the southeast. *Oligoneuron grandiflorus* (Raf.) Small—S .... var. *rigida*.

Leaves glabrous or nearly so except for the margins and often the midrib beneath; stem either glabrous or hairy. Southeastern phase, from s Ohio to NC (PP), Tenn (CU, RV, IP), Ga (RV, GC), Ala (GC), and Tex. *S. jacksonii* (Kuntze) Fern.—F. *Oligoneuron jacksonii* (Kuntze) Small—S. .... var. *glabrata* L. Braun.

53. S. ptarmicoides (Nees) Boivin. Stems 1–7 dm tall from a branched caudex, scabrous at least above. Leaves firm, glabrous or scabrous, entire or with a few remote salient teeth, tending to be trinerved, 3–20 cm × 1.5–10 mm, the lower linear-oblanceolate and petiolate, sometimes tufted, persistent, and larger than those above, sometimes smaller and deciduous, the others becoming sessile upward and linear or nearly so. Heads mostly 3–60 in an open, minutely bracteate, corymbiform inflorescence; involucre (4) 5–7 mm high, glabrous, its bracts imbricate, firm, greenish above but scarcely herbaceous, often with strongly thickened midrib; rays 10–25, white, 5–9 mm long; disk flowers numerous, white. Achenes glabrous. Pappus copious, many of the bristles clavellate-thickened toward the tip. (n = 9) Late summer–fall. Prairies and other open, usually dry, commonly calcareous places; Vt, NY, and w Que to Sask and Wyo, s to Colo, Mo, Ind, and rarely and locally to Ark (OZ), NC (PP), and Ga (RV). *Aster ptarmicoides* (Nees) T. & G.—F, G, S.

67. BRINTONIA Greene

1. **B. discoidea** (Elliott) Greene. Fibrous-rooted perennial herb 4–15 dm tall from a short, stout rhizome, the stem loosely and rather softly spreading-hairy. Leaves alternate, toothed or the upper entire, pinnately veined, rather numerous, somewhat hairy on both sides, the basal and lower cauline ones the largest, with ovate or subcordate to broadly elliptic blade 4–10 × 3–8 cm on a petiole 2–8 cm long, the others gradually reduced (and less petiolate) upward. Heads more or less numerous in an elongate, terminal, leafy-bracteate, often loosely thyrsoid inflorescence, rather small, discoid, white or chloroleucous or anthocyanic; involucre 4–6 mm high, its bracts rather narrow, firm, green-tipped, imbricate, at least the outer with loose or spreading tip; receptacle small, flattish, naked; flowers all tubular and perfect; style branches with ventromarginal stigmatic lines and elongate (nearly 1 mm), slender appendage, this externally hispidulous at the base, otherwise merely papillate. Achenes turgid, several-nerved, glabrous or nearly so. Pappus of numerous white or anthocyanic capillary bristles. (n = 9) Late summer–fall. Rich, sometimes swampy woods; GC from sw Ga and w Fla to La, and n to Cullman Co. (CU) in Ala.

68. CHRYSOMA Nutt.

1. **C. pauciflosculosa** (Michx.) Greene. Glabrous, glutinous, evergreen shrubs up to 1 m tall. Leaves alternate, firm, finely tessellate-reticulate on both sides with impressed reticulum and irregularly subisodiametric areoles (as seen in herbarium specimens), not veiny, entire, oblanceolate to linear-elliptic, mostly 2–6 cm × 2–10 mm, sessile or tapering to a short petiole. Heads small, numerous, short-pedunculate in dense, cymose clusters terminating the often elongate branches, yellow, radiate, with 1–2 (3) pistillate and fertile rays 4–6 mm long and often 2 mm wide, or some heads discoid; involucre 5–6 mm high, narrow, strongly glutinous, its small, chartaceous, stramineous, narrowly hyaline-margined bracts imbricate in several series, tending to be more or less aligned in vertical rows; receptacle small, minutely conic, naked, somewhat alveolate; disk flowers (2) 3–4 (5), perfect and fertile; style branches flattened, with ventromarginal stigmatic lines and lanceolate, externally papillate-puberulent appendage. Achenes several-nerved, strigose-sericeous. Pappus of numerous white or dingy capillary bristles. (n = 9) Late summer and fall, or all year southward. Sand hills and other sandy places on CP, especially near the coast; NC to n Fla, w to Miss. *Solidago pauciflosculosa* Michx.—R.

69. EUTHAMIA Nutt. FLAT-TOPPED GOLDENROD

Rhizomatous perennial herbs. Leaves alternate, resinous-punctate, numerous, narrow, commonly linear or nearly so, entire, sessile or subsessile, often evidently 3-nerved, mainly or wholly cauline, the lower generally deciduous. Heads small, more or less numerous, pedunculate or often sessile in small glomerules, forming a terminal, flat-topped, corymbiform inflorescence, yellow, radiate, with 7–25 (35) pistillate and fertile rays, these mostly less than 3 mm long; involucre more or less glutinous, its small, chartaceous, yellowish or green-tipped bracts imbricate in several series;

receptacle small, commonly more or less fimbrillate; disk flowers 2–12 (20), generally fewer than the rays, perfect and fertile; style branches flattened, with ventromarginal stigmatic lines and lanceolate, externally short-hairy appendage. Achenes several-nerved, short-hairy. Pappus of numerous white, capillary bristles. (x = 9) All spp. bloom in late summer and fall. Included in *Solidago* by F, G, R.

Euthamia has often been included within the genus *Solidago* as a section. All of the technical distinctions between the two genera are subject to exception, but *Euthamia* has such a distinctive aspect that there need never be any confusion between them. *Euthamia* may actually be more closely related to genera such as *Gutierrezia*, which have similar leaves and involucre, than to *Solidago*.

1 Leaves evidently 3-nerved, the larger ones ordinarily with 1 or 2 additional pairs of fainter lateral nerves; heads relatively broad, with mostly 15–25 or more rays and 20–35 or more flowers in all .... 1. *E. graminifolia*.
1 Leaves 1-nerved, or obscurely to sometimes evidently 3-nerved, without any additional lateral nerves; heads relatively narrow, with mostly 7–16 rays and 11–21 flowers in all.
2 Leaves mostly 3–6 mm wide and (7) 10–20 times as long as wide; involucre 4.5–6.5 mm high; plants not very resinous .... 2. *E. leptocephala*.
2 Leaves 1–3 (4) mm wide, mostly (15) 20–50 times as long as wide; involucre 3–4.5 (5) mm high; plants evidently resinous.
3 Disk flowers 5–7 (–9) and rays (8–) 10–16 .... 3. *E. tenuifolia*.
3 Disk flowers 3–4 (5) and rays 7–11 (–13) .... 4. *E. minor*.

1. E. graminifolia (L.) Nutt. Plant 3–15 dm tall, glabrous to densely spreading-hirtellous. Leaves mostly 4–13 cm × 3–10 mm and 10–20 times as long as wide, evidently 3-nerved, the larger ones ordinarily with 1 or 2 additional pairs of fainter lateral nerves. Heads mostly sessile or subsessile in small glomerules, mostly 20–35 (–45)-flowered, with 15–25 (35) rays, and (4) 5–10 (13) disk flowers; involucre turbinate, 3–5 mm high; rays minute and often scarcely spreading, ca. 1 mm long. (n = 9) Open, usually moist ground; widespread in n US and adj Can, s to Va, Ky, and Tenn (mainly IP), and in the mts. to NC; also in Forrest Co., Miss, where perhaps only intro. Our plants are var. *graminifolia*. *Solidago graminifolia* (L.) Salisb.—F, G, R.

2. E. leptocephala (T. & G.) Greene. Plants 3–10 dm tall, glabrous, or nearly so, less resinous than other spp. Leaves mostly 4–8 cm × 3–6 mm, 10–20 times as long as wide, more or less evidently 3-nerved, but without any additional lateral nerves, only sparsely and obscurely punctate, or sometimes more closely and evidently so, but the puncta then with scanty resin and appearing somewhat pustulate. Heads sessile in small glomerules or somewhat pedunculate, mostly 14–19-flowered, with 10–13 rays and 3–6 disk flowers; involucre narrowly turbinate, 4.5–6.5 mm high; rays minute and often scarcely spreading, ca. 1 mm long. (n = 18, 27) Open, often moist and sandy places, and thin woods; ME, OU, and OZ of s Mo, c Tenn (IP), and Ark, w to Tex, s to GC of La and Miss, thence e to Fla panhandle. *Solidago leptocephala* T. & G.—F, G.

3. E. tenuifolia (Pursh) Greene. Much like *E. minor*, but somewhat more robust. Larger leaves mostly 2–3 (4) mm wide, often with a pair of weak lateral nerves, with or without axillary fascicles. Heads pedicellate or some of them glomerate; rays mostly (8) 10–16 and disk flowers 5–7 (9), the flowers 17–21 in all. (n = 9) Blooming later than *E. minor*, where their ranges overlap. Open, sandy places, especially near the coast; CP from NS and Mass to Va, and irregularly to Ga and n Fla. *Solidago tenuifolia* Pursh—F, G. R. Intergrades with *E. minor* morphologically, but retains its populational identity at least at some localities within the area where the ranges overlap.

4. E. minor (Michx.) Greene. Plants 3–10 dm tall, glabrous or somewhat scabro-hirtellous. Leaves numerous, relatively thin and lax, linear or nearly filiform, 1-nerved or seldom also with a faint pair of lateral nerves, evidently resinous-punctate, the larger ones 3–6 cm × 1–2 (3) mm, 20–50 times as long as wide, often with axillary fascicles. Many or most of the heads commonly pedunculate; involucre 3–4.5 (5) mm high, evidently resinous; rays mostly 7–11 (13), ca. 2 mm long, spreading; disk flowers 3–4 (5). (n = 9) Open, sandy places, especially near the coast; CP from Md and Va to s Fla, w to s Ala, s Miss, and se La. Our most common sp. of the genus. *Solidago microcephala* (Greene) Bush—F, G, R. Perhaps better reduced to a var. of *E. tenuifolia* (the older specific epithet).

70. BIGELOWIA DC., nom. conserv.

Fibrous-rooted perennial herbs, glabrous and somewhat glutinous or viscid, especially in the inflorescence. Leaves alternate, entire, basally disposed, minutely glandular-punctate. Heads numerous in a terminal, flat-topped, corymbiform inflorescence, yellow with an unusual greenish cast, discoid; involucre subcylindric, its bracts weakly keeled, with chartaceous base and green or yellow-green tip, imbricate, tending to be aligned in vertical ranks; receptacle naked, either alveolate or with a central cusp; flowers 2–6, all tubular and perfect; style branches flattened, with ventro-marginal stigmatic lines and a lance-triangular, externally short-hairy appendage. Achenes turbinate or subcylindric, several-nerved, short-hairy. Pappus of more or less numerous capillary bristles. *Chondrophora* Raf. Anderson, L. C. 1970. Studies on *Bigelowia* (Astereae, Compositae). I. Morphology and taxonomy. Sida 3: 451–465.

1 Plants becoming colonial or loosely matted by means of rather short creeping rhizomes; leaves all linear, mostly 1–2 mm wide .... 1. *B. nuttallii.*
1 Plants with the stems clustered or solitary on a short rhizome or crown, the clump sometimes enlarging by offsets; basal leaves linear-oblanceolate to broadly oblanceolate, mostly 2–10 (14) mm wide .... 2. *B. nudata.*

1. B. nuttallii L. C. Anderson. Plants 2–8 dm tall, becoming colonial or loosely matted by means of rather short creeping rhizomes. Leaves linear, the basal ones clustered, mostly 6–13 cm × 1–2 mm, without distinction of blade and petiole, the cauline ones relatively numerous (mostly 12–23), progressively smaller. Involucres 6–9 mm high; flowers 3–5, the corolla 4–5 mm long; style appendage nearly as long as the stigmatic portion. Achenes 3–3.5 mm long. (n = 9, 18, 27) Late summer–fall. Dry prairies and thin soil on exposed sandstone or granite rocks and in dry pine woods and prairies; CP and PP of Ga; Washington Co., Fla; CU of Ala; c and w La and e Tex. *Chondrophora virgata* (Nutt.) Greene—S, misapplied.

2. B. nudata (Michx.) DC. Plants 2–8 dm tall, the stems clustered or solitary on a short rhizome or crown, the clump sometimes enlarging by offsets (habit in var. *australis* sometimes approaching that of *B. nuttallii*). Basal leaves narrowly linear-oblanceolate to broadly oblanceolate, tapering to a more or less distinctly petiolar base, mostly 2–10 (14) mm wide; cauline leaves relatively few (mostly 6–15), linear or nearly so, progressively smaller. Style appendage obviously shorter than the stigmatic portion. Achenes 1–2 mm long. (n = 9) Late summer and fall, or all year southward. Wet meadows, margins of ponds, and other moist (seldom dry), low places, often in sandy soil; CP from NC to s Fla, and w to se La. *Chondrophora nudata* (Michx.) Britton—R, S. Two vars.:

Basal leaves oblanceolate, mostly 5–10 (14) cm (petiole included) × 4–10 (14) mm; heads relatively small, the involucre 4.5–6 mm high, the 2–5 disk corollas 3–4 mm long. NC to n Fla, w to se La. . . . . var. *nudata*.

Basal leaves linear-oblanceolate, mostly 10–15 (20) cm × 2–4.5 mm; heads larger, the involucre 6–7.5 mm high, the 3–6 disk corollas 4–5 mm long. C and s pen Fla. Intermediate in some respects between *B. nudata* var. *nudata* and *B. nuttallii*, but morphologically and geographically confluent with the former and not the latter. . . . . var. *australis* (L. C. Anderson) Shinners.

71. GUTIERREZIA Lag.

1. G. dracunculoides (DC.) Blake. Taprooted, glabrous annual, mostly 3–8 dm tall, bushy-branched above. Leaves numerous, alternate, finely glandular-punctate, linear, up to 6 cm × 3 mm. Heads numerous, terminating the branchlets, yellow, radiate with 6–10 pistillate and fertile rays mostly 3–5 mm long; involucre 3–6 mm high, campanulate to hemispheric, glutinous, its bracts firm, imbricate, with stramineous base and abrupt green tip, the inner broader than the outer, receptacle small, flattish, naked; disk flowers fairly numerous, sterile, with obsolete ovary. Achenes multinerved, densely strigose. Pappus of the rays a minute, toothed crown, or of several short, more or less concrescent scales, that of the disk of 5–9 awns or very narrow scales, slightly dilated above, more or less united at the base, about as long as the corolla. (n = 5) Late summer–fall. Dry soil on plains and prairies, especially on limestone; sp. chiefly of Mo and Kans to Ark, Okla, Tex, and NM, but disjunct in CU of Ala and IP of Tenn, and to be expected elsewhere in the w part of our range. *Amphiachyris dracunculoides* (DC.) Nutt.—S.

72. ASTER L. ASTER

Fibrous-rooted perennial herbs from a rhizome, caudex, or crown, rarely (*A. subulatus*) taprooted annuals. Leaves alternate, simple. Heads solitary to more often several or numerous in various sorts of inflorescences, hemispheric to subcylindric, radiate, with 3 to numerous pistillate and fertile, white to anthocyanic rays, in a few species the rays reduced and inconspicuous, the corolla of the pistillate flowers then consisting mainly of the elongate, slender tube; involucral bracts in 2 or more series, equal or more often imbricate, usually more or less herbaceous at the tip and chartaceous below, sometimes herbaceous or chartaceous throughout; receptacle naked, flat or a little convex; disk flowers few to usually numerous, perfect and fertile, red or purple to yellow or occasionally white; style branches flattened, with mostly narrow and acute or acuminate, externally hairy appendage. Achenes several-nerved, glabrous or hairy, or seldom glandular. Pappus of numerous capillary bristles, sometimes with an additional short outer series. (x mostly = 5 or 9, seldom 7 or 8)

As in *Solidago*, there are numerous hybrids, often between species that do not seem closely allied. In the field these are generally less abundant than the parents and often can be readily recognized; their recognition in the herbarium is more difficult and uncertain.

The term leaves *basally disposed*, and leaves *chiefly cauline*, as used in the key and descriptions, are elucidated in the discussion under *Solidago*.

An effort has been made to juxtapose related species in the linear sequence here presented. In furtherance of that aim, our most nearly primitive species are placed somewhat after the beginning of the sequence, in Group III. The first 37 species (Groups I–VII) in the sequence form a fairly coherent major group, in which the inflorescence is often more or less elongate or paniculiform, and the involucral bracts tend to be definitely herbaceous-tipped. The last 28 species (Groups VIII–XVI) form several different alliances, some of which have often been recognized as distinct sections or genera. In most of the species of these several alliances the inflorescence tends to be corymbiform and flat- or round-topped, and the involucral bracts tend to be relatively firm and more chartaceous or scarious than in the first major group, even the usually green tip not very herbaceous. These differences are unfortunately not sufficiently constant to provide the basis for the fundamental organization of the key, which in order to be useful is necessarily to a considerable extent artificial. Some of the established names of species groups, sections, and segregate genera that are here included in *Aster* are indicated in parentheses in the key.

Reported chromosome counts for several species of *Aster* (notably in the *A. corelifolius* group) based on $x = 9$ have recently been challenged. It is possible that some or all of these species have $x = 8$, or both base numbers.

There is a considerable group of specimens from central and southern Florida that do not fit well into any of the species here recognized, but that are too diverse among themselves to be treated with any confidence as a single distinct species. Many of them approach *A. dumosus*, but differ in having narrower and more pointed involucral bracts, or broader lower leaves, or shorter or less conspicuously bracteate peduncles. Some of them are robust, more than 1 m tall; others are very slender and small, only about 2 dm tall, and all intermediate sizes exist. These specimens, which have often collectively been called *A. simmondsii* Small, may possibly reflect hybridization among *A. dumosus*, *A. elliottii*, and *A. bracei*. The type sheet of *A. simmondsii* consists of small, slender, few-headed plants that might conceivably represent a hybrid between *A. dumosus* and *A. bracei*. No attempt is made to provide for the *A. simmondsii* materials in the following treatment.

Key to the Groups of Species

1 At least the basal or lower cauline leaves cordate or subcordate at the base and evidently petioled.

2 Inflorescence corymbiform, commonly flat or round-topped, occasionally more elongate, with few and often leafy bracts; involucral bracts relatively broad and firm, the outer often more than 1 mm wide and seldom more than 2.5 (3) times as long as wide; plants (except *A. commixtus*) ordinarily becoming colonial by creeping rhizomes, often glandular or with white rays (Biotia) .. Group X.

2 Inflorescence more paniculiform, often elongate, its bracts either very narrow or very small or both, often numerous; involucral bracts relatively narrow, the outer not more than 1 mm wide and usually at least 3 times as long as wide; plants rarely becoming colonial, eglandular, and (except *A. sagittifolius*) seldom with white rays (Heterophylli) .... Group I.

1 None of the leaves at once cordate and evidently petioled.

3 Leaves silvery-silky on both sides (sometimes glabrate in age), entire, not more than ca. 5 cm long .... Group VI.

3 Leaves glabrous or hairy, but not at all silvery-silky, variously large or small and entire or toothed.
4 Plants coarse, freely branched, nearly leafless, broomlike or rushlike, often with some of the short axillary branches modified into thorns 1–2 cm long (Leucosyris) .... Group XV.
4 Plants distinctly otherwise, not broomlike or rushlike, never thorny.
5 Leaves very numerous and small, the largest ones (within 1 dm of the base) not more than ca. 3 cm long, the others rarely more than 1.5 cm long (Brachyphylli) .... Group V.
5 Leaves larger and usually less numerous.
6 Plants evidently glandular on the involucres, or on the peduncles, or both .... Group II.
6 Plants not glandular on the involucres and peduncles.
7 Pappus evidently (in *A. reticulatus* sometimes obscurely) double, the inner of long, slender, capillary bristles, the outer of short bristles ca. 1 mm long or less (Ianthe and Doellingeria) .... Group XIII.
7 Pappus simple, not divided into distinct outer and inner series (but the uppermost hairs of the achene sometimes simulating an outer pappus in species of group XII).
8 Leaves evidently auriculate-clasping (or even cordate-clasping) at the base .... Group III.
8 Leaves variously sessile or petiolate, but not evidently auriculate-clasping.
9 Plants annual, taprooted (*Oxytripolium*) .... Group XVI.
9 Plants perennial, fibrous-rooted from a rhizome, caudex, or crown.
10 Plants with succulent, narrow leaves seldom as much as 1 cm wide, strictly glabrous; involucral bracts scarcely herbaceous Group XIV.
10 Plants otherwise, the leaves only rarely somewhat fleshy, and then mostly well over 1 cm wide; involucral bracts often (though not always) with an evidently green and herbaceous tip; plants glabrous or more or less hairy.
11 Leaves basally disposed, elongate and narrow, firm and more or less grasslike, parallel-veined or scarcely veiny, the principal ones mostly more than (10) 15 times as long as wide and often with distinctly sheathing base, rarely more than 1 cm wide; pappus often (not always) coarse, with the larger bristles clavellate and flattened (Heleastrum) .... Group IX.
11 Leaves otherwise; pappus fine and soft except in some species of Group XII.
12 Rays few, mostly 3–8 (Sericocarpus) .... Group XII.
12 Rays more numerous, mostly (8) 9–40.
13 Achenes copiously glandular .... Group XI.
13 Achenes glabrous or hairy, but not glandular.
14 Involucral bracts, or some of them with subulate, marginally inrolled green tip; perennial from a caudex or very short and stout rhizome .... Group VII.
14 Involucral bracts flat, not with subulate, marginally inrolled green tip (but the tip spinulose-mucronate in *A. ericoides*); habit various.
15 Heads smaller, the involucre 2.5–6 (7) mm high; rays in most species white; inflorescence tending to be paniculiform .... Group IV.
15 Heads larger, the involucre 6–12 mm high; rays in most species anthocyanic; inflorescence in most species corymbiform .... Group VIII.

Group I

1 Involucral bracts, or many of them, with spreading or reflexed tip; rays often more than 25.
2 Leaves essentially glabrous .... 18. *A. curtisii.*
2 Leaves evidently scabrous or hairy on both sides .... 9. *A. anomalus.*
1 Involucral bracts with erect, often appressed tip; rays mostly 10–25.
3 Cauline leaves, or some of them, either sessile and cordate-clasping or with conspicuously auriculate-clasping petiole .... 8. *A. undulatus.*

3 Cauline leaves not at all clasping.
4 Leaves ordinarily entire or subentire; involucral bracts with short, broad, diamond-shaped green tip.
5 Plants with only the lower leaves cordate or subcordate; involucral bracts glabrous except for the sometimes ciliolate margins .... 1. *A. azureus*.
5 Plants with nearly all the leaves below the inflorescence cordate or subcordate; involucral bracts usually minutely hairy on the back .... 2. *A. shortii*.
4 Leaves evidently toothed (at least the lower ones); bracts various.
6 Plants with one or both sides of the leaves, or the stem, or both, more or less hairy or scabrous.
7 Involucral bracts obtuse or merely acute, the tip short and broad, its green color usually partly or even wholly replaced or obscured by anthocyanin; leaves usually sharply and deeply toothed, and with scarcely or obscurely winged petiole .... 3. *A. cordifolius*.
7 Involucral bracts strongly acute to acuminate, the green tip more or less elongate and narrow, ordinarily not at all purple to the naked eye; leaves usually shallowly and often bluntly toothed, the cauline ones often with strongly winged petiole.
8 Branches of the inflorescence with long, copiously and minutely bracteate, nonfloriferous base, or the peduncles mostly well over 1 cm long and copiously bracteate, or both .... 7. *A. texanus*.
8 Branches of the inflorescence shorter and more densely floriferous, either leafy or more floriferous toward the base, the peduncles seldom as much as 1 cm long.
9 Plants relatively densely hairy (see text); green tips of the involucral bracts tending to be elongate-rhombic; rays commonly bright blue 6. *A. drummondii*.
9 Plants relatively thinly hairy (see text); green tips of the involucral bracts very narrow; rays pale bluish to lilac or white .... 5. *A. sagittifolius*.
6 Plants essentially glabrous, or sometimes slightly puberulent in the inflorescence.
10 Green tips of the involucral bracts very narrow and elongate, the bracts acuminate or very acute; inflorescence narrow, with mostly strongly ascending branches and short, crowded peduncles .... 5. *A. sagittifolius*.
10 Green tips of the involucral bracts relatively short and broad, more or less diamond-shaped, the bracts obtuse or barely acute; inflorescence broader and more open, with longer peduncles .... 4. *A. lowrieanus*.

Group II

1 Leaves with strongly cordate-clasping base .... 10. *A. patens*.
1 Leaves with merely auriculate-clasping base, or not clasping.
2 Leaves chiefly cauline (see commentary under generic description).
3 Rays relatively few, mostly 15–40; stems scattered on elongate, creeping rhizomes.
4 Heads small, the involucre 5–8 mm high .... 11. *A. oblongifolius*.
4 Heads larger, the involucre 10–15 mm high .... 12. *A. grandiflorus*.
3 Rays relatively numerous, mostly 45–100; stems clustered on a caudex or very short rhizome .... 13. *A. novae-angliae*.
2 Leaves basally disposed.
5 Plants evidently glandular on the involucre or peduncles or both .... 41. *A. spectabilis*.
5 Plants only obscurely or inconspicuously glandular .... 42. *A. surculosus*.

Group III

1 Leaves with strongly cordate-clasping base .... 10. *A. patens*.
1 Leaves with merely auriculate-clasping base.
2 Scrambling climbers (habit unique in the genus) .... 20. *A. carolinianus*.
2 Plants more or less erect, not climbing.
3 Involucral bracts, at least the inner ones, long-acuminate or attenuate.
4 Perennial from a short, stout rhizome or caudex, occasionally with short stolons as well; stem uniformly hairy at least under the heads, densely spreading-hispid to occasionally glabrous below the inflorescence; leaves of the inflorescence not conspicuously crowded .... 14. *A. puniceus*.
4 Perennial from long, creeping rhizomes; stem and branches pubescent in lines above, glabrous or sparingly hispid below the inflorescence; leaves, at least those of the inflorescence, conspicuously crowded .... 15. *A. lucidulus*.

3 Involucral bracts obtuse to merely acute or occasionally acuminate.
5 Involucral bracts regularly imbricate and appressed, with short, diamond-shaped green tip mostly 1–2 mm long; plants mostly with a short, stout rhizome or branched caudex, sometimes with slender creeping rhizomes as well; herbage often or usually glaucous.
6 Principal leaves evidently or strongly auriculate-clasping, variable in shape, sometimes narrow, but often broader, then more than 2.5 cm wide and less than 5 times as long as wide .... 21. *A. laevis*.
6 Principal leaves only slightly auriculate-clasping, always relatively narrow, seldom more than 2.5 cm wide and seldom less than 5 times as long as wide (length measured to include petiole, if present).
7 Leaves chiefly cauline .... 22. *A. concinnus*.
7 Leaves basally disposed .... 23. *A. attenuatus*.
5 Involucral bracts evidently to more often only slightly or scarcely imbricate, with looser, often spreading tip, the green part not diamond-shaped, that of at least the larger bracts generally more than 2 mm long; plants with long rhizomes, and without a caudex, the herbage not glaucous.
8 Principal leaves with ovate or lanceolate, serrate blade abruptly contracted or occasionally more tapering into a relatively long, generally entire, strongly auriculate-clasping petiolar base .... 17. *A. prenanthoides*.
8 Leaves otherwise, commonly sessile or nearly so .... 19. *A. novi-belgii*.

Group IV

1 Involucral bracts, at least the outer, with loose or squarrose, minutely spinulose-mucronate tip; leaves entire, less than 1 cm wide .... 30. *A. ericoides*.
1 Involucral bracts appressed or a little loose, not spinulose-mucronate; leaves various.
2 Disk corollas deeply lobed, the lobes comprising 45–75 percent of the limb;* leaves usually hairy beneath, at least along the midrib, only seldom glabrous.
3 Plants without creeping rhizomes; leaves hairy only along the midrib beneath, or seldom glabrous .... 28. *A. lateriflorus*.
3 Plants with creeping rhizomes; leaves generally hairy over the surface beneath .... 27. *A. ontarionis*.
2 Disk corollas more shallowly lobed, the lobes comprising 17–45 percent of the limb; leaves glabrous beneath except in forms of *A. praealtus*.
4 Plants strongly colonial by creeping rhizomes, often robust and more than 1 m tall; peduncles either all short (less than 2 cm long), or sparsely bracteate, or the bracts leaflike.
5 Heads very small (involucre 2.5–3.5 or seldom 4 mm high, rays 3–6 mm long), numerous, the branches of the inflorescence often recurved and unilaterally racemiform .... 26. *A. vimineus*.
5 Heads larger (involucre 4–7 mm high, rays mostly 6–15 mm long) often numerous, but the branches of the inflorescence generally not recurved and unilaterally racemiform.
6 Veinlets of the leaf forming an obscure reticulum, or if the reticulum is evident, then the areolae clearly longer than broad; lobes of the disk corollas comprising 30–45 percent of the limb .... 25. *A. simplex*.
6 Veinlets of the leaf forming a conspicuous reticulum beneath, the areolae mostly nearly isodiametric; lobes of the disk corollas comprising 17–25 (30) percent of the limb .... 24. *A. praealtus*.
4 Plants (except often in *A. dumosus*) with short or no rhizomes, not strongly colonial, seldom more than 1 m tall; peduncles (except sometimes in *A. attenuatus*) conspicuously long, copiously bracteolate, usually at least some of them 2 cm long or more.
7 Leaves basally disposed .... 23. *A. attenuatus*.
7 Leaves chiefly cauline.
8 Leaves narrow, seldom as much as 1 cm wide; involucral bracts tending to be obtuse .... 29. *A. dumosus*.
8 Leaves wider, the larger ones generally 1 cm wide or more; involucral bracts generally sharply acute .... 22. *A. concinnus*.

*The disk corollas consist of a slender tube and a more swollen limb; the lobes are part of the limb.

Group V

1 Involucral bracts regularly imbricate, all less than 1.5 mm wide.
  2 Leaves (aside from those near the base) mostly squarrose-spreading or reflexed; involucral bracts glabrous except for the sometimes minutely ciliolate margins .... 31. *A. walteri*.
  2 Leaves (aside from those near the base) appressed or closely ascending; involucral bracts shortly hairy or scabrous on the back .... 32. *A. adnatus*.
1 Involucral bracts not much imbricate, some of the outer ones more or less enlarged and foliaceous, (1.5) 2–5 or 6 mm wide .... 33. *A. phyllolepis*.

Group VI

1 Achenes glabrous; inflorescence loosely corymbiform or corymbose-paniculiform, not virgate.
  2 Involucral bracts essentially glabrous on the back, but evidently ciliate .... 33. *A. phyllolepis*.
  2 Involucral bracts finely and densely sericeous on the back, usually not ciliate 34. *A. sericeus*.
1 Achenes densely sericeous; inflorescence mostly elongate-racemiform and more or less virgate .... 35. *A. concolor*.

Group VII

1 Involucre broadly urn-shaped, with mostly 40–100 flowers; plants often more than 5 dm tall .... 36. *A. pilosus*.
1 Involucre narrower, mostly narrowly obconic when pressed, with 16–32 flowers; plants only 1–5 dm tall .... 37. *A. depauperatus*.

Group VIII

1 Involucral bracts long-attenuate, not much imbricate, very narrow, the green tip generally well under 1 mm wide; rays commonly pink .... 16. *A. elliottii*.
1 Involucral bracts rounded or obtuse to strongly acute, but not attenuate, often strongly imbricate; rays mostly blue or purple, except in *A. commixtus*.
  2 Involucral bracts with loose, herbaceous tip, some of them spreading or reflexed, at least the larger ones usually 1–2 (3.5) mm wide, the scarious basal part of the bract relatively short and inconspicuous in relation to the herbaceous tip; herbage glabrous or nearly so (extreme forms of *A. novi-belgii* might be sought here) .... 18. *A. curtisii*.
  2 Involucral bracts either appressed, or, if with spreading or reflexed tip, then the bracts largely scarious, even the usually green tip not strongly herbaceous; herbage variously glabrous or hairy.
    3 Outer involucral bracts much narrower than the inner and very small, passing into the minute peduncular bracts .... 38. *A. turbinellus*.
    3 Outer involucral bracts not much if at all narrower than the inner, usually well differentiated from the generally more leaflike peduncular ones.
      4 Inflorescence densely leafy, usually elongate, the heads more or less numerous .... 24. *A. praealtus*.
      4 Inflorescence sparsely leafy, corymbiform, more or less flat-topped, with usually few heads.
        5 Leaves evidently veiny beneath, usually toothed, more or less hairy beneath, or seldom glabrous.
          6 Leaves basally disposed, the lower ones much the largest, mostly 3–15 cm wide, the cauline ones progressively reduced; plants often more than 1 m tall .... 39. *A. tataricus*.
          6 Leaves wholly cauline, the lower ones reduced and deciduous; plants seldom more than 1 m tall.
            7 Larger leaves commonly more than 3 cm wide; plants not strongly colonial, the rhizome short and stout; rays white to lavender .... 50. *A. commixtus*.
            7 Larger leaves up to 2.5 (3) cm wide; plants colonial by long, creeping rhizomes; rays violet .... 40. *A. radula*.
        5 Leaves only very obscurely veined beneath, usually glabrous and entire or nearly so.
          8 Involucre narrowly obconic or turbinate; rays mostly 9–14 .... 43. *A. gracilis*.
          8 Involucre broader, more campanulate; rays mostly 15–30 .... 42. *A. surculosus*.

Group IX

1 Inflorescence corymbiform, or the head sometimes solitary.
  2 Pappus relatively soft and fine; rays flowers often fewer than 15.
    3 Stem shortly hirtellous-puberulent above; involucral bracts often with loose or spreading tip .... 44. *A. avitus*.
    3 Stem glabrous; involucral bracts appressed .... 49. *A. chapmanii*.
  2 Pappus coarse, the larger bristles flattened and clavellate; rays seldom as few as 15.
    4 Involucral bracts tapering gradually from the base to the slender, spinulose-mucronate tip; heads broadly hemispheric, with very numerous disk flowers .... 47. *A. eryngiifolius*.
    4 Involucral bracts carrying their width well toward the summit, more abruptly pointed; heads campanulate-hemispheric, with less numerous disk flowers .... 45. *A. paludosus*.
1 Inflorescence spiciform or racemiform (or of spiciform or racemiform branches), or the head sometimes solitary.
  5 Heads campanulate-hemispheric to broadly hemispheric, with mostly 15–50 rays; involucral bracts often with loose or spreading tip.
    6 Stem essentially glabrous, or short-hairy only just beneath the heads; involucral bracts carrying their width well toward the summit, often rather abruptly pointed; heads campanulate-hemispheric, with only moderately numerous disk flowers .... 46. *A. hemisphericus*.
    6 Stem evidently spreading-villous throughout; involucral bracts tapering gradually from the base to the slender, spinulose-mucronate tip; heads broadly hemispheric, with very numerous disk flowers .... 47. *A. eryngiifolius*.
  5 Heads turbinate, relatively few-flowered, with mostly 8–15 rays; involucral bracts a little loose, but scarcely spreading .... 48. *A. spinulosus*.

Group X

1 Involucral bracts, or many of them, with distinctly squarrose or reflexed tip; plants mostly with rather short rhizomes, not strongly colonial .... 50. *A. commixtus*.
1 Involucral bracts appressed or sometimes a little loose; plants vigorously rhizomatous and colonial.
  2 Plants glandular in the inflorescence; rays tinged with lilac or purple .... 51. *A. macrophyllus*.
  2 Plants not glandular; rays usually white.
    3 Plants ordinarily with well-developed tufts of basal leaves on separate short shoots .... 52. *A. schreberi*.
    3 Plants ordinarily without well-developed tufts of basal leaves .... 53. *A. divaricatus*.

Group XI

A single species .... 54. *A. acuminatus*.

Group XII

1 Leaves, or some of them, generally toothed, more or less basally disposed, the larger ones 1–4.5 cm wide; disk flowers 9–20 .... 55. *A. paternus*.
1 Leaves all entire or nearly so, well distributed along the stem, up to 2 cm wide; disk flowers 5–10 (13).
  2 Herbage obviously rough-puberulent; leaves relatively broad, mostly 1.5–4 (5) times as long as wide .... 56. *A. tortifolius*.
  2 Herbage glabrous or nearly so; leaves narrow, mostly 5.5–20 times as long as wide .... 57. *A. solidagineus*.

Group XIII

1 Inner pappus bristles clavate; leaves veiny, scarcely rigid, mostly (0.7) 1 cm wide or more; rays white (Doellingeria).
  2 Leaves glabrous beneath, or shortly hairy on the midrib and main veins, rarely sparsely so across the surface; rays short, mostly 5–10 mm long.
    3 Heads relatively many-flowered, the disk flowers 16–40, the rays often more than 7.
      4 Achenes more or less strigose or puberulent; plants with creeping rhizomes .... 58. *A. umbellatus*.

4 Achenes glabrous; plants without creeping rhizomes .... 60. *A. infirmus*.
3 Heads relatively few-flowered, the disk flowers 4–14, the rays 2–7 .... 59. *A. sericocarpoides*.
2 Leaves densely and uniformly downy beneath; rays elongate, mostly 10–16 mm long .... 61. *A. reticulatus*.
1 Inner pappus bristles not clavate; leaves rigid, 1-nerved, only 1–4 mm wide; rays mostly violet (Ianthe) .... 62. *A. linariifolius*.

Group XIV

1 Stems arising singly from creeping rhizomes; heads larger, with mostly 15–25 rays, the involucre mostly 6–9 mm high .... 63. *A. tenuifolius*.
1 Stems clustered on a short, densely fibrous-rooted crown; heads smaller, with mostly 10–15 rays, the involucre mostly 4–6 mm high .... 64. *A. bracei*.

Group XV

A single species .... 65. *A. spinosus*.

Group XVI

A single species .... 66. *A. subulatus*.

1. A. azureus Lindl. Plants with a branched caudex or short rhizome; stems 2–15 dm tall, scabrous-puberulent to occasionally subglabrous. Leaves basally disposed, thick and firm, entire or occasionally shallowly serrate, scabrous-hispid above, the hairs on the lower surface softer, and usually longer and looser; basal and usually also some of the lowest cauline leaves long-petiolate, cordate (usually shallowly so) or subcordate, lanceolate or ovate in outline, 4–13 × 1.2–6 cm, some of the basal ones sometimes smaller and with more tapering base, those above more or less abruptly smaller, narrower, less petiolate, and generally not at all cordate, the upper sessile and lanceolate or linear. Inflorescence open-paniculiform, copiously and narrowly bracteate; involucre 4.5–8 mm high, its bracts well imbricate, obtusish to sharply acute, with a diamond-shaped green tip shorter than the chartaceous base, glabrous except for the often ciliolate margins; rays 10–25, blue (pink), 5–12 mm long. Achenes glabrous or nearly so. (n = 18) Late summer–fall. Prairies and dry, open woods; La and e Tex, n through Ark (OU, OZ), nw Miss, w Tenn (ME), Okla, Kans, Mo, and Iowa to s Minn, and thence e to w NY and s Ont. *A. poaceus* Burgess—S.

2. A. shortii Lindl. Plants with a short, branched caudex; stems 3–12 dm tall, glabrous or nearly so below, spreading-hirtellous above. Leaves entire or occasionally few-toothed, glabrous or scaberulous above, spreading-hirtellous beneath, lanceolate or rather narrowly ovate, acute or acuminate, petiolate, nearly all of those below the inflorescence cordate or subcordate, the lower cauline ones usually deeply so, 6–15 × 2–6 cm, the basal ones, if persistent, often shorter, broader, more rounded, and somewhat toothed, the middle and upper cauline ones only gradually reduced. Inflorescence open, paniculiform, with numerous usually narrow bracts, the bracteate peduncles often very long; involucre 4–6 mm high, its bracts narrow, acute or the outer obtusish, well imbricate, usually minutely pubescent, with small, more or less diamond-shaped green tip shorter than the chartaceous base; rays 10–20, blue, or rarely rose-red or white, 5–14 mm long. Achenes glabrous. (n = 9, 18) Fall. Woods; sw Pa and adj. WVa and nw Md (RV) to Ky (CU, IP), Tenn (RV, CU, IP), w Ga (CU, RV, PP), and ec Ala (PP), w to ne Miss, ne Iowa, and se Minn, and isolated in OU of Ark; an old, isolated station in DC may have been temporary. *A. camptosorus* Small—S.

3. A. cordifolius L. Plants with a branched caudex or short rhizome, occasionally with creeping rhizomes as well; stems 2–12 dm tall, glabrous below the inflorescence or occasionally loosely hairy with the hairs chiefly in lines. Leaves rather thin, sharply toothed, acuminate, more or less scabrous or scaberulous above, at least toward the margins, and sparsely to densely hirsute beneath with mostly long, flattened hairs, narrowly to broadly ovate, the larger ones 3.5–15 × 2.5–7.5 cm, all but the reduced ones of the inflorescence cordate (usually deeply so) and petiolate, the petioles progressively shorter upward, but only slightly if at all winged. Inflorescence paniculiform, with loosely ascending to widely spreading branches and often very numerous heads, the peduncles usually well developed and more or less bracteolate, sometimes up to 1.5 cm long; involucre 3–6 mm high, glabrous, its narrow bracts well imbricate, obtuse or merely acute, the green tip short and broad (or in the innermost more elongate), the green usually partly or sometimes wholly replaced or obscured by anthocyanin (visible to the naked eye in some of them); rays 8–20, blue or purple (white), 5–10 mm long. Achenes glabrous. (n = 9, 18) Fall. Woods and clearings; NS to Minn, s to Va, n Ga (BR, RV, CU), nw Ala, and Mo, mostly avoiding CP s of Va.

4. A. lowrieanus T. C. Porter. Resembling *A. cordifolius*, but glabrous except for the sometimes puberulent peduncles, the leaves very smooth to the touch, averaging narrower and more elongate, the middle and upper ones generally only shallowly cordate or merely subcordate or abruptly narrowed to the commonly conspicuously winged petiole, or even sessile and lanceolate; involucral bracts often lacking anthocyanin. (n = 18) Fall. Woods; s Ont and s Que, s through NY, NJ, and Pa to the mt. provinces of Md, WVa, Va, NC, Tenn, and Ga. Included in *A. cordifolius* by R. Perhaps originating by hybridization between *A. cordifolius* and *A. laevis*, but now self-perpetuating and distinct.

5. A. sagittifolius Willd. Plants with a branched caudex or short rhizome; stems 4–12 dm tall, glabrous or nearly so below the inflorescence, or the upper part occasionally villous-puberulent in lines. Leaves rather thick, shallowly toothed, glabrous or scabrous above, glabrous or more or less villous-hirsute with flattened hairs beneath, the lowest ones ovate or more often lance-ovate, acuminate, cordate, 6–15 × 2–6 cm, long-petiolate, those above progressively less cordate or merely abruptly narrowed to the shorter, often broadly winged petiole, or the upper tapering and sessile. Inflorescence paniculiform, elongate, with more or less strongly ascending, narrowly bracteate branches, the heads often very numerous, on narrowly bracteate peduncles rarely over 1 cm long and usually less than 5 mm long, thus appearing crowded; involucre 4–6 mm high, its imbricate bracts glabrous except for the sometimes ciliolate margins, very slender and long-pointed, with elongate, narrow green tip, often minutely purple-tipped under a lens; rays 8–20, usually pale blue or lilac, sometimes white, 4–8 mm long. Achenes glabrous. (n = 9, 18) Midsummer–fall. Stream banks, woods, and less often in open places, often in more mesic habitats than *A. drummondii*, with which it hybridizes extensively; sw Vt to Minn, s to Ga (GC), the Fla panhandle, Ala (GC), Miss, and Mo, mostly avoiding AC s of Va. *A. plumarius* Burgess—S.

6. A. drummondii Lindl. Plants with a branched caudex or short rhizomes; stems 4–12 dm tall, stouter than in *A. sagittifolius*, usually densely hairy at least above the middle with minute, stiffly spreading hairs. Leaves relatively firm, shallowly toothed, scabrous above, densely pubescent with short spreading hairs beneath, the lowest ones ovate or lance-ovate, acuminate, cordate, 6–14 × 2.5–6.5 cm, long-petiolate, those above progressively less cordate (or the upper merely broadly rounded) and with shorter, usually broadly winged petiole. Inflorescence paniculiform, with spreading or ascending, bracteate branches, the heads often numerous, on bracteate peduncles usually well under 1 cm long; involucre glabrous or puberulent, 4.5–7 mm high, its bracts firm, imbricate, sharply acute or acuminate but with broader and proportionately longer chartaceous base than in *A. sagittifolius*, the green tip tending to be elongate-rhombic; rays 10–20, bright blue, 5–10 mm long. Achenes sparsely short-hairy or glabrous. (n = 18) Fall. Typically in clearings and open woods; s Ohio to Minn, s to w Ky (ME), Miss, La, Kans, and Tex.

7. A. texanus Burgess. Resembling *A. drummondii*, but with more open inflorescence, the branches commonly very long, with copiously and minutely bracteate nonfloriferous base, or in more compact specimens merely the peduncles conspicuously elongate (mostly well over 1 cm long) and copiously bracteate; plants sometimes fully as hairy as *A. drummondii*, sometimes less so. (n = 9) Fall. Bottomlands and open woods; essentially Ozarkian species, from w Ky (IP and ME) to e Kans, s through Mo and Ark (all) to s Miss, nw La, e Okla, and e Tex. *A. trigonicus* Burgess—S.

8. A. undulatus L. Plants with a branched caudex or short rhizomes; stems 3–12 dm tall, densely pubescent with short, spreading hairs, varying to occasionally subglabrous below the inflorescence. Leaves entire or toothed, scabrous to glabrous above, usually shortly and rather loosely hairy beneath, at least the lower ones cordate or subcordate at the base and petioled, lance-ovate to ovate, 3.5–14 × 1.5–7 cm, those above extremely variable in size and shape, but always at least some of them either sessile and cordate-clasping or with the petiole enlarged and auriculate-clasping at the base. Inflorescence open, paniculiform, the branches and mostly well-developed peduncles more or less spreading and bracteate; involucre 4–7 mm high, usually minutely puberulent, its bracts imbricate, sharply acute or acuminate, sometimes very slender, more often a little wider as in *A. drummondii*, often minutely purple-tipped; rays 10–20, blue or lilac. Achenes minutely hairy at least above. (n = 9, 18) Late summer–fall. Mostly in dry, open woods and clearings; Me to c Fla, w to Ohio, se Ind, e Tenn, and c Miss, and se La; all provinces except OU and OZ, but not common on GC. *A. asperifolius* Burgess—S; *A. claviger* Burgess—S; *A. corrigiatus* Burgess—S; *A. gracilescens* Burgess—S; *A. linguiformis* Burgess—S; *A. loriformis* Burgess—S; *A. mohrii* Burgess—S; *A. proteus* Burgess—S; *A. sylvestris* Burgess—S; *A. triangularis* Burgess—S; *A. truellius* Burgess—S.

9. A. anomalus Engelm. Plants with a short, stout rhizome or branched caudex; stem 2–10 dm tall, pubescent with short, spreading, often coarse hairs, or merely scabrous. Leaves thick and firm, scabrous-hirsute above, more loosely and softly hairy beneath, entire or nearly so, petiolate, the basal and lower cauline ones ovate or lanceolate, deeply cordate, 4–9 × 2–5.5 cm, the middle and upper ones gradually or abruptly reduced. Inflorescence open, paniculiform, with numerous narrow bracts, the bracteate peduncles often very long; involucre 6–10 mm high, short-hairy or rarely subglabrous, the elongate, narrow, well-imbricate bracts with reflexed green tip; rays 20–45, bright blue, 7–15 mm long. Achenes glabrous. (n = 9) Fall. Dry woods, usually associated with limestone. OU and OZ of Ark, to Okla, Kans, Mo, and s Ill.

10. A. patens Aiton. Plants with a short caudex, sometimes with creeping rhizomes as well; stems rather slender and brittle, 2–15 dm tall, shortly and loosely hairy. Leaves more or less hairy or scabrous, at least beneath, sessile and conspicuously cordate-clasping, broadly ovate to oblong, entire, 2.5–15 × 0.8–4.5 cm, the lower soon deciduous. Heads few to rather numerous in an open, divaricately branched inflorescence; involucre 5–9 (–12) mm high, its bracts well imbricate, mostly acute, more or less glandular or short-hairy or both; rays 15–25 (30), blue or rarely pink, 8–15 mm long. Achenes shortly sericeous. (n = 5, 10, 20) Fall. Woods and dry, open places; Mass and s NH to Mo and Kans, s to n Fla, s La, and Tex. Divisible, with some difficulty, into 5 vars. with a degree of ecogeographic coherence:

Involucre more or less strongly glandular, often also hairy, but scarcely sericeous-strigose; plants occurring chiefly (but not entirely) e of the Mississippi R.
 Heads relatively large, the involucre 9–12 mm high; n = 25; Ga and SC, especially on and near PP; *A. georgianus* Alexander—S. .... var. *georgianus* (Alexander) Cronq.
 Heads smaller, the involucre 5–9 mm high; n = 5, 10.
  Leaves relatively thin, often constricted above the large, clasping base, the middle and upper ones mostly 3.5–6 times as long as wide; pubescence mostly a little softer than in the 2 following vars.; woods, mainly in and near the mt. provinces, from NY and Ohio, s to Ga and Ala, and also w to w Ky and n Miss; *A. phlogifolius* Muhl.—S. .... var. *phlogifolius* (Muhl.) Nees.
  Leaves relatively thick and firm, rarely constricted above the base, the middle and upper ones mostly 1.5–3.5 times as long as wide; in drier or more open places, on the average, than var. *phlogifolius*.
   Branches very long, conspicuously and minutely bracteate; plants averaging smaller and more slender than in var. *patens*; Va to n Fla, Miss, La, w Tenn (ME), and Tex, mainly on CP; *A. tenuicaulis* (C. Mohr) Burgess—S. .... var. *gracilis* Hook.
   Branches shorter, with fewer and larger bracts; widespread especially e of Mississippi R, but not commonly on CP s of Va .... var. *patens*.
Involucre only slightly or scarcely glandular, more or less strongly sericeous-strigose; involucral bracts broader and blunter than in the other vars.; habit varying from essentially like that of var. *patens* to more often like that of var. *gracilis*; essentially Ozarkian, ranging from Ark (all) and Mo to Kans, La, Miss, Okla, and Tex; *A. continuus* Small—S. .... var. *patentissimus* (Lindl.) T. & G.

11. A. oblongifolius Nutt. Plants rhizomatous, sometimes also with a short caudex; stem 1–10 dm tall, rigid, brittle, usually more or less branched, glandular upward, commonly also more or less hairy or scabrous. Leaves firm, entire, sessile and obscurely to evidently auriculate-clasping, narrowly to broadly oblong or lance-oblong, up to 8 × 2 cm, scabrous or short-hirsute, or sometimes glabrous except the margins, the lower ones soon deciduous, those of the branches numerous and reduced, becoming mere spreading bracts. Heads several or many, terminating the branches; involucre densely glandular, 5–8 mm high, its bracts in several series but not much imbricate, firm, with chartaceous base and long, green, loose or spreading, acute or acuminate tip; rays 15–40, blue or purple (rose), 1–1.5 cm long. Achenes strigose or finely sericeous. Fall. Dry, usually open places; DC and Pa, s through BR to s NC, and w to ND, Wyo, n Ala (IP), n Miss, Ark (OU, OZ), Tex, and NM. The var. *oblongifolius*, occurring largely to the w of our range, but entering Ark, is low and slender, usually much branched, seldom more than 4 dm tall, the upper part of the stem and branches merely glandular, or with short, mostly appressed or ascending hairs. Some of the Ark plants and all the remainder of our material belong to the more eastern var. *angustatus* Shinners, which is taller, up to 1 m, usually less intricately branched, and often with larger leaves, the upper part of the stem and branches with long, spreading hairs as well as glandular.

12. A. grandiflorus L. Similar to *A. oblongifolius* var. *angustatus*, but more openly branched, with smaller leaves, and with conspicuously larger heads, the involucre 10–15 mm high, its bracts wider and often more imbricate, the outer sometimes blunter; rays 12–25 mm long; stem rather sparsely or sometimes scarcely pubescent with usually spreading hairs, as well as glandular upward. Fall. Sandy places, often in pine woods or in old fields; Va (largely AC) and NC (AC and more abundantly on PP).

13. A. novae-angliae L. Plants with a stout caudex or short thick rhizome, occasionally with creeping rhizomes as well; stems clustered, 3–20 dm tall, usually strongly spreading-hirsute at least above, and becoming glandular upward as well. Leaves lanceolate, entire, 3–12 cm × 6–20 mm, sessile and conspicuously auriculate-clasping, scabrous or stiffly appressed-hairy above, more softly hairy beneath, or the upper leaves becoming glandular; lower leaves similar to the others but soon deciduous. Heads several or many in a leafy, usually short inflorescence, the involucre and peduncles densely glandular; involucre 6–10 mm high, its numerous slender bracts about equal, often purplish, with chartaceous base and loose or spreading attenuate tip, the outer sometimes a little broader and more foliaceous; rays 45–100, usually bright reddish purple or rosy, 1–2 cm long. Achenes densely sericeous or appressed-hirsute, their nerves obscure. (n = 5) Fall. Moist, open or sometimes wooded places; Mass and Vt to ND and Wyo, s generally to DC, Tenn, Ark (OU, OZ), and NM, and in the mt. provinces to Va, NC, and Ala; also irregularly on prairies on GC and ME of Ala and Miss.

14. A. puniceus L. Plants with a short, stout rhizome or caudex, sometimes with short, thick stolons as well; stem stout, 4–25 dm tall, simple or much branched above, uniformly spreading-hairy at least under the heads, conspicuously spreading-hispid to occasionally glabrous below. Leaves sessile (or the deciduous lower ones subpetiolate), auriculate-clasping, rather distantly serrate to occasionally entire, scabrous to subglabrous above, glabrous or spreading-hairy along the midrib beneath, lanceolate to oblong or elliptic-oblong, 7–20 cm × 12–40 mm. Heads few to many in a leafy inflorescence; involucre 6–12 mm high, its bracts slender and loose, scarcely or not at all imbricate, at least the inner ones long-acuminate to attenuate, often some of the outer enlarged and leafy, but still narrow; rays 30–60, blue (rose or white), 7–18 mm long. Achenes glabrous or nearly so. (n = 8) Late summer, fall. Swamps and other moist places; Nf to ND, s to Va (all), WVa, and Ill, and on PP and in mt. provinces to NC, Tenn, SC, Ga, and Ala.

15. A. lucidulus (A. Gray). Wieg. Similar to *A. puniceus*, but colonial by long, creeping rhizomes; stem and branches puberulent in lines above, glabrous or sparingly hispid below the inflorescence; leaves more crowded, especially upward, firmer, often shining, entire or nearly so; involucral bracts often less attenuate; rays blue or lavender (mostly rather pale) to occasionally white. Late summer, fall. Moist places; chiefly midwestern species, barely entering our range in WVa, and with apparently outlying stations in Morgan Co., WVa, and Buncombe Co., NC. *A. conduplicatus* Burgess—S.

16. A. elliottii T. & G. Plants colonial from well-developed creeping rhizomes, 6–16 dm tall, tending to be somewhat succulent; stem puberulent in lines, at least above. Leaves glabrous beneath, scaberulous above, serrate, the lowermost ones enlarged, long-petiolate, with elliptic blade up to 25 × 5 cm, but sometimes deciduous before anthesis, the cauline ones more or less reduced upward, becoming sessile or nearly so with more or less sheathing but not auriculate base. Inflorescence corymbiform or paniculiform; involucre 8–11 mm high, its bracts not much if at all imbricate, glabrous except for the sometimes ciliolate margins, narrow with long-attenuate, loose or somewhat squarrose green tip (this well under 1 mm wide), or the inner purplish instead of green; rays 25–45, pink or sometimes lavender, 7–12 mm long. Achenes glabrous or sparsely hairy. Fall. Swamps and other moist, low places; outer CP from se Va to s peninsular Fla, and w to the Fla panhandle and s Ala (CP & PP).

17. A. prenanthoides Muhl. Perennial from elongate rhizomes; stems 2–10 dm tall, often zigzag, pubescent in lines, becoming uniformly so under the heads and glabrate toward the base. Leaves scabrous to glabrous above, glabrous or loosely hairy on the midrib beneath, the main ones 6–20 × 1–5 cm, with ovate or lanceolate, serrate, acuminate blade abruptly contracted into a relatively long, generally entire, broadly winged, strongly auriculate-clasping petiolar base, the lower similar but reduced and deciduous, the reduced upper ones of the inflorescence becoming sessile and entire. Heads several or many in an open, sparsely leafy-bracteate inflorescence; involucre glabrous, 5–7 mm high, its bracts acute or obtusish, seldom much imbricate, rather loose, often with spreading tip, the green portion more elongate and relatively narrower than in *A. laevis* and its allies, often more than 2 mm long, the outer bracts sometimes wholly herbaceous; rays mostly 20–35, blue or pale purple (white), 7–15 mm long. Achenes strigose; pappus yellowish. (n = 9) Fall. Streambanks, meadows, and moist woods; NY to DC and in the mts. to NC and Tenn, w to WVa, Ky (IP), Ind, Iowa, and Minn.

18. A. curtisii T. & G. Stems 4–10 dm tall, scattered or few together on compact or shortly creeping rhizomes (filiform rhizomes seldom present as well); plants wholly glabrous, or often inconspicuously puberulent in lines in the inflorescence. Leaves entire or often toothed, the basal and lower cauline ones distinctly petiolate, with lance-ovate or rather narrowly ovate, generally basally rounded (seldom somewhat cordate) blade 4–11 × 1.5–4 cm, but often deciduous before anthesis, the others tapering to a broadly wing-petiolate or shortly subpetiolate or sessile but scarcely or not at all clasping base, up to 15 cm long overall and 2.5 cm wide. Heads several or rather many in a corymbiform to more often narrowly paniculiform or almost racemiform inflorescence; involucre (6) 7–10 mm high, its bracts usually more or less strongly imbricate (but sometimes the outer enlarged and foliaceous), loosely herbaceous-tipped, some or all of the green tips spreading or reflexed, the wider ones mostly 1–2 (3.5) mm wide, seldom narrower; rays mostly 13–30, showy, blue, 1–1.5 cm long. Achenes sparsely hairy or glabrous. Late summer, fall. From moist meadows to dry open woods, at middle and upper altitudes in BR of NC, Tenn, SC, and Ga.

19. A. novi-belgii L. Perennial from elongate rhizomes, large and stout to sometimes small and slender; stems 2–14 dm tall, sometimes puberulent in lines, sometimes glabrous except just under the heads. Leaves lanceolate to elliptic or lance-linear, sessile and more or less auriculate-clasping (scarcely so in extreme forms), though often narrowed toward the base, sharply serrate to entire, glabrous except for the scabrous-ciliate margins, often thick and firm, 4–17 cm × 4–25 mm, the lower similar to those above but reduced and deciduous. Heads several or many in an open or more often leafy-bracteate inflorescence; involucre glabrous, 5–10 mm high, its obtuse to sometimes sharply acute bracts subequal or more or less imbricate, generally at least some of them with rather loose or spreading tip; rays mostly 20–50, blue (rose or white), 6–14 mm long. Achenes strigose or subglabrous. (n = 9, 24, 27) Fall. Moist places, often in salt marshes; Nf to SC, chiefly along or near the coast, in our range only on AC. *A. elodes* T. & G.—S.

20. A. carolinianus Walter. CLIMBING ASTER. Slender, branching, somewhat woody, scrambling plants, climbing to ca. 4 m, spreading-villous or -villosulous to velvety-puberulent throughout. Leaves well distributed along the stem and branches, the principal ones broadly lanceolate and acuminate to elliptic-ovate and merely acutish, 3–7 (10) × 0.7–2 (3) cm, sessile or subsessile or often with an immediately suprabasal, shortly subpetiolar constriction, the base itself small but distinctly auriculate-clasping. Heads terminating the branches or forming small corymbiform inflorescences; involucre 8–11 mm high, short-hairy, its bracts more or less strongly imbricate, the outer commonly with loose or squarrose, often expanded green tip; rays ca. 40–60, pink or purple, 1–2 cm long. Achenes glabrous. Fall. Woods and thickets, from swamps and salt marshes to hammocks and dry, sandy pine barrens; nearly throughout Fla, n irregularly on AC to NC.

21. A. laevis L. Plants 3–10 dm tall from a short, stout rhizome or branched caudex, sometimes with short, slender, creeping red rhizomes as well; herbage glabrous except occasionally for some puberulent lines in the inflorescence, commonly somewhat glaucous. Leaves mainly cauline, thick and firm, highly variable in size and shape, but the larger ones more than 1 cm (often more than 2.5 cm) wide and often less than 5 times as long as wide, entire or sometimes toothed, sessile and more or less strongly auriculate-clasping, the lower tapering to a winged petiole and scarcely clasping, those of the inflorescence reduced and often bractlike, broadest at the clasping or subclasping base. Heads several or many in an open inflorescence; involucre 5–9 mm high, its firm, appressed, acute bracts conspicuously imbricate in several series, with short, diamond-shaped green tip mostly 1–2 mm long; rays mostly 15–30, blue or purple, 8–15 mm long. Achenes glabrous or nearly so; pappus usually reddish. (n = 24, 27) Late summer, fall. Open, usually dry places; Me to BC, s to n Ga, n Ala, Ark, and NM, but s of Va mostly not on AC and PP. Eastern Am plants, as described here, are var. *laevis*.

22. A. concinnus Willd. Much like *A. laevis*, but with narrower, often less evidently glaucous leaves, the principal ones only slightly or not at all clasping, often petiolate, seldom more than 2.5 cm wide and seldom less than 5 times as long as wide (petiole included in the length); inflorescence on the average more diffusely branched, with more numerous and smaller bracts, the heads mostly long-pedunculate. (n = 23, 24) Fall. Dry woods and open places; Pa to Ga and Ala, chiefly in mt. provinces and on PP; disjunct(?) in c Miss. Hardly distinct from *A. laevis*, at one extreme, and from *A. attenuatus*, at the other, but occupying a coherent geographic area largely separate from that of *A. attenuatus*, and more restricted than that of *A. laevis*. *A. laevis* var. *concinnus* (Willd.) House—R; *A. falcidens* Burgess—S; *A. purpuratus* Nees—S; *A. steeleorum* Shinners—F.

23. A. attenuatus Lindl. Like *A. concinnus*, but the leaves more or less basally disposed, the basal and lower cauline ones the largest and generally persistent, up to 20 cm long and 2 or 2.5 cm wide, the others progressively reduced, often to mere bracts; leaf margins often strongly scabrous; inflorescence sometimes diffuse as in *A. concinnus*, but sometimes more compactly branched, with short-pedunculate heads. Fall. Open woods, prairies, and other dry, open places; n Ga to Ala, La, Ark, and Tex, chiefly on and near CP. *A. ursinus* Burgess—S.

24. A. praealtus Poiret. Colonial by long rhizomes; stems 6–15 dm tall, puberulent in lines above or uniformly puberulent nearly throughout. Leaves all cauline, lanceolate, sessile but scarcely auriculate, or tapering to a narrow, subpetiolar base, thick and firm, conspicuously reticulate beneath with nearly isodiametric areolae, entire or nearly so, scabrous to subglabrous above, glabrous or puberulent beneath, the main ones 7–13 cm × 8–18 mm. Heads more or less numerous in a mostly elongate and leafy inflorescence; involucre 5–7 (8) mm high, its bracts narrow, sharply acute to acutish, glabrous except for the often ciliolate margins, evidently imbricate in several series, with narrow or elongate-rhombic appressed green tip; rays 6–15 mm long, bluish-purple or rarely white; lobes of the disk corollas comprising 17–25 (30) percent of the limb. Fall. Moist, low ground; Nebr. to Ariz, Tex, and n Mex, e to Mich, Ky (IP), n Ga (RV), and La, and disjunct(?) in the Fla panhandle. Most of our material belongs to the widespread var. *praealtus*, with the stem puberulent only in lines above, and with the lower surfaces of the leaves scantily scabrous-puberulent to subglabrous. The more strictly western var. *nebraskensis* (Britt.) Wieg. with the stem and often also the lower leaf-surfaces uniformly and conspicuously spreading-puberulent, occurs with us in Ark and La. *A. salicifolius* Lam.—S, a preoccupied name.

25. A. simplex Willd. Colonial by long rhizomes; stems 6–15 dm tall, puberulent in lines above (at least the branchlets and peduncles). Leaves all cauline, lanceolate or elliptic-lanceolate, serrate or occasionally entire, glabrous or somewhat scabrous above, sessile or tapering to a shortly subpetiolar base, sometimes a little clasping, but scarcely auriculate, the main ones mostly 8–15 × 1–3.5 cm, mostly not strongly reticulate, the areolae, if visible, generally irregular and longer than wide. Heads more or less numerous in an elongate, leafy inflorescence; involucre mostly 4–6 mm high, its bracts narrow, sharply acute to acutish, glabrous except for the often ciliolate margins, more or less strongly imbricate, with elongate, usually appressed green tip; rays 20–40, white or occasionally lavender or blue, mostly 6–12 mm long; lobes of the disk corollas comprising 30–45 percent of the limb. (n = 8, 16, 24, 32, 36) Fall. Moist, low places; NS to ND, s to NC, Tenn, Ark, La, and Tex. Our plants are var. *simplex*. *A. lamarckianus* Nees—S; *A. tradescanti* L.—S, misapplied.

26. A. vimineus Lam. Colonial by long rhizomes; stems 4–15 dm tall, glabrous or more or less puberulent in lines. Leaves all cauline, glabrous, or slightly scabrous above, linear to narrowly lanceolate, acute, tapering to the sessile base, entire or slightly toothed, up to 11 × 1 cm, those of the branches becoming much reduced. Heads mostly numerous in an open, ample inflorescence with long, divaricate, divergently bracteate, often recurved branches that tend to be secund, the shortly bracteate peduncles short or up to 1.5 cm long; involucre 2.5–3.5 (4) mm high, glabrous, its bracts imbricate, with mostly elongate green tip; rays 15–30, white or seldom purplish, 3–6 mm long; lobes of the disk corollas comprising ca. 40 percent of the limb. Fall. Mostly in moist, open places, and in floodplain forests; Me to Fla, chiefly near the coast, thence along CP to the Mississippi R valley and upstream chiefly along river-bottoms to w Tenn, s Ohio and w Mo. *A. brachypholis* Small—S; *A. racemosus* Ell.—F, R, S.

27. A. ontarionis Wieg. Similar to *A. lateriflorus*, but with long, creeping rhizomes, and often verging toward *A. simplex* in habit; stem uniformly spreading-puberulent, at least above. Leaves uniformly (and sometimes sparsely) puberulent beneath, or sometimes more densely so and almost villous along the midrib, more scabrous or subglabrous above. Involucre glabrous or sometimes puberulent; lobes of the disk corollas comprising 45–65 percent of the limb. (n = 16) Fall. River bottomlands; n NY to Ky (IP), Tenn (IP), and n Miss, w to Minn, SD, Ark (OU), and Okla. *A. missouriensis* Britton—S, a preoccupied name.

28. A. lateriflorus (L.) Britton. Stems 3–12 dm tall from a branched caudex or short stout rhizome, more or less curly-villous to glabrous. Leaves scabrous or subglabrous above, apparently glabrous beneath except for the usually villous or puberulent midrib; basal and lower cauline leaves mostly soon deciduous, or the basal ones sometimes persistent, petiolate, with obovate to elliptic or subrotund blade up to 8 × 4 cm, those above sessile or nearly so, broadly linear to more often lanceolate, lance-elliptic, or subrhombic, tending to taper from the middle to both ends, entire or serrate, the main ones 5–15 × 0.5–3 cm, those of the branches often abruptly reduced. Heads more or less numerous in a widely branched or occasionally more simple inflorescence, commonly subracemiform on the branches; involucre glabrous, 4–5.5 mm high, its bracts imbricate in few series, obtuse or acute, with evident, fairly broad green tip, often suffused with purple upward; rays 9–14, white or slightly purplish, 4–6.5 mm long; lobes of the disk corollas recurved, comprising 50–75 percent of the limb. (n = 8, 16, 24) Fall. Various habitats, most commonly in open woods, dry, open places, and on beaches; Magdalen Isl. (Que) to c peninsular Fla, w to Minn, Mo, and Tex. *A. agrostifolius* Burgess—S; *A. hirsuticaulis* Lindl.—S; *A. spatelliformis* Burgess—S.

29. A. dumosus L. Plants with creeping rhizomes or sometimes a short stout rhizome or caudex, the stems 3–10 dm tall, glabrous or puberulent upward. Leaves linear to lance-linear or linear-elliptic; sessile, entire or nearly so, more or less scabrous above, glabrous beneath, all or nearly all cauline, the principal ones 3–11 cm × 3–10 mm, those of the branches much reduced. Inflorescence open, usually ample, often diffuse, the branches with numerous spreading or ascending, oblong or spatulate bracts; heads more or less numerous, conspicuously long-pedunculate, the peduncles bracteate like the branches of the inflorescence, generally at least some of them 2 cm long or more, sometimes up to 15 cm; involucre 4–6 (7) mm high, glabrous, its bracts strongly imbricate, with short broad green tip, tending to be obtuse, often some of them dilated upward; rays 13–30, pale lavender to bluish or sometimes white, 5–9 mm long; lobes of the disk corollas comprising 20–35 percent of the limb. Fall. Dry or moist, often sandy places; Mass to s Fla, w to Mich, Ark, and La, most common on or near CP. *A. coridifolius* Michx.—S; *A. gracilipes* (Wieg.) Alexander—S; *A. pinifolius* Alexander—S.

A. fontinalis Alexander, from s Fla, may prove to be a distinct var. or sp. It differs from characteristic *A. dumosus* in its harsher and more copious pubescence, and in that the leaves tend to be somewhat clasping. These features suggest genetic infiltration from *A. patens*, but that sp. apparently reaches its s limit in n Fla.

30. A. ericoides L. Plants 3–10 dm tall, hairy, colonial by elongate creeping rhizomes, the stems arising singly. Leaves numerous, linear, sessile, up to 6 cm × 7 mm, the lower and often also the middle ones soon deciduous, those of the branches reduced and divaricate. Heads numerous, small, commonly somewhat secund on the divergent or recurved branches; involucre 3–5 mm high, its bracts more or less strongly imbricate, the outer obtuse or acutish, spinulose-mucronate, and more or less squarrose, some or all of the bracts coarsely ciliolate-margined, and usually also shortly hairy on the back; rays 8–20, white (blue or pink) 3–6 mm long; disk flowers 5–13 (15). Achenes sericeous-strigose. (n = 5) Late summer–fall. Dry, open places; s Me to s Sask, s to Del, n Va, Tenn, s Ill, Ark (mainly OU and OZ), Tex, n Mex, and se Ariz.

31. A. walteri Alexander. Plants rhizomatous, often from a thickened base, 1–8 dm tall, slender, freely and openly long-branched when well developed, rather sparsely to copiously scabrous-hispidulous or partly glabrous. Leaves very numerous and very small, mostly or all cauline, somewhat fleshy-thickened; lower leaves oblong or oblong-oblanceolate, up to ca. 3 cm long and 1 cm wide, but often soon deciduous, the others more ovate or lance-oblong, seldom more than 1.5 cm long, broadly rounded at the auriculate-clasping and often shortly adnate-decurrent base, squarrose-spreading or reflexed, the rameal ones numerous, progressively smaller, often only 1–2 mm long. Heads mostly terminating the branches; involucre 4.5–7 mm high, its bracts glabrous or in part minutely ciliolate-margined, seldom short-hairy as in *A. adnatus*, well imbricate in several series, the outer rounded or broadly obtuse; rays 10–21, lilac to blue-purple, 7–15 mm long. Achenes short-hairy. Fall (winter). Mostly in dry, sandy pine barrens; AC of NC to Fla (more common in Fla), and irregularly w to Apalachicola. *A. squarrosus* Walt., not All. Intergrades to some extent with *A. adnatus* in Fla.

32. A. adnatus Nutt. Similar to *A. walteri*, often a little more densely short-hairy; leaves (except those near the base) erect or closely ascending, basally a little narrower and distinctly adnate-decurrent, the adnate part often as long as or longer than the free part; involucral bracts shortly scabrous-hispidulous to finely scaberulous over the surface. Fall–winter. Mostly in sandy pine barrens; Bahama Isl and s Fla, n generally to ca. 29 degrees, thence mainly on GC to sw Ga and w on GC to Miss.

33. A. phyllolepis T. & G. Much like *A. sericeus*, but averaging less densely and less softly hairy, the leaves sometimes merely strigillose and soon glabrate; involucre mostly 8–13 mm high, the enlarged outer involucral bracts averaging wider, sometimes as much as 5–6 mm wide, essentially glabrous on the back but evidently ciliolate on the margins; rays mostly 20–30. Fall. Dry, open, often sandy woods and prairies; La and Tex.

34. A. sericeus Vent. Stems clustered on a short, branched caudex, brittle, wiry, 3–7 dm tall, more or less branched upward, thinly sericeous, or glabrate below. Leaves sericeous, entire, the basal oblanceolate and petiolate, but these and the ones on the lower half of the stem soon deciduous, the others sessile but only slightly or not at all clasping, lanceolate or lance-ovate to oblong or elliptic, up to 4 × 1 cm. Heads several or many in a widely branched, corymbiform or corymbose-paniculiform inflorescence, often clustered at the ends of the branches; involucre 6–10 mm high, its broad, acute bracts finely and densely sericeous on the back, usually not ciliate, borne in several series but not much imbricate, often partly anthocyanic, the outer tending to be somewhat enlarged, foliaceous-tipped, and often loose or spreading, the larger ones mostly 1.5–3 mm wide; rays mostly 15–25, deep violet to rose-purple (white), 8–15 mm long. Achenes glabrous, closely 8- to 12-nerved. Fall. Dry prairies and other open places; Mich to SD, s to Mo and Tex, and irregularly e in Tenn to CU; records from farther e are in error.

35. A. concolor L. Plants with a short caudex or crown, often with creeping rhizomes as well; stems slender, 3–10 dm tall, simple or sparingly branched, thinly sericeous or sometimes merely strigose, rarely spreading-villous, glabrate below. Leaves sericeous, sometimes glabrate in age, entire, the lower broadly elliptic, up to 5 × 1.5 cm. Inflorescence narrow and racemiform or occasionally with racemiform branches, the peduncles minutely bracteate; involucre 5–9 mm high, densely and finely sericeous, its bracts evidently imbricate in several series, narrow, all less than 1.5 mm wide, sometimes partly anthocyanic, green-tipped but not foliaceous-expanded, sometimes with loose or spreading tip; rays 8–16, blue or sometimes pink, 7–12 mm long. Achenes densely sericeous, the pubescence obscuring the nerves. Fall. Dry, sandy places, often among pines; coastal states (all prov.) from Mass to s Fla and w to

La, and up the ME to sw Tenn, less commonly inland to the mts. of Ky and Tenn (CU). *A. plumosus* Small—S; *A. simulatus* Small—S.

36. A. pilosus Willd. Stems 1–15 dm tall from a stout caudex. Basal and lower cauline leaves soon deciduous, or the basal persistent, oblanceolate, and petiolate, those above sessile or nearly so, entire or slightly toothed, linear to lance-elliptic, seldom more than 10 × 1 (2) cm, the upper and rameal ones numerous and reduced, often subulate. Heads few to usually numerous in an often diffuse inflorescence, sometimes secund on the branches, their copiously subulate-bracteate to nearly naked peduncles 3–40 mm long; involucre broadly urn-shaped, constricted above the middle, then flaring, glabrous, 3.5–8 mm high, its bracts imbricate or subequal, with loose, subulate, marginally inrolled green tip; heads mostly 40–100-flowered, the rays 16–35, 5–10 mm long. (n = 16, 24) Summer–fall. Open, rather dry places, often in sandy soil; common and widespread from Me and NS to Ga and the Fla panhandle, w to Wis, Iowa, Kans, Ark, and La. Four vars.:

Rays white, rarely pink or purple, 5–10 mm long; involucre 3.5–5.5 mm high, rarely more; heads 40- to 70-flowered.
  Stem and often also the leaves sparsely to more often densely spreading-hirsute; plants generally robust and many-headed; the common var., occurring nearly throughout the range of the species, except at the ne; *A. pilosus* var. *platyphyllus* (T. & G.) Blake—F. .... var. *pilosus*.
  Stem and leaves nearly or quite glabrous.
    Similar to var. *pilosus*, although the leaves average narrower; chiefly in the coastal states, especially on CP, but also scattered far inland; possibly better considered as a phase of var. *pilosus*; *A. juniperinus* Burgess—S; *A. ramosissimus* Mill.—S. .... var. *demotus* Blake.
    Smaller, seldom more than 5 dm tall, with few, mostly less than 50 heads; northern and especially northeastern variety, occasionally extending s to the n fringe of our range, and to the mts. of NC and Tenn; *A. faxoni* Porter—S (in large part) .... var. *pringlei* (A. Gray) Blake.
Rays purple to pink or lavender, 8–15 mm long; involucre 5–8 mm high, seldom much imbricate; flowers up to 100 or perhaps more; herbage glabrous, or the stem sparsely spreading-hairy; Ky (IP) and Tenn (IP, CU) to Ga (RV, PP) and Ala (PP); *A. priceae* Britton—S. .... var. *priceae* (Britton) Cronq.

37. A. depauperatus (T. C. Porter) Fern. Like a slender, wiry, depauperate form of *A. pilosus*, only 1–5 dm tall, glabrous throughout, all the cauline leaves, except the numerous, much reduced, subulate ones of the inflorescence, commonly deciduous by flowering time, heads narrow, 16- to 32-flowered, the 9–16 rays mostly 2–5 mm long. Midsummer–fall. On serpentine in WVa, Md, Pa, and Del.

38. A. turbinellus Lindl. Stems 4–12 dm tall from a branched caudex, glabrous or more or less spreading-hirsute. Leaves entire, firm, glabrous except for the scabrous-ciliolate margins and sometimes some coarse hairs on the midrib beneath, the lower soon deciduous, the others broadly linear to oblong or lance-elliptic, sessile or subpetiolate, but not clasping, 6–10 cm × 8–20 mm, the upper more reduced, those of the often long and wiry branches becoming bractlike. Heads usually more or less numerous, turbinate, conspicuously long-pedunculate, the peduncles beset with minute, mostly appressed bracts; involucre glabrous, 7–12 mm high, its bracts obtuse or rounded, multiseriate, the outer narrow and small, passing into those of the peduncles, the inner much broader, shortly green-tipped; rays 15–20, violet, 8–12 mm long. Achenes strigillose and minutely punctate; pappus tawny-rufescent. (n = 48–50) Fall. Mostly in dry open places and in open woods; c Ill to Kans, Ark (mainly OU and OZ), and La.

39. A. tataricus L. f. Coarse, rough-hairy plants, 5–20 dm tall from a stout caudex. Lower leaves long-petiolate, with large, elliptic, conspicuously toothed blade 8–40 × 2.5–15 cm, the middle and upper smaller, sessile or nearly so, and mostly entire. Inflorescence corymbiform, flat-topped; involucre 7–10 mm high, strigose-puberulent or subglabrous, its bracts not much imbricate, the larger ones mostly 1–2 mm wide; rays mostly 15–20, purple or blue, 1–2 cm long. (n = 27) Fall. Native of s Siberia, casually escaped from cult. in e US, as in Va, NC, Tenn, and Ala.

40. A. radula Aiton. Plants rhizomatous, 1–12 dm tall, the stems glabrous except for some puberulence beneath the heads. Leaves veiny, with several pairs of evidently divergent primary lateral veins, more or less serrate, sessile or nearly so, scabrous above, hairy or occasionally glabrous beneath, the lower ones much reduced and soon deciduous, the others elliptic to oblong or lance-oblong, 3–10 × 0.6–2.5 (3) cm. Heads several in a short and broad, sparsely leafy-bracteate, corymbiform inflorescence, or seldom solitary; involucre 6–11 mm high, its relatively broad, firm bracts more or less imbricate, rounded to acute, puberulent or more often glabrous except for the ciliate or fimbriate margins, often squarrose-tipped, the green tip often ill-defined or wanting; rays 15–40, violet, 8–15 mm long. Achenes glabrous. Late summer, early fall. Bogs, streamsides, and other moist places; Del, Md, and WVa to Nf and Lab.

41. A. spectabilis Aiton. Plants rhizomatous; stems 1–9 dm tall, usually densely glandular, at least above, or occasionally merely spreading-villous. Leaves firm, scabrous, especially above, or glabrous, entire or remotely and shallowly toothed, the basal and lowest cauline ones well developed and usually persistent, with elliptic blade 2–16 cm × 8–40 mm tapering to the well-developed petiole, those above more oblong, becoming sessile and somewhat reduced, well spaced. Heads few to rather many in an open, corymbiform, sparsely leafy-bracteate inflorescence; involucre 8–16 mm high, its bracts broad and firm, more or less imbricate, with chartaceous base and loose or spreading green tip, densely glandular (at least the inner), the outer sometimes viscid-villous as well; rays 15–35, rich violet-purple, 1–2.5 cm long. Achenes short-hairy. (n = 36) Fall. Dry, sandy soil, often among pines; Mass to SC, near the coast, and scattered across PP of NC to BR, and in Jackson and DeKalb Cos., Ala. *A. smallii* Alexander—S.

42. A. surculosus Michx. Similar to *A. spectabilis*, but not glandular, or only slightly and inconspicuously so; rhizome sometimes with nodular-thickened, woody portions; stem spreading-hirtellous (or merely strigose below), often slightly viscidulous, but scarcely glandular, or sometimes slightly so on the peduncles; heads averaging smaller, the involucre 7–12 mm high, its bracts averaging narrower and blunter, only obscurely or not at all glandular, otherwise glabrous except for the often ciliolate margins; rays paler, more bluish. Late summer–fall. Various habitats, especially in sandy soil; mt. provinces of Ky, Tenn, NC, SC, and Ga, encroaching onto IP in Tenn.

43. A. gracilis Nutt. Plants with a thickened, hard, cormlike base, sometimes rhizomatous as well, the several puberulent or hirtellous to occasionally subglabrous stems 1.5–5 dm tall. Leaves thick and firm, obscurely veined except for the evident midrib and sometimes a single pair of slowly divergent laterals, entire or nearly so, not evidently sheathing below, the basal with elliptic blade 2–6 cm × 8–20 mm, shorter than the petiole, often deciduous, the cauline narrower and most of them sessile, 1.5–9 cm × 2.5–14 mm, 4–12 (15) times as long as wide. Heads several or rather many in a short and broad, usually sparsely bracteate, corymbiform inflorescence, narrow, the involucre narrowly obconic or turbinate, 7–12 mm high, its relatively broad, firm bracts glabrous or obscurely puberulent, imbricate, the outer shortly green-tipped, the inner scarcely so or merely purple-margined, commonly some or all of them shortly squarrose; rays 8–14, blue-violet to rose-purple, often rather pale, 5–8 mm long. Achenes thinly strigillose or glabrate. (n = 9) Midsummer, early fall. Dry, sandy places, often among pines; CP from NJ to SC and Burke Co., Ga. Approaches *A. surculosus*, on one hand, and *A. paludosus* and *A. avitus*, on the other.

44. A. avitus Alexander. Perennial from shortly creeping rhizomes, 3–8 dm tall; stems glabrous or nearly so below, shortly hirtellous-puberulent above. Leaves firm, glabrous, somewhat grasslike, linear to lance-linear or oblong-linear, not veiny, entire or often some of them remotely spinulose-toothed, up to ca. 15 cm long and 1 cm wide, somewhat basally disposed but the lowermost ones tending to be deciduous, the lower tapering to a more or less petiolar, sheathing base, the others progressively reduced and sessile. Heads several to rather many in a flat-topped, sparsely bracteate, corymbiform inflorescence, relatively narrow and few-flowered; involucre turbinate, 7–9 mm high, its bracts well imbricate, firm, linear-oblong, with conspicuously chartaceous base and short, often slightly expanded and often somewhat loose or spreading green tip, the outer generally more acute than the inner; rays mostly 8–13, lavender, 5–10 mm long; disk flowers mostly 15–20. Achenes glabrous; pappus of slender bristles. Late summer, early fall. On Stone Mt., Ga, and on granite flatrocks in Ga and Pickens Co., SC.

45. A. paludosus Aiton. Perennial from scaly, often nodular-woody creeping rhizomes, 2–8 dm tall; stem glabrous below, hirtellous-puberulent to often loosely villous-hirsute above, at least on the peduncles. Leaves firm, glabrous or nearly so, grasslike, linear or nearly so, not veiny, only the midrib evident, entire or occasionally some of them remotely spinulose-toothed, up to ca. 20 cm long overall and ca. 1 cm wide, more or less basally disposed but the lowermost ones sometimes deciduous, the lower tapering to a more or less petiolar, sheathing base, the others progressively reduced and sessile. Heads (1–) several in an open, corymbiform inflorescence; involucre campanulate or campanulate-hemispheric, 9–12 mm high, glabrous or short-hairy, its bracts seldom much imbricate, linear-oblong, loose and often distally spreading, firm with chartaceous base and green tip, or the outer nearly wholly green, the outer often more acute than the inner, the larger ones mostly 1.5–2.5 mm wide; rays mostly 15–35, deep lavender or purple, 1–2 cm long; disk flowers mostly more than 30. Achenes glabrous or slightly strigose; pappus coarse and firm, the larger bristles flattened and often slightly clavellate. (n = 9) Fall. Moist savannas, margins of pools and swamps, and low pinelands, seldom on sand hills; AC of NC, SC, and Ga.

46. A. hemisphericus Alexander. Similar to *A. paludosus*, but the inflorescence elongate-racemiform (or of spiciform or racemiform branches) instead of corymbiform; stem commonly glabrous or nearly so, or short-hairy only just beneath the heads; involucre glabrous. Late summer, fall. Prairies and open woods, less commonly in moist, low ground; Mo, Kans, and Tex to Ark (all), Tenn (mainly IP and ME), La, Miss, Ala (all), Ga (RV), and the Fla panhandle. *A. paludosus* subsp. *hemisphericus* (Alexander) Cronq.—G; *A. gattingeri* Alexander—S; *A. pedionomus* Alexander—S; *A. verutifolius* Alexander—S. Sometimes difficult to separate from *A. paludosus*, but geographically distinct.

47. A. eryngiifolius T. & G. Rather coarse perennial from a very short, stout rhizome or a mere crown, 3–7 dm tall, the stem spreading-villous. Leaves glabrous or nearly so, narrow and elongate, all linear or nearly so, firmly grasslike, finely parallel-veined, only the midrib prominent, usually some of them with a few distant spinulose teeth, strongly basally disposed, the basal and lowest cauline ones mostly 7–35 cm × 3–8 mm, the petiole sometimes scarcely differentiated from the blade; cauline leaves fairly numerous but progressively reduced, mostly sessile. Heads 1–6, relatively large, the disk broadly hemispheric, mostly 1.5–3 cm wide; inflorescence often corymbiform especially when the heads are only 2–3, otherwise generally racemiform; involucre 9–12 mm high, its bracts several-seriate but not much imbricate, very firm, greenish at least above but scarcely herbaceous, often scabrous-ciliolate, otherwise glabrous or nearly so, tapering gradually from the base to the very slender, minutely spinulose tip, distally loose or spreading; rays mostly 25–50, white or pinkish, 1–2 cm long. Achenes glabrous; pappus reddish-tinged, coarse, the larger bristles clavellate and flattened. Late spring–midsummer. Low flatwoods and swampy places; Fla panhandle and adj. Ga and Ala.

48. A. spinulosus Chapman. Perennial from a very short, stout rhizome or a mere crown; stems clustered, 2–7 dm tall, glabrous or rather coarsely spreading-hairy. Leaves essentially glabrous except for the often ciliate and sometimes remotely spinulose-toothed margins, narrow and elongate, firmly grasslike and somewhat fleshy, finely or obscurely parallel-veined (only the midrib prominent), strongly basally disposed, the basal and lowest cauline ones mostly 10–30 cm × 1–5 mm, the petiole sometimes scarcely differentiated from the blade; cauline leaves fairly numerous but progressively reduced. Heads 3–15 in a spiciform or racemiform inflorescence, relatively narrow and few-flowered, more or less turbinate, the disk mostly 1–1.5 cm wide as pressed; involucre 6–9 mm high, its bracts several-seriate but not much imbricate, very firm, greenish above but scarcely herbaceous, a little loose but scarcely spreading, tapering from above the middle or near the base to the minutely spinulose tip; rays mostly 8–15, anthocyanic, 1–1.5 or 2 cm long. Achenes glabrous or nearly so; pappus reddish-tinged, coarse, the larger bristles clavellate and flattened. Midsummer. Moist or dry pinelands and swamps in the Apalachicola Valley, Fla.

49. A. chapmanii T. & G. Perennial from a very short, stout rhizome or a mere crown, 3–8 dm tall; herbage glabrous. Leaves narrow and elongate, all linear or nearly so, entire, fleshy-firm, not veiny, only the midrib evident, strongly basally disposed, the basal and lowest cauline ones 10–30 cm × 2–7 mm, the slender blade tapering gradually into the long petiole; cauline leaves fairly numerous but strongly reduced upward, mostly linear and sessile. Heads several in an open-corymbiform inflorescence, terminating the slender, more or less elongate, slender-bracteate peduncular branches; involucre 7–9 mm high, its bracts well imbricate, appressed, firm, not strongly green-tipped, often partly anthocyanic, ciliolate on the margins or somewhat puberulent on the back, or both; rays ca. 8 to ca. 21, purple or blue-lavender, 1–2 cm long. Achenes glabrous; pappus bristles fine and soft, somewhat yellowish. Fall. Wet savannas and swampy pinelands in the Apalachicola Valley of Fla, and w to se Ala; also disjunct in St. Lucie Co., Fla.

50. A. commixtus (Nees) Kuntze. Plants 3–12 dm tall from a short, coarse rhizome, not strongly colonial, with or without some separate clusters of well-developed, cordate-based radical leaves; usually at least some of the basal and lower cauline leaves long-petiolate, with broad, basally cordate or subcordate, toothed blade mostly 7–20 × 5–12 cm, the other leaves also relatively large and broad, but sessile or short-petiolate and not cordate, seldom none of the leaves cordate; herbage and inflorescence somewhat hairy, especially upward, but scarcely or not at all glandular.

Inflorescence corymbiform, tending to be flat-topped, its bracts few and broad; involucre 7–12 mm high, its bracts well imbricate, largely chartaceous, and sometimes not very noticeably green-tipped, many or all of them with conspicuously squarrose or reflexed tip, the outer ones mostly 0.7–1.2 mm wide and up to ca. 3 times as long; rays 7–17, white to lavender, 1–1.5 cm long. Late summer–early fall. Deciduous woods; PP (down to the fall line) from s NC to Ga and ec Ala. *A. mirabilis* T. & G.—S.

51. A. macrophyllus L. Rhizomatous and colonial, 2–12 dm tall, with abundant clusters of radical leaves; glandular at least in the inflorescence, otherwise glabrous or rough-hairy. Leaves thick and firm, crenate or serrate with usually mucronate teeth, the basal and lower cauline ones cordate, 4–30 × 3–20 cm, commonly short-acuminate to obtuse, long-petiolate, the middle and upper gradually or abruptly reduced, becoming sessile and ovate to lanceolate or elliptic. Inflorescence corymbiform, its bracts few and broad; involucre 7–11 mm high, generally glandular, sometimes also short-hairy, its bracts firm, well imbricate, appressed, rounded to sharply acute, the green tip sometimes obscure, the outer 1–2.5 mm wide and not more than 2.5 times as long as wide; rays 9–20, 7–15 mm long, more or less tinged with lilac or purple. (n = 36) Late midsummer, early fall. Mountain woods; NB and Que to Minn, s to Pa and Ind, and in BR to n Ga, extending into RV and AP at least in Md and WVa. *A. multiformis* Burgess—S; *A. riciniatus* Burgess—S.

52. A. schreberi Nees. Similar to *A. macrophyllus*, but eglandular, with merely puberulent inflorescence, and normally with white rays; basal leaves tending to have a rectangular sinus; involucre 5–10 mm high, often very narrow, its bracts averaging narrower than in *A. macrophyllus* and sometimes a little loose (but not squarrose-tipped), those of the inner row much the longest; rays 6–14. (n = 9) Mostly late summer, early fall. Woods, less common than nos. 51 and 53; NH to Wis, s to Del, Md, WVa, Va (BR), Ky (CU), Tenn (RV), and Ala (CU or RV).

53. A. divaricatus L. Rhizomatous and colonial, 2–10 dm tall, usually without tufts of radical leaves; stem puberulent at least in the inflorescence, not glandular. Leaves rather thin, sharply serrate with usually mucronulate teeth, glabrous or with some long, mostly appressed hairs, especially along the main veins beneath; lower leaves ovate (often narrowly so), with cordate base, conspicuously acuminate, 4–20 × 2–10 cm, long petiolate, the lowest often smaller than those above and commonly deciduous, the middle and upper ones progressively less cordate, less petiolate, and more or less reduced. Inflorescence corymbiform, occasionally becoming elongate, its bracts few and often broad; involucre 5–10 mm high, its bracts firm, well imbricate, rounded to acute, very shortly green-tipped, otherwise mostly whitish, the outer 0.7–1.5 mm wide (or a little wider in var. *chlorolepis*) and seldom more than 2.5 times as long; rays 5–16 (20), white, or in var. *chlorolepis* sometimes lilac-tinged. (n in var. *divaricatus* = 9, 10) Late summer, fall. Woods, common and highly variable; NH to Del, DC, Pa, and s Ohio, and s chiefly in mt. provinces and on PP to SC, Ga, and Ala. Two vars.: Var. *chlorolepis* (Burgess) Ahles—R, with relatively large heads, the involucre 7–10 mm high, its bracts averaging broader and with not so pale chartaceous part as in the next var., the inner ones, as in no. 51, often purplish distally, the rays (10) 12–16 (20), 10–20 mm long. Woods and forest openings at high elev.; BR of sw Va, NC, Tenn, and Ga. *A. chlorolepis* Burgess—G, S. Var. *divaricatus*, with smaller heads, the involucre 5–8 mm high, the rays 5–10 (12), 5–15 mm long; widespread, but mostly not in the habitat of the previous var. *A. boykinii* Burgess—S; *A. castaneus* Burgess—

S; *A. excavatus* Burgess—S; *A. flexilis* Burgess—S; *A. stilletiformis* Burgess—S; *A. tenebrosus* Burgess—S. *A. glomeratus* Bernh.—F.

54. A. acuminatus Michx. Plants with slender creeping rhizomes, these often apically enlarged and scaly; stem 2–8 dm tall, loosely villous or puberulent and commonly somewhat viscidulous. Leaves mostly 10–22 below the inflorescence, thin, glabrous or slightly scabrous above, viscidulous-puberulent beneath, at least along the larger veins, the lower much reduced and soon deciduous, the others elliptic or obovate, acuminate, more or less sharply and saliently few-toothed, tapering to a sessile or shortly petiolar base, often forming a pseudowhorl beneath the inflorescence, the largest ones 6–17 × (1.5) 2–6 cm. Heads several or many in an open, slenderly branched, sparsely subulate-bracteate, mostly corymbiform inflorescence; involucre 6–9 mm high, its slender, sharply pointed bracts well imbricate, thin, pale stramineous to faintly greenish or somewhat purple-tinged, scarcely herbaceous, the outer ciliate-margined, otherwise glabrous; rays 10–21, white or faintly pinkish, 9–15 mm long. Achenes copiously glandular. (n = 9) Late summer, early fall. Woods; Nf and Que to Va, e WVa, NC, Tenn, and n Ga, in our range restricted to the mt. provinces, especially BR.

55. A. paternus Cronq. Stems 1.5–6 dm tall from a branched caudex, generally scabrous-puberulent in the inflorescence. Leaves ciliate-margined and sometimes hairy over the surface as well, at least some of them evidently toothed, the basal and lower cauline ones generally enlarged and persistent, broadly oblanceolate to obovate, elliptic, or even subrotund, petiolate, the blade 1.5–10 × 1–4.5 cm; cauline leaves becoming sessile upward, otherwise scarcely to strongly reduced. Inflorescence corymbiform, flat-topped, the heads commonly in small glomerules; involucre glabrous, narrow, 5–9 mm high, its bracts well imbricate, broad, with short spreading green tip, or the inner wholly chartaceous; rays 4–8, white (pink), 4–8 mm long; disk flowers 9–20, 4–5.5 mm long (dry), white or ochroleucous, or seldom lavender. Achenes densely sericeous, the upper hairs simulating an outer pappus; pappus bristles usually reddish, obscurely clavellate above. Summer. Dry woods; Me and Vt, s throughout all the coastal states to Ga (all), and w to s Ohio, WVa, e Ky (CU, IP), e Tenn (BR, RV, CU, encroaching onto IP), e Ala (CU, GC, RV, PP), and reputedly to Miss and the Fla panhandle. *Sericocarpus asteroides* (L.) B.S.P.—F, S.

56. A. tortifolius Michx. Stems 3–10 dm tall, few or solitary from a fibrous-rooted crown; herbage densely, uniformly, and rather coarsely spreading-puberulent throughout, or the upper surfaces of the leaves merely scaberulous-puberulent, the leaves commonly also minutely atomiferous-glandular and punctate (as seen at 20×). Leaves all cauline, numerous, small but relatively broad, mostly spatulate-obovate to broadly elliptic-oblanceolate and tapering to a narrow, sessile or shortly subpetiolar base, 1.5–4 × 0.6–1.5 (2) cm, 1.5–4 (5) times as long as wide. Inflorescence corymbiform, flat-topped, the heads individually short-pedunculate or often in small glomerules; involucre evidently puberulent, narrow, 6–8 mm high, its bracts well imbricate, firm, shortly green-tipped but scarcely herbaceous; rays (3) 4–5 (–7), white, 4–8 mm long; disk flowers 6–9 (–13), white or ochroleucous, mostly 6–8 mm long (dry). Achenes densely sericeous-strigose; pappus bristles white, obscurely clavellate above. Spring–summer. Sandy, moist or dry places on CP, often in pine woods; throughout Fla, n to NC, and w to La. *Sericocarpus bifoliatus* (Walt.) Porter—S; *S. acutisquamosus* (Nash) Small—S.

57. A. solidagineus Michx. Essentially glabrous, 2–6 dm tall from a short stout caudex. Leaves entire, linear or narrowly oblong (or the lower oblong-oblanceolate), 2–8 cm × 2–12 mm, more than 5 times as long as wide, sessile, or the lower petiolate, the lower only slightly if at all larger than those above. Inflorescence flat-topped, the heads tending to be in small glomerules; involucre cylindric, 4–7 mm high, its bracts well imbricate, broad and firm, the outer shortly green-tipped, the inner chartaceous and erose-margined; rays 3–6, white, 5–10 mm long; disk flowers 5–10, 4–5.5 mm long (dry), white or ochroleucous. Achenes more or less densely sericeous, pappus white, some of the bristles obscurely clavellate above. Summer. Dry woods and open ground; Mass, s throughout all the coastal states to Ga (all), w to WVa, s Ind, w Ky, w Tenn, and La. *Sericocarpus linifolius* (L.) B.S.P.—F, S.

58. A. umbellatus Miller. Plants (4) 10–20 dm tall from well-developed creeping rhizomes; stem generally glabrous or nearly so below the inflorescence. Leaves entire, scabrous or subglabrous above, glabrous (and often glaucous) beneath or short-hairy on the midrib and main veins only, finely reticulate beneath and with well-developed primary lateral veins, the lower reduced and soon deciduous, the others mostly 4–16 cm × 7–35 mm, sessile or nearly so and tapering to both ends, mostly rather narrowly elliptic or lance-elliptic and 4–6 times as long as wide, but sometimes shaped like the next sp. Heads more or less numerous (30–300 or more) in a corymbiform, generally dense inflorescence that tends to be flat-topped; involucre 3–5 mm high, glabrous or somewhat puberulent, its bracts well imbricate, greenish but scarcely herbaceous, relatively thin and slender, seldom any of them as much as 1 mm wide, acute or obtuse; flowers 23–54 per head, the rays (6) 7–14, white, 5–8 mm long, the disk flowers 16–40, ochroleucous. Achenes sparsely to evidently strigose or puberulent; pappus double, the inner of firm, rather sordid bristles, the larger ones more or less clavellate-thickened toward the tip, the outer less than 1 mm long. ($n = 9$) Late summer, early fall. Moist, low places; Nf to Minn, s generally to Va and Ky, and along the mts. to NC, Tenn, ne Ala, and n Ga, also extending onto IP in Tenn and PP in Ga. *Doellingeria umbellata* (Mill.) Nees—S.

59. A. sericocarpoides (Small) K. Schum. Rather similar to *A. umbellatus*; stems arising singly from a short rhizome, but the plants at least sometimes with long, slender, stoloniform rhizomes as well; leaves wider, up to 5 cm wide, mostly 1.5–4 times as long as wide, the principal ones mostly 3–11 × 1.5–4 cm; heads narrower and fewer-flowered, with 2–7 rays, 4–14 disk flowers, and 6–21 flowers in all; involucre 4–7 mm high, its bracts wider, the larger ones generally ca. 1 mm wide or a bit more, up to 1.5 mm; achenes glabrous or slightly hairy. Fall. In bogs and other wet, low places, especially in seepage areas in pine woods; on and near CP, from NC to n Fla, w to La and reputedly Tex. *A. umbellatus* var. *brevisquamus* Fern.—R, misapplied; *A. umbellatus* var. *latifolius* A. Gray; *Doellingeria humilis* (Willd.) Britt.—S, misapplied.

60. A. infirmus Michx. Plants 4–11 dm tall; stems mostly solitary from a short, fibrous-rooted crown, glabrous. Leaves entire, reticulate-veiny, glabrous or scabrous above, glabrous or more often short-hirsute on the midrib and main veins beneath, seldom sparsely short-hairy over the whole lower surface, the principal ones elliptic or elliptic-ovate, (3) 6–13 × 1.5–5 cm, 2–4 (5) times as long as wide, acute or acuminate, with shortly petiolar or subsessile base, the reduced lower ones obovate or broadly oblanceolate and rounded or obtuse, the lowest soon deciduous. Heads (2) 5–35 (75) in an open, sparsely or scarcely leafy, corymbiform inflorescence; involucre 4.5–7 mm high, glabrous or puberulent, its bracts well imbricate, relatively firm and broad, the larger ones commonly ca. 1 mm wide or a bit more, often longitudinally striate; heads 24–45-flowered, the rays 5–9, white, 6–10 mm long, broad and showy, the disk flowers ca. 18–36, ochroleucous. Achenes glabrous, pappus double, the inner of rather firm, sordid bristles, most of which are clavellate-thickened above, the outer less than 1 mm long. Midsummer, early fall. Woods; Mass to Va (all), WVa, and Ky (CU), and s mainly in the mt. provinces and on PP to Tenn, Ga, and Ala. *Doellingeria infirma* (Michx.) Greene—S.

61. A. reticulatus Pursh. Stems numerous, densely clustered on a branching caudex, 4–9 dm tall; plants densely downy-puberulent throughout, especially on the lower surfaces of the leaves. Leaves entire or remotely toothed, essentially cauline, the lowermost ones reduced, the principal ones mostly elliptic or elliptic-oblong to somewhat obovate, sessile or nearly so but hardly clasping, mostly 2.5–8 × 1–3.5 cm, 2–3.5 times as long as wide. Heads mostly 5–30 in a corymbiform inflorescence; involucre 5–10 mm high, evidently puberulent, its bracts slightly to strongly imbricate, acute or acuminate, with greenish but scarcely herbaceous midstrip; rays mostly 8–20, white or ochroleucous, narrow and elongate, commonly 10–15 mm long, often irregularly cleft at the tip; disk flowers numerous, yellow or reddish. Achenes densely hairy; pappus double, the inner of firm but rather slender bristles, not clavellate, the outer ca. 1 mm long or less and often inconspicuous. Spring–midsummer. Moist to rather dry, usually sandy places, often in pine woods; nearly throughout Fla, n onto CP of Ga and s SC. *Doellingeria reticulata* (Pursh) Greene—S.

62. A. linariifolius L. Plants with a short caudex, rarely with creeping rhizomes as well; stems several, wiry, 1–5 (7) dm tall, finely puberulent, becoming tomentose-puberulent upward. Leaves numerous and similar, firm, linear or nearly so, entire, minutely scaberulous, scabrous-ciliate on the margins, nerveless except for the prominent midrib, 1.2–4 cm × 1.2–4 mm, the lowest ones soon deciduous. Heads solitary, or more often several in a mostly corymbiform inflorescence; involucre 6–9 mm high, its bracts strongly imbricate, firm, keeled, greenish upward but scarcely herbaceous, finely scaberulous like the leaves, acute to broadly rounded, the inner more or less fringed-ciliate upward and usually purple-margined; rays 10–20, violet (white), broad and showy, 7–12 mm long; disk yellow or anthocyanic. Achenes copiously long-hairy; pappus double, the inner bristles elongate, firm, and tawny, the outer ca. 1 mm long or less. ($n = 9$) Late summer, fall. Dry ground and open woods, especially in sandy soil; Me and Que to Ga (CP) and n Fla, w to Wis, Mo, Ark, La, and Tex. *Ionactis linariifolius* (L.) Greene—S.

63. A. tenuifolius L. PERENNIAL SALT-MARSH ASTER. Glabrous perennial 2–7 dm tall, the stems often zigzag, arising singly from slender creeping rhizomes. Leaves few, fleshy, linear or nearly so, 4–15 cm × 1–5 (12) mm, the lower soon deciduous, the upper reduced and often bractlike. Heads several or many in an open inflorescence, or occasionally solitary, the peduncles subulate-bracteate; involucre 6–9 mm high, its bracts more or less imbricate, firm, chartaceous, sometimes greenish upward but scarcely herbaceous, acuminate to sharply acute, the inner often purple-margined; rays mostly 15–25, blue to pink or occasionally whitish, 4–7 mm long, sometimes circinately outrolled. Fall. Salt marshes along the coast; Mass to ne Fla, and along the Gulf Coast from c Fla to La.

64. A. bracei Britton. Much like *A. tenuifolius*, but not rhizomatous, the stems clustered on a short, densely fibrous-rooted crown; leaves averaging narrower, up to ca. 3 mm wide, the upper ones more often strongly reduced; heads relatively small, the involucre mostly 4–6 mm high, the rays mostly 10–15, 3–6 mm long, very often circinately outrolled. Winter. Coastal marshes and low pine woods; peninsular Fla, Bahama Isl., and Cuba.

65. A. spinosus Benth. Coarse, erect, freely branched, nearly leafless, broomlike or rushlike, glabrous, rhizomatous perennial mostly (0.6) 1–2.5 m tall, the stems striate, green and photosynthetic, some of the branches often modified into thorns 1–2 cm long. Foliage leaves, when produced, oblanceolate, up to 6 cm long and nearly 1 cm wide, but ephemeral, the leaves otherwise much reduced, linear or scalelike, and mostly less than 1 cm long. Heads terminating the slender branches; involucre 3.5–6 mm high, shorter than the disk, its bracts well imbricate, scarcely herbaceous, hyaline-margined and sometimes minutely erose-fringed; rays mostly 15–30, white, becoming circinately outrolled, 2–5 mm long. Achenes glabrous; pappus fine and soft, commonly equaling or surpassing the numerous disk flowers. (n = 9) Summer, early fall. Streambanks and bottomlands; La to s Calif, and n Mex. *Leucosyris spinosa* (Benth.) Greene—S.

66. A. subulatus Michx. Glabrous annual 1–15 dm tall from a short taproot. Leaves entire, linear or nearly so, up to 20 × 1 cm (lower leaves rarely broader and up to 3.5 cm in var. *cubensis*). Heads remotely solitary to usually several or very many in an open inflorescence; involucre 5–8 mm high, its usually well-imbricate bracts acuminate, somewhat greenish but scarcely herbaceous, frequently purplish toward the margins and tips, all except sometimes the outer with scarious or hyaline margins; rays mostly 15–50, anthocyanic to sometimes white, rather short, ascending or spreading at anthesis, but often becoming circinately outrolled at maturity or in drying. (n = 5, 10) Midsummer–fall, or into winter in Fla. Usually in moist, often maritime or otherwise saline habitats; along the coast from NB and s Me to Fla, w across the s part of our range (inland as well as coastal) to the Pacific, and s into S Am. A highly variable species, represented in our range by three vars.: var. *subulatus* and var. *ligulatus* both pass into var. *cubensis*.

Ligules very slender, mostly 0.2–0.5 mm wide (to 0.7 mm in var. *cubensis*), seldom more than 3 mm long, straight and spreading at anthesis, but at maturity (or in drying) commonly becoming circinately outrolled and not much if at all exceeding the pappus.

- Ligules at maturity (or when dried) very short, shorter than or about equaling the pappus, disk flowers relatively few, commonly 5–15; plants up to ca. 1 m tall, chiefly maritime and submaritime, growing in saline or sometimes nonsaline marshes and similarly wet places along and near the coast from s Me to c Fla (both coasts) and La (Annual salt marsh aster) .... var. *subulatus*.
- Ligules mostly a little longer, commonly equaling or more often a little surpassing the pappus when dry or mature; disk flowers few as in var. *subulatus*, or more numerous and up to ca. 22; plants up to ca. 1.5 m tall, mostly nonmaritime, often weedy, not confined to marshy habitats, not succulent; widespread in Latin Am, and extending into s US, as in Fla and s La. *A. inconspicuus* Less—S .... var. *cubensis* (DC.) Shinners.

Ligules broader, mostly (0.5) 0.7–1.5 mm wide, evidently surpassing the pappus, mostly (2) 3–6 mm long, straight or sometimes circinately outrolled at maturity; disk flowers variously few or up to ca. 40; plants up to ca. 1.5 m tall, mostly nonmaritime, often weedy, not confined to marshy habitats, not succulent; across s US from Fla to NM, and n through Ark and Okla into Mo, Kans, and s Neb. *A. exilis* Ell.—F, S, probably only misapplied .... var. *ligulatus* Shinners.

73. ERIGERON L. DAISY, FLEABANE

Herbs. Leaves alternate (sometimes all basal), entire to toothed or lobed. Heads 1–many, hemispheric to turbinate, radiate or seldom disciform or discoid, the rays pistillate and fertile, white or anthocyanic; involucral bracts narrow, herbaceous and equal to scarcely herbaceous and imbricate, but without *Aster*-like green tip; receptacle flat or nearly so, naked; disk flowers numerous, yellow, perfect and fertile; style branches flattened, with ventromarginal stigmatic lines and a short (in ours not more than ca. 0.3 mm), acute to more often obtuse, externally short-hairy appendage, or the appendage virtually obsolete in a few spp. such as *E. annuus* and *E. strigosus*. Achenes in our spp. 2- to 4-nerved. Pappus of several to many capillary bristles, with or without a short outer series of minute bristles or scales. Plants typically blooming in the spring and early summer (unlike *Aster*), but some weedy spp. continuing to bloom until fall. Cronquist, A. 1947. Revision of the North American species of *Erigeron*, north of Mexico. Brittonia 6: 121–302.

1 Pappus of the ray and disk flowers alike, of elongate capillary bristles, sometimes also with some short outer setae.
  2 Plants more or less erect (sometimes curved or decumbent at the base); stolons and rhizomes absent except in *E. pulchellus*; rays not circinately rolled.
    3 Plant bearing superficial rhizomes; disk corollas (fresh) mostly 4.5–6 mm long; rays ca. 1 mm wide, or more .... 1. *E. pulchellus*.
    3 Plant without rhizomes; disk corollas 1.5–4 mm long; rays (except in *E. vernus*) ca. 0.5 mm wide, or less.
      4 Achenes 4-nerved; rays mostly 25–40, 0.5–1.3 mm wide; plants subscapose .... 2. *E. vernus*.
      4 Achenes 2-nerved, rarely obscurely 4-nerved; rays mostly 60–400, ca. 0.5 mm wide or less; plants more or less leafy-stemmed, at least below the middle.
        5 Heads larger, the involucre 4–6 mm high, the rays mostly 150–400 and 5–10 mm long, deep-pink or rose-purple to white .... 3. *E. philadelphicus*.
        5 Heads smaller, the involucre 2.5–4 mm high, the rays mostly 60–250 and 2.5–5 (6) mm long, blue or less often white, seldom pink.
          6 Pappus simple or nearly so; stem spreading-hairy throughout or sometimes appressed-hairy above the middle (seldom the upper 2/3) .... 4. *E. quercifolius*.
          6 Pappus double, with some short outer setae in addition to the long, slender bristles; stem appressed-hairy at least above the middle, more spreading-hairy only toward the base (seldom up to 1/2 its length) .... 5. *E. tenuis*.
  2 Plants trailing or ascending, often with slender stolons or superficial rhizomes; rays usually circinately rolled distally, at least when dry .... 6. *E. myrionactis*.
1 Pappus of the ray and disk flowers unlike, that of the disk flowers composed of elongate, capillary bristles and very short, slender outer scales, that of the ray flowers lacking the capillary bristles; spp. commonly apomictic.
  7 Foliage ample; plants robust, mostly 6–15 dm tall; pubescence of the stem mostly long and spreading, or shorter and more appressed only near the top .... 7. *E. annuus*.
  7 Foliage sparse; plants more slender, mostly 3–7 (9) dm tall; pubescence of the stem mostly short and appressed, or short and spreading toward the base .... 8. *E. strigosus*.

1. E. pulchellus Michx. Biennial or short-lived perennial 1.5–6 dm tall from a short, fibrous-rooted caudex, and perennating by slender, stoloniform rhizomes. Basal leaves oblanceolate to suborbicular, 2–13 × 0.6–5 cm, commonly toothed; cauline leaves ovate to lanceolate or oblong, reduced upward. Heads solitary or few; involucre 5–7 mm high; disk 10–20 mm wide; rays 50–100, 6–10 mm long, ca. 1 mm wide or a little more, blue or occasionally pink, rarely white; disk corollas 4.5–6 mm long. Achenes 2-nerved, or seldom 4-nerved. Pappus simple. (n = 9) Spring. Woods, streambanks, and moist, open places; Me to Minn, s to s Ga, Miss, La, and Tex, more common northward and in the mts. Most of our plants belong to the widespread, evidently spreading-hairy var. *pulchellus*. The var. *brauniae* Fern., essentially glabrous except for the ciliate leaf-margins and glandular involucres, occurs locally in ne Ky (Lewis Co.) and adjacent Ohio.

2. E. vernus (L.) T. & G. Biennial or short-lived perennial 1.5–5 dm tall with a short, subsimple caudex and rosulate offsets, glabrous or the stem sparsely appressed-hairy. Basal leaves thick, oblanceolate to suborbicular, obtuse or rounded, 2–15 cm × 4–25 mm, denticulate; cauline leaves very few and small. Heads 1–many; involucre 3–4 mm high, glutinous or sometimes sparsely hairy; disk 5–11 mm wide; rays 25–40, white, 4–8 × 0.5–1.3 mm; disk corollas 2.5–3.8 mm long. Achenes 4-nerved. Pappus simple. (n = 9, 2n = ca. 27) Spring, or all year southward. Sandy or peaty places, sphagnum bogs, and pine barrens; CP from Va to s Fla and w to La.

3. E. philadelphicus L. Biennial or short-lived perennial 2–7 dm tall, pubescent with long, spreading hairs, varying to subglabrous. Basal leaves narrowly oblanceolate to obovate, coarsely crenate-toothed or lobed, rounded above, seldom more than 15 × 3 cm; cauline leaves clasping, mostly oblong or ovate, more or less reduced. Heads 1–many; involucre 4–6 mm high, more or less hirsute with flattened hairs, or subglabrous; disk 6–15 mm wide; rays very numerous, mostly 150–400, 5–10 mm long, 0.5 mm wide or less, deep pink or rose-purple to white; disk corollas 2.5–3.2 mm long. Achenes 2-nerved. Pappus simple, of ca. 20–30 bristles. (n = 9) Spring. Various habitats, somewhat weedy, varying in size with the habitat; widespread in N Am, and throughout our range except peninsular Fla.

4. E. quercifolius Lam. Resembling *E. philadelphicus*. Herbage more or less villous-hirsute, the hairs usually shorter, those of the upper part (up to 1/2, seldom 2/3) of the stem sometimes appressed. Stem slender, 1–4 (6) dm tall. Basal leaves oblanceolate to obovate, sinuately lobed to subentire, 1–14 cm × 5–40 mm; cauline leaves more or less reduced, sometimes clasping. Heads 1–many, the inflorescence more nearly naked than in no. 3; involucre 2.5–4 mm high, viscid-villous, the bracts darker than in no. 3; disk 5–10 mm wide; rays 100–250, up to 5 × 0.5 mm, blue or light blue-lavender, varying to sometimes white or seldom pinkish; disk corollas 1.5–2.5 mm long. Pappus of ca. 10–15 bristles, seldom with a few inconspicuous outer short setae. (n = 9, 18) Spring, sometimes continuing until fall, or all year southward. Moist, sandy places and pine woods; throughout Fla, n on AC to s Va and w on GC to Tex.

5. E. tenuis T. & G. Annual or biennial, 1–4 dm tall, averaging more slender and less hairy than no. 4, sometimes decumbent at the base. Stem moderately short-hirsute, the hairs of at least the upper half mostly appressed, those toward the base spreading. Basal leaves (sometimes soon deciduous) oblanceolate to suborbicular, with entire or irregularly toothed to irregularly lobulate blade mostly 1–3 × 0.5–1.5 cm, on a petiole 1–3 cm long; middle cauline leaves mostly narrowly oblong, sessile, entire or toothed, 1.5–4 × 0.3–1 cm. Heads 1–many; involucre 2.5–4 mm high, subglabrous to moderately hirsute with curved hairs, and sometimes finely glandular, its bracts often with somewhat expanded, very thin tip; disk 5–10 mm wide; rays 60–120, blue or white, 2.5–4.5 × 0.3–0.5 mm; disk corolla 2–2.7 mm long. Achenes 2-nerved. Pappus double, the inner of ca. 10–15 bristles, the outer (visible at 20×) of short setae. (n = 9, 18) Spring. Prairies, open woods, and sandy roadsides; common in Tex, extending e less commonly to La, w Miss, Ark, Mo, and Okla.

6. E. myrionactis Small. Trailing or ascending plants, often with slender stolons or superficial rhizomes; herbage more or less hirsute or villous-hirsute, the hairs of the leaves mostly more or less appressed, those of the stem and peduncles spreading or more commonly reflexed. Leaves obovate or spatulate, coarsely toothed or shallowly lobed, the blade mostly 1–4 × 0.5–2.5 cm, with cuneate, sessile base, or tapering to a petiole 1–4 cm long. Peduncles terminal or axillary, naked, erect, 0.5–2 dm long; involucre 6–8 mm high, villous-hirsute and more or less viscid; disk 8–13 mm wide; rays 150 or more, white or pinkish, 5–7 × ca. 0.3 mm, usually circinately rolled; disk corollas 3.5–4.5 mm long. Achenes 2-nerved. Pappus simple, of ca. 20–25 bristles. (n = 9) Spring, or continuing until fall. Sandy beaches and salt marshes along the coast; s Miss to La and s Tex.

7. E. annuus (L.) Pers. Annual or rarely biennial, 6–15 dm tall, amply leafy. Stem hirsute, the hairs spreading except near the top. Basal leaf blades elliptic to suborbicular, coarsely toothed, up to 10 × 7 cm, more or less abruptly long-petiolate; cauline leaves numerous, broadly lanceolate or broader, all except sometimes the uppermost usually sharply toothed. Heads several to very numerous; involucre 3–5 mm high, finely glandular and sparsely beset with long, flattened, transparent hairs; disk 6–10 mm wide; rays 80–125, white or rarely anthocyanic, 4–10 × 0.5–1.0 mm; disk corollas 2.0–2.8 mm long. Achenes 2-nerved. Pappus of the disk flowers double, with 10–15 bristles and several very short, slender scales (visible at 20×), that of the ray flowers of short scales only. (2n = 27, 54) Spring, continuing until fall, or all year southward. A weed in disturbed sites, especially in moist or fertile soil, throughout our range (and most of US and s Can), but more common northward.

8. E. strigosus Muhl ex Willd. Annual or rarely biennial, 3–7 (9) dm tall, sparsely leafy, rather shortly appressed-hairy, or the stem more spreading-hairy toward the base. Basal leaves mostly oblanceolate to elliptic, entire or toothed, not more than 15 cm (petiole included) × 2.5 cm; cauline leaves linear to lanceolate, mostly entire. Heads several to very many; involucre 2–4 mm high, obscurely glandular and more or less short-hairy; disk 5–10 mm wide; rays 50–100, white or seldom somewhat anthocyanic, up to 6 mm long, 0.4–1.0 mm wide; disk corollas 1.5–2.5 mm long. Achenes and pappus as in no. 7. (2n = 18, 27, 36, 54) Spring, often continuing until fall. A weed in disturbed sites, often in poorer or drier soil than no. 7, over most of the US and s Can, and found throughout our range. Two ill-defined but geographically significant vars. in our range:

Heads tiny, the involucre only 2–3 mm high; inflorescence diffuse and subnaked, the peduncles often flexuous; coastal states, chiefly on CP, from NJ to Fla and Tex .... var. *beyrichii* (Fischer & Meyer) A. Gray.

Heads averaging larger, the involucre (2.5) 3–4 mm high; inflorescence not diffuse, or if so, then somewhat leafy; widespread, but seldom on CP s of Va .... var. *strigosus*.

74. CONYZA L., nom. conserv.

Herbs, often weedy, ours all annual with a more or less well-developed taproot. Leaves alternate, in our spp. chiefly cauline, narrow, and entire or merely toothed or remotely lacinate-toothed. Heads several to numerous, mostly rather small, inconspicuously radiate or disciform; involucral bracts more or less imbricate, scarcely herbaceous; receptacle flat or nearly so, naked; pistillate flowers numerous, the corolla very slender, tubular, and rayless, or in some spp. (including all ours) with a very short, narrow, inconspicuous, white or purplish ray barely if at all exceeding the pappus, disk flowers few (in our spp. not more than ca. 21), tubular and perfect, yellowish or somewhat anthocyanic; style branches flattened, with ventromarginal stigmatic lines and short, blunt, externally short-hairy appendage. Achenes 1- to 2-nerved, or nerveless. Pappus of capillary bristles, sometimes with a short outer series.

1 Plants simple or nearly so up to the inflorescence, with a well-defined central axis, often more than 3 dm tall.
  2 Involucre 4–6 mm high, copiously short-hairy; pistillate flowers mostly (50) 70–200 or more .... 1. *C. bonariensis*.
  2 Involucre 3–4 mm high, glabrous or nearly so; pistillate flowers mostly 25–40 .... 2. *C. canadensis*.
1 Plants diffusely branched from near the base, without a central axis, mostly 1–3 dm tall .... 3. *C. ramosissima*.

1. C. bonariensis (L.) Cronq. Annual weed 1–10 dm tall or more, copiously and loosely hairy, habitally like no. 2, or often with some of the lateral branches elongate and overtopping the central axis. Lower leaves in robust plants sometimes up to 15 × 2 cm. Heads larger than in no. 2, the disk often more than 1 cm wide; involucre 4–6 mm high, copiously short-hairy; pistillate flowers (50–) 70–200 or more, with a very short or scarcely developed ligule up to 0.5 mm long, this generally surpassed by the style and equaling to more often surpassed by the often tawny to reddish pappus. (n = 27) Spring–summer. A weed in waste places, widespread in tropical Am, probably only intro. in our range, on CP n to Va, rarely elsewhere. *C. floribunda* H.B.K.—G; *Erigeron bonariensis* L.—F, R; *Leptilon bonariense* (L.) Small—S; *L. linifolium* (Willd.) Small—S.

2. C. canadensis (L.) Cronq. HORSEWEED. Annual weed 1–15 dm tall, simple or nearly so to the inflorescence, subglabrous to coarsely spreading-hirsute, the leaves commonly with some coarse marginal cilia toward the base even when the plant is otherwise glabrate. Leaves numerous, oblanceolate to linear, toothed (especially the lower) or entire, gradually reduced upward, the cauline ones up to 8 cm × 8 mm, the basal ones larger and broader but generally deciduous. Heads, except on depauperate plants, numerous in a long and open inflorescence; involucre 3–4 mm high, glabrous or with a few small, scattered hairs, the bracts strongly imbricate, brown or with brown midvein; pistillate flowers mostly 25–40 (commonly ca. 34), the white or sometimes pinkish ligule 0.5–1.0 mm long, equaling or shortly surpassing the style and pappus. (n = 9) Midsummer, fall, or all year southward. A weed in waste places throughout the US and s Can, s to tropical Am, and intro. elsewhere. We have two well-marked vars. Var. *pusilla* (Nutt.) Cronq., chiefly on and near CP, is subglabrous, and has some or all of the involucral bracts minutely purple-tipped. *Erigeron canadensis* var. *pusillus* (Nutt.) Ahles—R; *E. pusillus* Nutt.—F; *Leptilon pusillum* (Nutt.) Britton—S. Var. *canadensis*, widespread in e US but only sporadically on CP, has the stem coarsely spreading-hirsute and lacks the purple tips on the bracts. *Erigeron canadensis* L.—F; *E. canadensis* var. *canadensis*—R; *Leptilon canadensis* (L.) Britton—S.

3. C. ramosissima Cronq. Diffusely branched, slender, more or less hairy, 1–3 dm tall, with no well-defined central axis. Leaves narrowly linear, up to 4 cm × 2 mm, the uppermost reduced to mere bracts. Heads numerous, much like those of no. 2, often covering the broad, much-branched summit of the plant; involucre 3–4 mm high, the outer bracts short-hairy, the inner glabrous; rays minute, purplish, about equaling or slightly exceeding the pappus. (n = 9) Summer. A weed in waste places, especially in sandy soil or along streams; Ohio to ne Ala (CU), w to Minn, Ark (OZ), s La (GC), and Tex. *Erigeron divaricatus* Michx.—F; *Leptilon divaricatum* (Michx.) Raf.—S.

75. BACCHARIS L. GROUNDSEL TREE

Dioecious shrubs. Leaves alternate, simple. Fertile heads with more or less numerous pistillate flowers that have tubular-filiform, eligulate corolla; sterile heads with more or less numerous functionally staminate flowers, the ovary abortive, the style branches sometimes connate; flowers white to yellowish or greenish; involucral bracts subequal to strongly imbricate, chartaceous or subherbaceous; receptacle flat or merely convex, naked. Achenes usually somewhat compressed, several-nerved. Pappus of numerous capillary bristles, those of the sterile heads fewer and shorter than those of the fertile ones, often shorter than the corolla and somewhat fringed distally.

1 Leaves linear or nearly so, mostly 1–3 (4.5) mm wide . . . . 1. *B. angustifolia*.
1 Leaves broader, most of them at least (0.5) 1 cm wide.
  2 Leaves entire, spatulate-obovate, 1.5–3 (3.5) cm long (petiole included) . . . . . . 2. *B. dioica*.
  2 Leaves mostly with a few coarse teeth, larger, at least some of them generally more than 3.5 cm long overall (to 8 cm).
    3 Heads mostly or all in glomerules scattered along the leafy branches, not forming a distinctive terminal inflorescence . . . . 3. *B. glomeruliflora*.
    3 Heads in terminal, leafy-bracteate inflorescences along the major branches, individually sessile in small glomerules, or many of them pedunculate . . . 4. *B. halimifolia*.

1. B. angustifolia Michx. Much-branched, glutinous and shining, glabrous shrubs up to 4 m tall. Leaves numerous, succulent, linear, mostly 2–6 cm long and 1–3 (4.5) mm wide, 1-nerved or somewhat 3-nerved, entire or seldom with 1–few coarse, divergent teeth. Heads short-pedunculate or in small glomerules toward the branch tips; pistillate involucres 3.5–5 mm high, broadly rounded at the base, the bracts imbricate in several series; pistillate heads with the pappus much elongate at maturity, surpassing the involucre and the corollas. Late summer, fall. Salt marshes and dune hollows along the seacoast; NC to both coasts of peninsular Fla, and w to La.

2. B. dioica Vahl. Much-branched shrubs 0.5–3 m tall, glabrous and glutinous, or slightly scurfy. Leaves numerous, thick and firm, with obscure lateral veins, entire, obovate-spatulate, broadly rounded distally, varying to broadly obtuse, submucronate, or slightly retuse, tapering to a shortly petiolar base, 1.5–3 (3.5) cm long overall and 7–18 mm wide. Heads in small, pedunculate, axillary and terminal glomerules toward the branch tips, or some heads individually pedunculate; pistillate involucres 5–7 mm high, narrow-based, with strongly imbricate bracts in several series, the inner often recurved-spreading at maturity; pappus of the pistillate heads not notably elongate. Fall. Hammocks, and dune hollows near the shore; s Fla to WI.

3. B. glomeruliflora Pers. Much like no. 4, but the heads mostly or all in sessile or pedunculate axillary glomerules scattered along the branches, not forming a distinctive terminal inflorescence; upper leaves subtending the glomerules, often somewhat reduced and less toothed than those below, but still well developed; inner involucral bracts broader and blunter than those of no. 4. Mostly fall. Hammocks, moist woods, and swamps near the coast; se NC, s around Fla and n to the panhandle, and also inland in s Fla.

4. B. halimifolia L. GROUNDSEL TREE, SEA MYRTLE, MANGLIER. Freely branched shrubs mostly 1–3 (4) m tall, glabrous and somewhat glutinous, the branches and involucres sometimes minutely scurfy. Leaves thick and firm, puncticulate, pinnately veined or somewhat triplinerved, short-petiolate, the blade elliptic to broadly obovate, up to ca. 6 × 4 cm, coarsely few-toothed, especially distally, those of the inflorescence smaller, narrower, and mostly entire. Heads in numerous, small, pedunculate clusters or glomerules, or often many of them individually pedunculate, forming terminal, leafy-bracteate inflorescences on the major branches; pistillate involucres 4–6.5 mm high, the bracts strongly imbricate in several series, the inner ones narrow and more or less acute; pistillate heads with the pappus much elongate at maturity, conspicuously surpassing the involucre and corollas. (n = 9) Fall. Marshes, beaches, and hammocks, especially near the seashore; coastal states from Mass to Fla and Tex, chiefly on CP, but sometimes on PP, and inland to OU in Ark, [WI].

76. BOLTONIA L'Her.

Essentially glabrous, fibrous-rooted perennial herbs, often stoloniferous or rhizomatous. Leaves alternate, entire or nearly so, narrow, often more or less linear, mostly sessile, sometimes (not in ours) decurrent. Heads few to numerous, radiate, with mostly 20–60 pistillate and fertile, white to blue or pink rays; involucral bracts subequal to more often evidently imbricate, scarious-margined, with green or greenish midrib or tip; receptacle small, hemispheric or conic, naked; disk flowers numerous, perfect and fertile, yellow; style branches flattened, with ventromarginal stigmatic lines and a short, lanceolate, externally papillate hairy appendage. Achenes obovate, strongly flattened, evidently to obscurely wing-margined. Pappus of several minute bristles and 2 (4) somewhat longer awns, these commonly reduced or wanting in the ray achenes, sometimes also in the disk.

1 Inflorescence narrow and few-headed to broader and corymbiform, but not diffusely branched, always more or less leafy-bracteate, at least some of its leaves more than 1 cm long; heads larger, the rays mostly (5) 8–15 mm long, the disk mostly 6–10 mm wide; pappus awns often more than 1 mm long .... 1. *B. asteroides*.
1 Inflorescence diffusely branched, leafy-bracteate or not; heads mostly smaller, the rays mostly 5–8 mm long, the disk mostly 3–6 (8) mm wide; pappus awns less than 1 mm long, or wanting.
  2 Inflorescence more or less leafy-bracteate, at least some of its leaves more than 1 cm long; some of the leaves present at anthesis 1 cm wide or more .... 2. *B. caroliniana*.
  2 Inflorescence merely subulate-bracteate, its bracts commonly less than 1 cm long; leaves present at anthesis seldom more than 0.5 (1) cm wide .... 3. *B. diffusa*.

1. B. asteroides (L.) L'Herit. Plants 3–15 (20) dm tall. Leaves broadly linear to lanceolate or narrowly lance-elliptic, reduced upward, the larger ones 5–15 cm × 5–20 mm. Inflorescence commonly corymbiform, somewhat leafy, the leaves narrow, at least some of them more than 1 cm long; involucral bracts more or less imbricate; rays white to pink, purple, or blue, (5) 8–15 mm long; disk mostly 6–10 mm wide. Achenes evidently wing-margined. Pappus awns usually well developed (though short), (0.2)

0.6–2 mm long. (n = 9, 18) Midsummer–fall. Moist or wet places of various sorts; NJ to n Fla, w to ND, Okla, s La, and e Tex, but missing from much of the Appalachian region; escaped and locally established elsewhere. Three wholly intergradient geographic vars.:

Heads relatively few, seldom more than 25; plants averaging smaller and less leafy than the other vars.; coastal states from NJ to Fla and La, but not restricted to CP; var. *glastifolia* (Hill) Fern.—F. .... var. *asteroides*.
Heads more numerous, more than 25 except in occasional depauperate plants; mostly not in coastal states.
  Involucral bracts mostly linear or nearly so and acute or acutish; chiefly upper Miss valley (as broadly defined) and ne to w end of Lake Erie, only marginally in our range, as in Ky; *B. latisquama* var. *recognita* Fern. & Grisc.—F. .... var. *recognita* (Fern. & Grisc.) Cronq.
  Involucral bracts, except the inner, spatulate and rounded-obtuse; OU and OZ of w Ark and e Okla, n to Mo, Kans, and seldom beyond; *B. latisquama* A. Gray—F. .... var. *latisquama* (A. Gray) Cronq.

2. B. caroliniana (Walter) Fern. Plants mostly 8–20 dm tall. Leaves lanceolate to linear-oblanceolate, or the lower broader, tapering to both ends, the larger ones mostly 8–15 cm × 1–2.5 cm, some of those present at full anthesis generally at least 1 cm wide. Inflorescence diffusely branched, somewhat leafy, the leaves mostly narrow and more than 1 cm long; involucral bracts narrow, acute, subequal or slightly imbricate; rays white or sometimes lilac, mostly 5–8 mm long; disk mostly 3–6 (8) mm wide. Achenes scarcely or only very narrowly wing-margined. Pappus awns none or minute, less than 0.5 mm long. (n = 9) Late summer, fall. Moist lowlands; AC and PP of SC, NC, and s Va. *B. ravenelii* Fern. & Grisc.—F, with unusually large heads.

3. B. diffusa Elliott. DOLL'S DAISY. Plants mostly 5–15 dm tall. Leaves mostly linear or nearly so, the larger ones mostly 3–11 cm long and up to 0.5 or 1 (2) cm wide, but the lower leaves commonly deciduous, so that the leaves present in well-grown flowering plants are rarely more than 0.5 (1) cm wide. Inflorescence diffusely branched, merely subulate-bracteate, the bracts commonly less than 1 cm long; involucral bracts narrow, acute or acuminate, evidently imbricate; rays white or lilac, mostly 5–8 mm long; disk mostly 3–6 mm wide. Achenes evidently wing-margined. Pappus awns less than 1 mm long, or sometimes obsolete. (n = 9, 18) Midsummer–fall. Moist or wet to sometimes rather dry places; peninsular Fla, n through Ga (but only rarely to SC and NC) and Tenn to Ky, w to s Ill, Mo, Ark, e Okla, La, and e Tex. *B. diffusa* var. *interior* Fern. & Grisc.—F.

77. CHAETOPAPPA DC.

1. C. asteroides DC. Slender, freely branched, taprooted pubescent annual 5–30 cm tall. Leaves alternate, mostly linear or subulate and less than 1 cm long, the few basal ones a little larger and oblanceolate or spatulate. Heads terminating the branches, small, cylindric or narrowly campanulate, radiate, with 5–15 pistillate and fertile, white or pink rays 3–5 mm long; involucre 3–5 mm high, its bracts imbricate, thin, green, with scarious margins and tip; receptacle small, flat, naked; disk only 2–3 mm wide, its flowers few and perfect; style branches short, flattened, with ventromarginal stigmatic lines and a short, blunt, externally short-hairy appendage. Achenes slender, 5-nerved. Pappus of 5 thin, short scales, alternating with as many long, slender awns. (n = 8) Spring–midsummer. Dry, often sandy places; sw Mo and se Kans, s and w to La, Tex, and Mex.

78. APHANOSTEPHUS DC.

1. A. skirrhobasis (DC.) Trelease. Taprooted annual herb (seldom overwintering) 1–5 dm tall, simple (especially when first beginning to flower) to often much branched, shortly spreading-hairy throughout. Leaves alternate, entire or toothed to pinnatifid, the lower ones with bluntly oblanceolate to oblong-elliptic blade up to 6 cm long on a petiole often of equal length, the others sessile or nearly so and gradually reduced upward. Heads terminating the stem and branches, radiate; involucre mostly 6–8 mm high, its bracts with green or greenish midstrip and chartaceous or scarious margins, imbricate in several series; receptacle naked, nearly flat to broadly convex or low-conic; rays mostly 15–45, pistillate and fertile, white above, usually at least partly red or purple beneath; disk mostly 7–13 mm wide; disk flowers numerous, perfect and fertile, yellow, the base of the corolla becoming bulbous-thickened and indurated in fruit; style branches flattened, with ventromarginal stigmatic lines and a short, broadly triangular, externally short-hairy appendage. Achenes columnar, quadrate, evidently grooved, glabrous or nearly so. Pappus an irregular, scaly, lacerate crown, or of unequal, acute scales less than 1 mm long. (n = 3) Spring–fall. Open, often sandy places; Tex and NM to Kans, Mo, and possibly Ark, and along the Gulf Coast from n Fla to n Mex. Most or all of our material belongs to the morphologically poorly characterized but ecogeographically significant var. *thalassius* Shinners, a plant of beaches and dunes along the Gulf Coast, typically low and spreading, densely and softly pubescent, the pubescence of the involucre also fine and soft. The more strictly inland var. *skirrhobasis*, highly variable in habit but tending to be more coarsely pubescent, particularly on the involucre, is doubtfully reported from Washington Co., Ark; the locality may be in error. Shinners, L. H. 1946. Revision of the genus *Aphanostephus* DC. Wrightia 1: 95–121.

79. ASTRANTHIUM Nutt. DAISY

1. A. integrifolium (Michx.) Nutt. Caulescent, obscurely to evidently taprooted annual, 0.4–4.5 dm tall, sparingly to freely branched (often from near the base) or simple, sparsely hairy, more densely so at the base. Leaves alternate, simple, small, entire, up to 8 × 2 cm, the lower oblanceolate to spatulate, the others mostly linear or elliptic. Heads 1–many, mostly long-pedunculate, with 8–22 pistillate and fertile, mostly blue or purple rays 5–10 mm long; involucre 4–5 mm high, of 2–3 series of subequal bracts, these green with membranous or scarious margins; receptacle convex, cushionlike, naked; disk flowers numerous, perfect and fertile, yellow; style branches with short, ventromarginal stigmatic lines and well-developed, slender, externally papillate-hairy appendage. Achenes obovate, somewhat compressed, 2-nerved, minutely glandular-glochidiate. Pappus obsolete, represented only by a minute ring set well in from the lateral margins of the achene. (n = 8) Spring. Wooded, often sandy places, or on barrens or along roadsides; essentially Ozarkian sp. occurring from Tex to Kans, Mo, and Ark (OU, OZ), and e irregularly to Miss (ME), Ky (IP), Tenn (IP), and nw Ga (RV). DeJong, D. C. D. 1965. A systematic study of the genus *Astranthium* (Compositae, Astereae). Publ. Mus. Michigan State Univ. Biol. Ser. 2: 429–528.

80. BELLIS L. EUROPEAN or ENGLISH DAISY

1. **B. perennis** L. Fibrous-rooted, scapose perennial, more or less spreading-hairy. Leaves simple, basically alternate, elliptic to obovate or orbicular, the blade dentate or denticulate, up to ca. 4 cm long and 2 cm wide, narrowed to a petiole of equal or greater length. Scapes 5–15 (20) cm tall. Heads solitary, radiate, with numerous pistillate and fertile, white to pink rays ca. 1 cm long or less; involucral bracts commonly ca. 13, herbaceous, equal, essentially uniseriate, 4–6 mm long; receptacle naked, conic, elongating somewhat in age; disk 5–10 mm wide; disk flowers numerous, perfect and fertile, yellow; style branches with ventromarginal stigmatic lines and a short, ovate, externally papillate-hairy appendage scarcely as long as wide. Achenes compressed, 2-nerved, glochidiate-puberulent. Pappus obsolete. (n = 9) Spring, continuing more or less through the summer. Native of Europe, sometimes planted in lawns and flower borders, and more or less established as a weed in lawns and waste places across n US, in our range found only along the n, as in Va and the mts. of NC.

81. INULA L.

1. **I. helenium** L. ELECAMPANE. Coarse perennial herb, sometimes 2 m tall, the stem finely spreading-hairy. Leaves alternate, simple, irregularly and shallowly dentate, densely velvety beneath, sparsely spreading-hairy or subglabrous above, the lower long-petioled and elliptic, with blade sometimes 5 × 2 dm, the upper smaller, becoming ovate, sessile and cordate-clasping. Heads few, pedunculate, yellow, radiate, large, the disk broadly hemispheric, 3–5 cm wide, the involucre 2–2.5 cm high; outer bracts broad, herbaceous, densely short-hairy; inner bracts narrow, subscarious, glabrous; receptacle naked, flat or nearly so; rays numerous, pistillate and fertile, slender, mostly 1.5–2.5 cm long; disk flowers numerous, perfect, with tubular, 5-toothed corolla; anthers long-tailed at the base; style branches flattened, externally slightly papillate distally, with well-developed ventromarginal stigmatic lines extending all the way around the rounded tip. Achenes glabrous, columnar and more or less distinctly quadrangular, multistriate. Pappus a single series of numerous capillary bristles, united at the base. (n = 10) Spring–summer. Introduced from Europe; cult., escaped, and widely naturalized in fields and waste places in n US, s in our range to NC and Tenn.

Two related spp., *I. viscosa* (L.) Aiton and *Pulicaria arabica* (L.) Cass., both with much smaller heads and leaves, were collected many years ago as ballast weeds at Pensacola, Fla, but do not appear to have persisted. *I. viscosa* has a pappus of a single series of numerous, distinct, capillary bristles; *Pulicaria arabica* has a double pappus, the outer a short, lacerate crown, the inner a few firm capillary bristles. These two spp. appear in Small as *Cupularia viscosa* (Ait.) Godr. & Gren., and *Vicoa auriculata* Cass., respectively, the latter name misapplied.

82. FILAGO L., nom. conserv. COTTON ROSE

1. F. germanica (L.) Hudson. HERBA IMPIA. Slender, white-woolly annual 0.5–4 dm tall, simple below or branched from near the base, the stem terminated by a dense cluster of sessile heads, this usually subtended by several leafy branches terminating in similar clusters, and these often again proliferous. Leaves numerous and small, erect or nearly so, alternate, more or less linear, 1–2 cm × 1–3 mm. Heads disciform, white or whitish, rather narrow, mostly 3.5–4.5 mm high, woolly only toward the base; involucral bracts imbricate, with green midvein and yellowish-scarious or -hyaline margins, passing into the largely scarious and brownish-stramineous bracts of the shortly cylindric-obconic receptacle, the involucral bracts and the outer receptacular bracts each terminating in a short, shining bristle; pistillate flowers with tubular-filiform corolla, borne in several series, the outer ones epappose and each subtended by a boat-shaped bract, the inner ones bractless and with a pappus of capillary bristles; disk flowers few, often only 2, perfect and fertile, with a capillary pappus; anthers shortly tailed; style branches short and very slender, externally minutely papillate, with ventromarginal stigmatic lines extending to the rounded tip. Achenes tiny, 0.5 mm long, sparsely glandular-papillate, obscurely nerved. (n = 14) Summer. Native of Europe, intro. in disturbed habitats here and there from NY to Ohio, s to NC and reputedly Ga. *Gifola germanica* (L.) Dumort.—S.

83. EVAX Gaertn. RABBIT TOBACCO

Small, white-woolly annuals, often diffusely branched. Leaves small, alternate, simple, entire. Heads disciform, in small, terminal glomerules, so thoroughly embedded in wool that the involucre can be discerned only by careful dissection; involucre small, its bracts largely hyaline-scarious, in ours partly greenish; receptacle convex (in ours) to conic or columnar, chaffy throughout, its bracts in ours as long as or longer than the involucral bracts; pistillate flowers in several series, with short, filiform-tubular corolla and exserted, bifid style; disk flowers few (2–5 in our spp.), central, in ours and most other spp. functionally staminate, with undivided style and vestigial or no ovary; anthers shortly tailed at the base. Achenes small, nerveless, mostly glabrous, ours minutely papillate and slightly obcompressed. Pappus none.

1 Some of the bracteal leaves within the glomerule obviously protruding;
receptacular bracts mostly 2.5–3.5 mm long .... 1. *E. prolifera*.
1 None of the bracteal leaves obviously protruding from within the glomerule;
receptacular bracts mostly less than 2 mm long .... 2. *E. multicaulis*.

1. E. prolifera Nutt. ex DC. Slender annual up to 15 cm tall, simple below or branched from the base, and in well-developed plants commonly with 1–several slender branches arising at the base of the terminal cluster of heads and surpassing the main stem, these branches sometimes again branched in the same way ("Herba Impia" habit). Leaves numerous, ascending, oblong-spatulate or oblanceolate, 5–15 mm long. Heads in terminal glomerules, each glomerule commonly 7–20 mm thick, loosely subtended by oblong to lance-ovate leaves 6–12 mm long that collectively simulate an involucre; some of the leafy bracts subtending the heads within a glomerule obviously protruding from the main woolly mass; involucral bracts 1.5–2 mm long, soft, rather narrow, hyaline-scarious with a greenish, distally slightly expanded midvein; receptacular bracts, or most of them, larger and firmer than those of the involucre, mostly 2.5–3.5 mm long, smooth and shining below, woolly near the tip, collectively easily deciduous with their included flowers; staminate corolla borne on a definite stipe (vestigial ovary) ca. 0.5 mm long. Achenes planoconvex, thus with 2 more or less definite margins. Spring. Prairies and other open, dry or vernally moist places; Tex to Colo and SD, e to Ark, La, and the chalk prairies of Miss and Ala; intro. in SC.

2. E. multicaulis DC. Similar in aspect to no. 1, but the branching less consistently or less clearly of the Herba Impia pattern; leaves subtending the glomerules shorter, 3–10 mm long; bracteal leaves among the heads of a glomerule shorter, not protruding; involucral bracts broader and often shorter, often spatulate-obovate, with a large distal or subdistal green area; receptacular bracts not enlarged, not especially firm, not collectively deciduous, mostly less than 2 mm long; staminate flowers sessile, without an ovarian stipe; achenes elliptic in cross section, marginless. Spring. Open, dry or vernally moist places; sw US, e to La and irregularly intro. e to Ga and SC. *Filaginopsis nivea* Small—S.

84. PLUCHEA Cass. MARSH-FLEABANE

Aromatic herbs or shrubs. Leaves alternate, simple, entire or merely toothed. Heads generally more or less numerous in a flat-topped to elongate-paniculate inflorescence, disciform, the flowers white to yellow or pink-purple; involucre ovoid to broadly campanulate or hemispheric, its bracts firm, in several series, usually more or less imbricate; receptacle flat, naked; outer flowers pistillate, in several series, the filiform corolla shorter than the style; central flowers few, functionally staminate, often with undivided style; anthers filiform-tailed. Achenes tiny, 4–6-angled. Pappus a single series of capillary bristles. (x = 10) Godfrey, R. K. 1952. *Pluchea*, section *Stylimnus*, in North America. J. Elisha Mitchell Soc. 68. 238–271. Gillis, W. T. 1977. *Pluchea* revisited. Taxon 26: 587–591.

1 Shrubs .... 1. *P. symphytifolia.*
1 Herbs.
 2 Stem evidently winged by the decurrent leaf bases .... 2. *P. suaveolens.*
 2 Stem not winged, the leaf bases not decurrent.
  3 Leaves sessile and broad-based, commonly more or less clasping.
   4 Heads relatively large and broad-bracted, the involucre mostly 9–11 mm high, its middle bracts mostly 2–3 mm wide and up to 2 (2.5) times as long as wide .... 3. *P. longifolia.*
   4 Heads smaller and with narrower bracts, the involucre mostly 5–8 mm high, its middle bracts less than 2 mm wide and at least 2.5 times as long as wide.
    5 Corollas ochroleucous .... 4. *P. foetida.*
    5 Corollas pink-purple .... 5. *P. rosea.*
  3 Leaves petiolate, or tapering to a slender base.
   6 Involucre evidently pubescent with short, several-celled, glandular-viscid hairs; mainly in salt marshes .... 6. *P. odorata.*
   6 Involucre merely atomiferous-glandular, sometimes very sparsely so; not in salty places .... 7. *P. camphorata.*

1. P. symphytifolia (Miller) Gillis. Much branched shrub 1.5–4 m tall, shortly glandular-tomentose, or the leaves sometimes merely glandular, especially on the upper surface. Leaf blades firm, green above, pale beneath, elliptic to oblong-obovate, mostly 5–20 × 2–8 cm, entire or often with very small callous teeth; petioles 1–2.5 cm long. Heads in a dense, compound, corymbiform inflorescence; involucre 4.5–6 mm high, its middle and outer bracts glandular-tomentose, much wider than the subglabrous inner ones; disk broad in relation to its height, 5–10 mm wide; corollas pink-lavender. (n = 10) Late winter–spring. Intro. along roadsides and borders of hammocks in s Fla; native to Mex, WI, and n S Am. *P. odorata* Cass—S, misapplied.

2. P. suaveolens (Vell.) Kuntze. Coarse, fibrous-rooted, usually branching perennial herb 5–20 dm tall, finely hirtellous-strigose and glandular, or the younger parts almost sericeous-tomentulose; stems evidently winged by the long-decurrent leaf bases. Leaves numerous, finely callous-toothed, lanceolate to lance-elliptic or the lower more spatulate-oblanceolate, the better developed ones mostly 5–15 × 1–3 cm, tapering to the base. Heads numerous in broad, more or less flat-topped or rounded, corymbiform terminal cymes; involucre 4–7 mm high, its bracts finely glandular-hairy, not much imbricate; disk relatively broad and flat, mostly 8–15 mm wide; corollas white to rose-purple. (n = 10) Midsummer. Moist or wet, sunny places; adventive from S Am on ballast in Fla and Ala, but not recently collected. *P. quitoc* DC.—S.

3. P. longifolia Nash. Much like *P. foetida*, but more robust, with larger leaves and larger heads with wider bracts. Plants mostly 6–15 dm tall. Leaves numerous and crowded, coarsely and irregularly serrate, often glandular-villous beneath, the larger ones mostly 8–20 × 3–6 cm. Involucre mostly 9–11 mm high, evidently higher in relation to its width than in *P. foetida*, its middle and outer bracts notably broad, the middle ones commonly 2–3 mm wide and up to 2 (2.5) times as long as wide; disk mostly 8–10 mm wide. Midsummer–fall. Swamps, marshes, and lake shores; sc to ne Fla.

4. P. foetida (L.) DC. Coarse perennial herb from a fibrous-rooted crown or short rhizome, mostly 4–10 dm tall, glandular and often somewhat cobwebby-puberulent; stem and to some extent also the leaves tending to be somewhat anthocyanic in life. Leaves oblong to elliptic, lance-ovate, or ovate, mostly 4–10 (13) × 1–4 (4.5) cm, sharply callous-denticulate, reticulate-veiny, rounded to acute at the tip, closely sessile, broad-based and generally more or less clasping. Heads several or many in a short and broad, often flat-topped inflorescence; involucre mostly 5–8 mm high, glandular and often somewhat cobwebby or obscurely short-hairy, its bracts imbricate in several series, the outer much shorter than the inner, obtuse or merely acute or often abruptly apiculate, the middle ones broader than the inner but less than 2 mm wide and at least 2.5 times as long as wide; inner bracts stramineous or occasionally anthocyanic; disk 6–12 mm wide, broad in relation to its height; corollas ochroleucous. (n = 10) Late summer–fall, or all year southward. Permanently wet soil, in meadows, about ponds, in borrow pits, and in swampy woods; s NJ to s Fla (and Hispaniola), w to s Ark and e Tex, mainly on CP. *P. imbricata* (Kearney) Nash—S; *P. tenuifolia* Small—S.

5. P. rosea Godfrey. Much like *P. foetida*. Herbage often a little more hairy, not anthocyanic. Involucre not much if at all imbricate, the outer bracts often nearly or fully as long as the inner, more narrowly pointed, commonly acuminate or sharply acute, these and the middle bracts glandular-villous, more evidently hairy than in *P. foetida*; inner bracts generally more or less anthocyanic; disk mostly 5–9 mm wide; corollas pink purple. (n = 10) Late spring–fall, or all year southward. In open or shaded habitats with a fluctuating water table, alternately wet and dry; CP from NC to Fla, Tex, Mex, and WI. Ours is var. *rosea*.

6. **P. odorata** (L.) Cass. More or less fibrous-rooted annual up to 10 (15) dm tall, glandular-puberulent to occasionally subglabrate. Leaves ample, somewhat succulent in life, but drying thin, with lanceolate to elliptic or ovate blade mostly 4–15 × 1–7 cm, more or less serrate, acute or acuminate, evidently short-petiolate or sometimes (especially the upper) merely tapering to the narrow base. Heads numerous in a generally more or less flat-topped or layered inflorescence; involucre 4–7 mm high, evidently pubescent with short, multicellular, glandular-viscid hairs, the bracts imbricate in several series, commonly pink or purple at least distally; corollas rose-purplish. (n = 10) Late summer, fall, or all year southward. Salt or brackish (seldom freshwater) marshes; chiefly along the coast, from Mass to Fla, Tex, WI, and n S Am, and at scattered inland stations westward, as in Hempstead Co., Ark. Plants occurring along the coast from Mass to Md differ from the bulk of the species population in their combination of short stature (seldom more than 6 dm) and large heads (involucre 5.5–7 mm high, disk 5–9 mm wide). These may be distinguished as var. *succulenta* (Fern.) Cronq. comb. nov., based on *P. purpurascens* (Swartz) DC. var. *succulenta* Fern. Rhodora 44: 227. 1942. In Va and NC the var. *succulenta* passes into the widespread var. *odorata,* which is commonly more than 6 dm tall when well developed, and has involucres 4–5.5 mm high and disks 4–7 mm wide. Occasional plants of var. *odorata* have heads in the size range of var. *succulenta,* but are more robust. *P. camphorata* (L.) DC.—S, misapplied. *P. purpurascens* (Swartz) DC.—F, G, R.

7. **P. camphorata** (L.) DC. Similar to *P. odorata* in aspect, but averaging a little taller, up to 2 m, sometimes perennial, and more glabrate. Leaves thinner, on the average more serrate, and more evidently petiolate, averaging a little narrower and more acuminate. Inflorescence generally round-topped, often more or less elongate; involucre 4–6 mm high, sometimes purplish, but more often not, merely granular-glandular, or nearly glabrous; disk 3–6 mm wide. (n = 10) Late summer, fall, or all year southward. Wet or moist, nonsaline places; Del and Md to n Fla, w to s Ohio, s Ill, and e Okla and Tex. *P. petiolata* Cass.—S.

85. SACHSIA Griseb.

1. **S. polycephala** Griseb. Fibrous-rooted, scapose perennial 1–6 dm tall, finely glandular almost throughout, also somewhat sericeous-woolly at the crown and cobwebby-sericeous on the lower side of the leaves when young. Leaves rosulate, oblanceolate to obovate, 2.5–12 × 1–3.5 cm, obscurely to strongly callous-toothed. Heads disciform, several or numerous in an open-corymbiform inflorescence; involucre 5–8 mm high, its bracts slender, acute, largely scarious, well imbricate in several series, the inner inconspicuously fringed-ciliate toward the tip; receptacle flat, naked; flowers white, the pistillate ones numerous in several series, their slender, tubular-filiform corolla shorter than the style, about equaling the pappus, with a terminal, toothed, subligulate portion ca. 0.5 mm long; disk flowers fewer than the pistillate ones, functionally staminate but with a well-developed, bifid style, the branches slender, blunt, externally papillate, with poorly developed ventromarginal stigmatic lines; anthers sagittate at the base and very shortly but distinctly tailed. Achenes tiny, columnar, several-nerved, those of the disk nearly full-sized, but empty. Pappus a single series of ca. 20 slender, capillary bristles. Fall–spring. Pinelands; tropical Fla, Cuba, Hispaniola, and Bahama Isl. *S. bahamensis* Urban—S.

86. PTEROCAULON Elliott

1. **P. pycnostachyum** (Michx.) Elliott. BLACK ROOT. Perennial from a cluster of tuberous-thickened roots, 2–7 dm tall, densely and persistently gray- or rusty-tomentose throughout, except for the green, thinly tomentose or glabrate upper surface of the leaves; stem conspicuously winged by the long-decurrent leaf-bases. Leaves all cauline, alternate, simple, entire to repand-dentate, lanceolate to oblong or linear-oblong or rather narrowly elliptic (or the lower obovate), sessile, mostly 5–11 × (0.6) 1–3 (3.5) cm. Heads disciform, numerous in a very dense, terminal, spiciform, acropetally flowering inflorescence commonly 3–10 cm long and 1.3–2 cm thick, or in several such inflorescences, each on a short branch; involucre campanulate, ca. 5 mm high, with several series of slender, acuminate, stramineous, largely scarious, not much imbricate bracts; receptacle small, naked; flowers yellow, the pistillate ones numerous in several series, with a tubular-filiform corolla about equaling the pappus and a little shorter than the style, merely toothed at the summit; central flowers few, functionally staminate, the slender style shortly bifid, with externally papillate-hairy, astigmatic branches; disk corollas relatively slender, scarcely expanded upward, strongly 5-lobed; anthers sagittate at the base, the connate auricles shortly tailed. Achenes tiny, sericeous especially along the several ribs, glandular-tuberculate between the ribs, and with a smooth and shining basal annulus; disk achenes empty. Pappus of numerous capillary bristles. (n = 10) Spring–summer, or all year southward. Sandy pinelands and sandhill swales; CP of NC, SC, Ga, Fla, and se Ala. *P. undulatum* (Walt.) C. Mohr—S.

The tropical American species *P. virgatum* (L.) DC. has recently been collected in sw La. It is taller than *P. pycnostachyum*, up to 1.5 m, with narrower leaves, these mostly more than 7 times as long as wide, often all of them less than 1 cm wide. *P. pycnostachyum*, in contrast, has the principal leaves mostly 2–7 times as long as wide.

87. FACELIS Cass

1. **F. retusa** (Lam.) Schultz-Bip. Slender annual up to 3 dm tall, simple and erect, or more often with several basally decumbent branches from near the ground; stem loosely and often rather thinly white-woolly. Leaves numerous, ascending, alternate, simple, entire, linear-spatulate, with mucronate, otherwise subtruncately rounded to slightly retuse tip, mostly 7–20 × 1.5–4 mm. Heads disciform, sessile, closely subtended by ordinary or reduced leaves, crowded into a capitate-spiciform, terminal inflorescence or into several such; involucre turbinate-cylindric, 8–11 mm high, its bracts well imbricate in several series, largely hyaline-scarious, stramineous or partly greenish, the inner often anthocyanic near the tip; receptacle naked, flat with a depressed center; pistillate flowers in several series, with tubular-filiform corolla and exserted, bifid style; perfect flowers few, central, with slender, tubular, 5-toothed corolla; anthers sagittate and shortly tailed at the base; style branches very slender, externally minutely papillate-hairy, with ventromarginal stigmatic lines extending to the blunt tip. Achenes small, slightly compressed, narrowly obovate, obscurely 2-nerved, densely long-white-hairy. Pappus of strongly plumose capillary bristles well surpassing the corollas. Spring. S Am weed intro. along roadsides and in dooryards

and lawns and other disturbed sites in our coastal states from NC to n Fla and w to Tex. *F. apiculata* Cass.—S.

88. GNAPHALIUM L. CUDWEED, EVERLASTING

Woolly herbs, annual or perennial, not rhizomatous. Leaves alternate, simple, entire, commonly numerous and rather small or narrow. Heads in variously arranged glomerules, disciform, the flowers yellow or whitish, or in extralimital spp. sometimes pink; involucre ovoid or campanulate, its bracts more or less imbricate in several series, scarious at least toward the tip, often whitish; receptacle flat to subconic, naked; outer flowers pistillate, numerous in several series, with tubular-filiform corolla commonly about equaling the pappus; central flowers few, somewhat coarser, and perfect; anthers tailed at the base; style branches slender, minutely papillate outside, slightly flattened, with ventromarginal stigmatic lines extending to the truncate or slightly expanded tip. Achenes small, terete or slightly compressed, nerveless, in our spp. glabrous or merely papillate. Pappus of capillary bristles, these sometimes thickened at the summit, sometimes united into a ring at the base.

1 Pappus bristles distinct and falling separately, or merely cohering temporarily in small groups by means of tiny, interlocking basal hairs; inflorescence diverse, but not spiciform-thyrsoid.
  2 Involucre 4–7 mm high; plants when well-developed more than 2.5 dm tall; inflorescence obviously terminal, often ample.
    3 Leaves merely sessile, neither decurrent nor adnate-auriculate.
      4 Stem woolly, scarcely glandular except sometimes near the base .... 1. *G. obtusifolium*.
      4 Stem glandular-hairy, scarcely woolly except in the inflorescence .... 2. *G. helleri*.
    3 Leaves shortly but distinctly decurrent or adnate-auriculate at the base.
      5 Upper surface of the leaves coarsely glandular-hairy rather than tomentose .... 3. *G. viscosum*.
      5 Upper surface of the leaves loosely tomentose, not at all glandular .... 4. *G. chilense*.
  2 Involucre 2–3 mm high; plants up to 2.5 dm tall; inflorescence of numerous small axillary and terminal clusters overtopped by their subtending leaves .... 5. *G. uliginosum*.
1 Pappus deciduous as a unit, its bristles united into a ring at the base; inflorescence narrow, usually spiciform-thyrsoid .... 6. *G. purpureum*.

1. G. obtusifolium L. Erect, fragrant annual or winter-annual, (1–) 3–10 dm tall; stem thinly white-woolly, commonly becoming subglabrous (or even a little glandular) near the base. Leaves numerous, essentially all cauline, lance-linear, up to 10 × 1 cm, sessile but not decurrent, white-woolly beneath, green and from glabrous to slightly glandular or slightly woolly above. Inflorescence ample, branched, and many-headed in well-developed plants, flat or round-topped and often elongate; involucre yellowish-white or dingy, campanulate, woolly only near the base, 5–7 mm high, its bracts acutish to obtuse or somewhat rounded. Achenes glabrous. Pappus bristles distinct, falling separately, or sometimes temporarily coherent in small groups by means of tiny, interlocking basal hairs. ($n = 14$) Summer, fall, or all year in s Fla. A common native weed in open, often sandy places; widespread in N Am, and throughout our range.

2. G. helleri Britton. Similar to no. 1, but less common, and averaging a little smaller; stem glandular-hairy, becoming woolly in the inflorescence; leaves woolly beneath, more or less glandular-hairy above, the hairs shorter and sparser than those of the stem. Late summer, fall. Dry, commonly sandy soil, often in woods, sometimes with no. 1; irregularly from Me to Ga, w to Ind, Ark, and Tex. *G. obtusifolium* var. *helleri* (Britton) Blake—F; *G. obtusifolium* var. *micradenium* Weatherby—F.

3. G. viscosum H.B.K. Similar to no. 1; stem more or less glandular-hairy, becoming woolly in the inflorescence; rarely somewhat woolly near the base, as well as glandular; leaves distinctly (though rather shortly) decurrent at the base, the upper surface more or less glandular-hairy, the lower woolly or sometimes glandular-hairy; involucral bracts more or less sharply acute. Late summer, fall. Open places; widespread in w US and in Mex, and e occasionally to Que, Pa, WVa, and Tenn. *G. macounii* Greene—F, S.

4. G. chilense Sprengel. Erect annual or biennial, 2–8 dm tall, often several-stemmed; herbage loosely tomentose throughout, not glandular. Leaves numerous and narrow, mainly or wholly cauline, mostly 2–6 cm long and 2–5 (8) mm wide, sessile and evidently adnate-auriculate. Heads in (1–) several small, dense, terminal glomerules that often form a flat-topped cluster; involucre 4–6 mm high, woolly only at the base, its shining, yellowish-white bracts broadly rounded toward the often erose tip. Achenes glabrous. Pappus bristles distinct, falling separately, or often lightly coherent in small groups by means of tiny, interlocking basal hairs. Summer. In disturbed, open, usually moist sites; native to w US, and casually intro. eastward, as in NC and SC.

5. G. uliginosum L. Branching annual, 0.5–2.5 dm tall, generally diffuse; stem densely and often rather loosely white-woolly, the leaves sparsely so. Leaves numerous, mainly or wholly cauline, linear or oblanceolate, up to 4 cm × 5 mm. Heads glomerate in numerous small clusters in the axils and at the ends of the branches, overtopped by their subtending leaves; involucre 2–3 mm high, woolly at the base, its bracts greenish or brown, often paler at the tip, not much imbricate; acute, or the outer obtuse. Achenes papillate or smooth. Pappus bristles distinct, falling separately. ($n = 7$) Midsummer, fall. Streambanks and waste places, wet or dry; a European weed, thoroughly established in ne US and adj. Can, extending s to Va and WVa in our range.

6. G. purpureum L. Thinly woolly annual or biennial 1–4 (10) dm tall. Lowest leaves spatulate or oblanceolate, rounded to the generally micronate tip, up to 10 × 2 cm, often forming a persistent basal cluster or rosette. Heads numerous in a terminal, somewhat leafy-bracteate, spiciform-thyrsoid, seldom branched, sometimes interrupted inflorescence; involucre 3–5 mm high, woolly below, its bracts imbricate, mostly acute to acuminate, light brown, often tinged with anthocyanin. Achenes papillate. Pappus deciduous as a unit, its bristles united into a ring at the base. ($n = 7$, 14) Spring–summer. A widespread native Am weed, often in sandy soil, not entirely restricted to disturbed habitats. The polymorphic var. *purpureum*, occurring throughout our range and far beyond, has the leaves (except the uppermost) oblanceolate or spatulate and tending to be obviously greener and less hairy on the upper than on the lower surface. Var. *americanum* (Miller) Klatt—R; var. *spathulatum* (Lam.) Ahles—R; *G. spathulatum* Lam.—S; *G. peregrinum* Fern.—F. The more stable var. *falcatum* (T. & G.) Lam., occurring in the coastal states from Va to Fla and La, and s into tropical Am, has the leaves (except the lowermost) all linear or merely linear-oblanceolate and tending to be about equally hairy on both sides. *G. falcatum* Lam.—S; *G. calviceps* Fern.—F.

89. **ANAPHALIS** DC. PEARLY EVERLASTING

1. A. margaritacea (L.) Benth. & Hook. Erect, rhizomatous, polygamodioecious perennial, commonly 3–9 dm tall, loosely white-woolly, or the pubescence rusty in age. Leaves numerous, all cauline, alternate, simple, entire, lanceolate to linear, up to 12 × 2 cm, sessile, commonly less pubescent above than beneath, or glabrous above, the margins often revolute. Heads numerous and crowded in a short, broad inflorescence, but most of them individually short-pedunculate, ca. 1 cm wide or less; involucre 5–7 mm high, its bracts imbricate in several series, almost wholly dry and scarious, pearly-white, sometimes with a small basal dark spot, woolly only at the base; heads on male plants containing only tubular, functionally staminate flowers, the style shortly bifid, with truncate, astigmatic branches, the anthers distinctly tailed, on female plants containing numerous pistillate flowers with tubular-filiform corolla shorter than the pappus, and also a few central functionally staminate flowers. Achenes small, papillate. Pappus of distinct capillary bristles, neither clavellate nor conspicuously barbellate. (n = 14) Summer. Various habitats, chiefly dry and open; n N Am and e Asia, s in our range to Del and WVa.

90. **ANTENNARIA** Gaertn. EVERLASTING, PUSSY-TOES, LADIES' TOBACCO

Dioecious, woolly, perennial herbs, ours fibrous-rooted, stoloniferous, and mat-forming or colonial. Leaves alternate, simple, entire, in our spp. the largest ones basal and at the ends of the stolons. Heads rather small, unisexual, disciform or discoid, solitary to fairly numerous in a usually congested inflorescence; involucral bracts imbricate in several series, scarious at least at the tip, white or often colored; receptacle flat or convex, naked; anthers tailed at the base; staminate flowers with usually undivided style and scanty pappus, the bristles commonly barbellate or clavate; pistillate flowers with filiform-tubular corolla, bifid style, and copious pappus of capillary, naked bristles slightly united at the base. Achenes terete or nearly so, nerveless, glabrous or papillate. Absence of staminate plants from a local population indicates apomixis.

1 Basal leaves and those at the ends of the stolons relatively small, mostly less than 1.5 cm wide, 1-nerved or obscurely 3-nerved; heads several .... 1. *A. neglecta.*
1 Basal leaves and those at the ends of the stolons relatively large, prominently 3–5-nerved, the larger ones 1.5 cm wide or more.
  2 Heads several .... 2. *A. plantaginifolia.*
  2 Head solitary .... 3. *A. solitaria.*

1. A. neglecta Greene. Similar to no. 2, but averaging lower, and with consistently smaller leaves, the basal ones and those at the ends of the stolons mostly less than 1.5 cm wide, 1-nerved or rather obscurely 3-nerved. Forms connecting to no. 2 occur infrequently. (Diploids, tetraploids, and approximate hexaploids based on x = 14) Spring. Woods and open places; Nf and Que to Yukon, s to DC, Va, WVa, s Ind, Mo, and Calif, and to be expected in Ky. Four vars. in our range:

Basal leaves and those at the ends of the stolons glabrous above nearly or quite from the first; involucre 7–9 mm high; apomictic or sometimes sexual; widespread in ne US, and s reputedly to Va. *A. canadensis* Greene—F .... var. *randii* (Fern.) Cronq.
Basal leaves and those at the ends of the stolons only tardily glabrate.
 Stolons long, procumbent, with small and often few leaves; basal leaf blades tending to taper gradually to the base; involucre 7–10 mm high; diploids and polyploids, sexual or apomictic; widespread in n US, and s occasionally to Va and WVa. *A. petaloidea* Fern.—F .... var. *neglecta*.
 Stolons shorter, generally merely decumbent at the base, and more leafy; basal leaf blades tending to be more abruptly contracted and somewhat petiolate.
  Heads small, the pistillate involucre mostly 5–7 mm high; plants mostly smaller than in the next var., commonly diploid and sexual; shale barrens of Va and WVa. *A. virginica* Stebbins—F .... var. *argillicola* (Stebbins) Cronq.
  Heads larger, the pistillate involucres 7–10 mm high; plants sexual or more often apomictic, tetraploid so far as known; widespread in n US, and s occasionally to Va, WVa, and e Tenn (BR). *A. neodioica* Greene—F, *A. brainerdii* Fern.—F .... var. *attenuata* (Fern.) Cronq.

2. A. plantaginifolia (L.) Richards. Plants 1–4 dm tall, stoloniferous, the stolons sparsely leafy or merely bracteate. Basal leaves and those at the ends of the stolon densely and persistently tomentose beneath, only sparsely so (and eventually glabrate) above, or the upper side glabrous or nearly so from the first, relatively large, 3–5-nerved, evidently petiolate, the blade ovate to elliptic or obovate, mucronate, the larger ones 3–6 × 1.5–5 cm, often persisting throughout the winter; cauline leaves reduced, mostly linear or lanceolate. Heads several in a generally subcapitate cyme; pistillate involucres 5–10 mm high, the bracts white-tipped (often pinkish toward the base), striate, staminate involucres averaging smaller, with broader and more conspicuous white tips to the bracts; styles often crimson. (Diploids, presumably tetraploids, and hexaploids based on $x = 14$) Spring. Open woods and dry ground; Que to n Fla, w to Minn and Tex. Three vars. in our range:

Pistillate involucres mostly 5–7 mm high; plants usually or always sexual and diploid; basal leaves and those at the ends of the stolons tardily glabrate; common and widespread. *A. caroliniana* Rydb.—S. .... var. *plantaginifolia*.
Pistillate involucres mostly 7–10 mm high; plants usually apomictic and polyploid, but occasionally sexual and presumably diploid; less common, widespread northward, but only irregularly distributed s of NC and Mo.
 Basal leaves and those at the ends of the stolons tardily glabrate. *A. munda* Fern.—F; *A. fallax* Greene—F, S; *A. calophylla* Greene—S .... var. *ambigens* (Greene) Cronq.
 Basal leaves and those at the ends of the stolons green and glabrous above (or with a few coarse, glandular hairs) nearly or quite from the first. *A. parlinii* Fern.—F. .... var. *arnoglossa* (Greene) Cronq.

3. A. solitaria Rydb. Similar to no. 2, but differing sharply in its solitary heads; stems 1–2.5 dm tall, nearly naked; basal leaves averaging longer (to 7.5 cm) and somewhat narrower in shape; stolons nearly naked; involucre 8–10 mm high; plants commonly sexual. ($n = 14$) Spring. Woods; Va, WVa, and sw Pa to s Ind, s to Ga and La.

91. EUPATORIUM L.

Annual to more often perennial herbs (most of our species fibrous-rooted perennial herbs from a rhizome or crown or short caudex), or especially in extralimital species often shrubs. Leaves opposite or less commonly whorled, or sometimes the upper ones or most of them alternate, often glandular-punctate, simple and usually toothed or seldom lobulate, or in a few species evidently pinnatifid or trifid, then usually with slender segments. Heads small to medium-sized, mostly in basically cymose, corymbiform to paniculiform inflorescences, seldom in diffuse panicles, strictly discoid, the flowers all tubular and perfect; involucre cylindric to campanulate or hemispheric, its bracts variously green and subherbaceous to chartaceous or coriaceous, seldom distally hyaline-scarious and somewhat petaloid, equal or subequal to strongly imbricate; receptacle naked, flat or seldom conic, in a few of our species the portion bearing the involucral bracts elongate and more or less stout-columnar or columnar-obconic; flowers blue or lavender to purple or pink, or often white; anthers with a small, hyaline, apical appendage, minutely rounded-auriculate at the base; style branches papillate, elongate, linear or often somewhat clavate, obtuse, with ventromarginal stigmatic lines near the base. Achenes mostly 5 (–8)-angled and -nerved, glabrous, or inconspicuously hairy along the nerves, or in many of our species atomiferous-glandular; pappus a single series of capillary bristles. (x most commonly = 10, less often 17 or other numbers) Fryor, W. R., 1964. Natural hybridization between two perennial species of Eupatorium (Compositae). M.S. thesis, Florida State University. Marushat, H. D. 1969. Natural hybridization in the dog-fennels (Eupatorium spp., Compositae). M.S. thesis, Florida State University. Montgomery, J. D., and D. E. Fairbrothers, 1970. A biosystematic study of the Eupatorium rotundifolium complex (Compositae). Brittonia 22: 134–150. Sullivan, V. I. 1972. Investigations of the breeding systems, formation of auto- and alloploids and the reticulate pattern of hybridization in North American Eupatorium (Compositae). Ph.D. thesis, Florida State University. I wish to acknowledge also the advice and counsel of R. K. Godfrey, who directed the three Florida State theses.

Our species all have a basic chromosome number of x = 10, except for species 11–14, which are diploids with n = 17. Species 1–4 form a closely related, hybridizing, mainly diploid group, and species 11–14 form another. Species 15–33 form a large, intricately reticulate complex of diploids, autopolyploids, established or temporary allopolyploids, and temporary hybrids. The diploids are largely or wholly sexual and outcrossing; the polyploids, in contrast, are largely or wholly apomictic. *E. pinnatifidum* Ell., for convenience assigned a binomial here, is a temporary diploid hybrid of strikingly distinct morphology. Many other less conspicuous hybrids are not formally treated here. Some of the species recognized here consist wholly of diploids. Others consist of both diploid and polyploid elements, the polyploids being associated, for taxonomic purposes, with the diploids that they most resemble. Only *E. anomalum* Nash appears to consist wholly of polyploids that cannot usefully be associated taxonomically with diploids. In addition to the taxa recognized here, there are scattered small populations representing diverse sorts of allopolyploids that may not be permanently established. It does not seem useful to provide these with formal names.

The large genus *Eupatorium* is divided by some authors into numerous much smaller genera. Some of these are indicated in parentheses in the key.

1 Leaves all or mostly in whorls of 3–7, generally at least 2 cm wide (*Eupatoriadelphus*).
  2 Leaves more or less strongly triplinerved, rather abruptly contracted to the petiole, thick and firm, up to 12 (15) cm long; flowers (4) 6–9 (–12) per head .... 2. *E. dubium*.

2 Leaves otherwise, either more gradually narrowed to the petiole, or not at all triplinerved, or commonly with both of these differences, also often more than 15 cm long.

3 Flowers mostly 9–22 per head; inflorescence or its segments flat-topped .... 1. *E. maculatum*.

3 Flowers mostly 4–7 per head; inflorescence convex.

4 Stem purplish only or chiefly at the nodes, solid or sometimes eventually developing a slender central cavity .... 3. *E. purpureum*.

4 Stem more or less purplish essentially throughout, hollow with a large central cavity .... 4. *E. fistulosum*.

1 Leaves mostly opposite, sometimes some of them alternate, ternate in rare individuals of some species, or, if regularly whorled, then well under 2 cm wide.

5 Flowers more or less numerous, usually at least 9 per head, white or often anthocyanic; leaves entire or merely toothed.

6 Receptacle conic; flowers mostly 35–70, commonly blue (*Conocephalum*) .... 5. *E. coelestinum*.

6 Receptacle (at least the portion bearing the flowers) flat or merely convex; flowers seldom more than 35 (to 40 in *C. ivaefolium*), blue to pink, lavender, or white.

7 Involucral bracts strongly imbricate in several series, smooth or merely somewhat glandular on the back, thick and very firm (or some of them with thin, anthocyanic, petaloid tip) (*Osmia*).

8 Inner (and to some extent the middle) involucral bracts with somewhat petaloid, loose and slightly expanded, hyaline-scarious, anthocyanic tip; achenes ca. 2 (–3) mm long; leaves tapering to a sessile or shortly petiolar base .... 6. *E. ivaefolium*.

8 Inner involucral bracts, like the others, neither petaloid or distally expanded; achenes 3–4.5 mm long; leaves more or less abruptly contracted to a definite petiole.

9 Heads smaller, the involucre 5.5–7.5 (8) mm high; leaves small, the larger ones mostly 1.5–4 × 0.7–2 cm .... 7. *E. frustratum*.

9 Heads larger, the involucre (8) 8.5–11 mm high; leaves larger, the larger ones mostly (3.5) 5–10 × (1.5) 2–7 cm .... 8. *E. odoratum*.

7 Involucral bracts otherwise, either not strongly imbricate, or if so then obviously hairy or of distinctly thinner texture or both, never with anthocyanic, petaloid tip.

10 Shrub .... 9. *E. villosum*.

10 Herbs.

11 Flowers anthocyanic, commonly pink or purplish; leaves evidently petiolate .... 10. *E. incarnatum*.

11 Flowers normally white; leaves petiolate or sessile.

12 Involucral bracts not strongly imbricate, the principal ones subequal and subbiseriate, a few shorter outer ones sometimes also present (*Ageratina*).

13 Leaf blades relatively thick and firm, crenate or crenate-serrate, small, up to ca. 7 (10) cm long.

14 Leaves short-petiolate, the blade mostly more than 4 times as long as the petiole .... 11. *E. aromaticum*.

14 Leaves with longer petiole, the blade only (1.5) 2–4 times as long .... 12. *E. jucundum*.

13 Leaf blades relatively thin and soft, serrate or dentate, usually coarsely and sharply so, often well over 7 cm long.

15 Leaf blade obviously longer than the petiole, up to ca. 5 times as long, not extremely thin and delicate .... 13. *E. rugosum*.

15 Leaf blade about as long as the petiole, extremely thin and delicate .... 14. *E. luciae-brauniae*.

12 Involucral bracts evidently imbricate in 3 or more series, the outer less than half as long as the inner (*Uncasia*).

16 Leaves evidently petiolate .... 15. *E. serotinum*.

16 Leaves sessile.

17 Leaves neither especially broad at the base nor at all connate, seldom as much as 1.5 cm wide .... 16. *E. resinosum*.

17 Leaves broad-based and nearly always connate-perfoliate, the larger ones seldom less than 1.5 cm wide .... 17. *E. perfoliatum*.

5 Flowers mostly 5 in each head, rarely 3–7 (to 9 in *E. pinnatifidum*, with pinnatifid leaves).

18 Principal leaves pinnatifid or ternate or pinnately dissected (*Traganthes*).
19 Principal leaves irregularly pinnatifid or bipinnatifid, often with fairly broad segments and midstrip, the segments not elongate and linear or filiform as in the following species; inflorescence terminal, corymbose-paniculiform, not diffuse .... 18. *E. pinnatifidum*.
19 Principal leaves with narrow midstrip and elongate, narrow, linear or filiform segments; inflorescence diffuse or paniculate.
20 Stem evidently puberulent, at least above; inflorescence diffusely paniculate, but the branches neither recurved nor secund.
21 Leaves or leaf segments linear, at least the wider ones commonly 1–2.5 (4) mm wide .... 19. *E. compositifolium*.
21 Leaves or leaf segments filiform, commonly ca. 0.5 (–1) mm wide .. 20. *E. capillifolium*.
20 Stem and branches glabrous even in the inflorescence; inflorescence paniculate, with elongate, strongly recurved and secund main branches; leaves much as in *E. capillifolium* .... 21. *E. leptophyllum*.
18 Leaves all entire to deeply toothed or seldom lobulate, not at all pinnatifid (*Uncasia*).
22 Leaves with an evident slender petiole mostly (0.5) 1–2.5 cm long ...... 22. *E. mikanioides*.
22 Leaves sessile or nearly so, sometimes tapering to a shortly subpetiolar base well under 1 cm long.
23 Involucral bracts acuminate to attenuate (or the inner rounded and strongly mucronate in forms of *E. album*).
24 Larger leaves mostly 1.5–3 cm wide; involucre 8–11 mm high ...... 23. *E. album*.
24 Larger leaves up to 1 (1.3) cm wide; involucre 5–7 mm high .... 24. *E. leucolepis*.
23 Involucral bracts rounded to acute, but not acuminate.
25 Leaves broadly cuneate or subtruncate to rounded or subcordate at the base.
26 Plants glabrous below the inflorescence; leaves acuminate or gradually and narrowly acute, pinnately veined, not trinerved or triplinerved .... 25. *E. sessilifolium*.
26 Plants evidently hairy below as well as in the inflorescence; leaves usually more or less evidently trinerved or triplinerved from at or near the base .... 26. *E. rotundifolium*.
25 Leaves tapering to a narrow base.
27 Plants with conspicuously tuberous-thickened short rhizomes.
28 Leaves mostly 10–20 mm wide, usually spreading or ascending .... 27. *E. anomalum*.
28 Leaves up to 10 (12) mm wide, tending to be recurved or deflexed .... 28. *E. mohrii*.
27 Plants without tuberous-thickened rhizomes, mostly from a short crown or caudex or caudexlike rhizome instead.
29 Plants generally branched at or near the ground level; slender sterile axillary shoots of the middle and lower leaves commonly elongating .... 29. *E. cuneifolium*.
29 Plants generally simple below the middle, the stems generally distinct down to the caudex or rhizome, the middle and lower axillary shoots generally not elongating.
30 Leaves mostly 2.5–7 times as long as wide, the larger ones seldom less than 1 cm wide, all opposite, or the upper ones alternate.
31 Involucre 2.5–4.5 mm high; leaves obtuse to acute, but tending to carry their width well above the middle, finely and densely puberulent to subglabrous, triplinerved, the principal pair of lateral veins emerging as evident branches from the midvein (or in *E. lancifolium* sometimes trinerved as in *E. altissimum*).
32 Leaves finely and densely puberulent on both sides, sometimes more shortly so above than beneath .... 30. *E. semiserratum*.
32 Leaves subglabrous, often with the hairs confined chiefly to the main veins beneath .... 31. *E. lancifolium*.
31 Involucre 4.5–7 mm high; leaves gradually and very narrowly acute or acuminate, rather coarsely hirsute-puberulent at least beneath, trinerved, the principal pair of lateral veins distinct from the midrib all the way to the base of the leaf .... 32. *E. altissimum*.

30 Leaves mostly 6–40 times as long as wide, seldom more than 1 cm wide, mostly ternate or quaternate, but sometimes merely opposite, or even alternate above .... 33. *E. hyssopifolium*.

1. E. maculatum L. JOE-PYE WEED. Stems 6–20 dm tall, speckled or sometimes more evenly purplish, glabrous or often glutinous-puberulent, especially upward, only seldom glaucous. Leaves mostly in whorls of 4 or 5, lance-elliptic to lanceolate or lance-ovate, relatively gradually narrowed to the short petiole, 6–20 × 2–9 cm, seldom evidently triplinerved, sharply serrate. Inflorescence or its divisions more or less flat-topped in life; involucre 6.5–9 mm high, often purplish, its bracts well imbricate in several series, often 3–5-nerved, obtuse, essentially glabrous, or the outer often inconspicuously short-hairy; flowers purple to rather pale lavender (8) 9–22 per head. (n = 10) Summer. Moist places especially in calcareous soils; Nf to BC, s to Md, WVa, Ill, and NM, and along the higher mts. to sw NC and adj. Tenn. Ours is the e Am var. *maculatum*.

2. E. dubium Willd. JOE-PYE WEED. Stems 4–10 (15) dm tall, generally purple-speckled, viscid-puberulent at least near the summit, scarcely or not at all glaucous. Leaves mostly in whorls of 3 or 4, thick and firm, often somewhat rugose, ovate or lance-ovate, relatively abruptly contracted to the short petiole, 5–12 (15) × 2–7 cm, coarsely serrate, more or less strongly triplinerved. Inflorescence dense, slightly to strongly convex, not large; involucre 6.5–9 mm high, often purplish, its bracts well imbricate in several series, obtuse, often 3–5-nerved, essentially glabrous, or the outer often inconspicuously short-hairy; flowers purple, (4) 6–9 (12) per head. (n = 10) Late summer, fall. Moist places, especially in sandy or gravelly, acid soil; near the coast from NS and s NH to SC; doubtfully in the Apalachicola region of Fla. *E. purpureum* L.—S, misapplied.

3. E. purpureum L. JOE-PYE WEED. Stems 6–20 dm tall, slightly glaucous, usually purple only or chiefly at the nodes, otherwise greenish, usually solid, the pith remaining intact or sometimes ultimately developing a slender central cavity. Leaves mostly in whorls of 3 or 4 (seldom some or nearly all of them merely paired), lanceolate to ovate or elliptic, 8–30 × 2.5–15 cm, gradually or sometimes rather abruptly narrowed to the short petiole, pinnately veined, usually sharply toothed. Inflorescence convex, involucre imbricate, 6.5–9 mm high, its bracts well imbricate in several series, commonly 3-nerved, obtuse or acutish, essentially glabrous, or the outer often inconspicuously short hairy; flowers mostly 4–7 per head; corolla generally very pale pinkish or purplish, but variable, 4.5–7.5 mm long. (n = 10, 20) Summer. Thickets and open woods, often in drier habitats than related spp.; s NH and Wis to Iowa, s to Va (all), NC and SC (chiefly mt. provinces and PP) and the mt. provinces of Ga, and occasionally s to CP of sw Ga and the Apalachicola region of Fla, in the w to Tenn (all), Ark (OZ, n ME), and Okla. *E. trifoliatum* L.—S.

4. E. fistulosum Barratt. JOE-PYE WEED. Similar to *E. purpureum*, often more robust, up to 3 m tall; stem strongly glaucous, usually purplish throughout, hollow with a large central cavity. Leaves mostly in whorls of 4–7, more elliptic, generally narrowly so, with mostly finer, more rounded, blunter teeth. Corolla generally bright-pink-purple. (n = 10) Summer. Bottomlands and moist woods; s Me to Iowa, s to c Fla, Ala, Miss, La, and Tex, especially common southward. *E. maculatum* Justineus—S, misapplied.

5. E. coelestinum L. MIST FLOWER. Plants rhizomatous; stems 3–9 dm tall, puberulent. Leaves opposite, petiolate, the blade crenate or crenate-serrate, deltoid-ovate, sometimes narrowly so, 3–10 × 2–5 cm, trinerved or triplinerved, sparsely appressed-hairy or subglabrous, often atomiferous-glandular. Involucre 3–5 mm high, its bracts narrow, firm, long-pointed, more or less imbricate; receptacle conic; flowers 35–70 per head, bright blue or violet (rarely reddish-purple), often purplish when dry. Achenes ca. (1) 1.5 mm long; pappus scanty. (n = 10) Midsummer–fall, or all year southward. Woods, streambanks, meadows, and fields; throughout our range, n and w to NJ, Ohio, Ill, Kans, and Tex, and in WI. *Conoclinium coelestinum* (L.) DC.—S.

6. E. ivaefolium L. Herbs; stem solitary, 3–20 dm tall, loosely spreading-hairy and often glandular, freely branched above when well-developed. Leaves opposite, very often with axillary fascicles, often deflexed, lanceolate or narrowly elliptic, tapering to a sessile or shortly petiolar base, evidently trinerved, the larger ones mostly 3–7 × 0.5–1.5 (2) cm, the lower surface hairy (commonly rough-strigose) along the main veins and with scattered glands, the upper surface more scabrous. Heads in open or compact, corymbiform clusters, forming a diffuse inflorescence in well-developed plants; involucre 5–7.5 mm high, short-cylindric, its bracts conspicuously imbricate in several series, strongly 3 (5)–nerved, the outer broadly obtuse and sometimes apiculate, bearing some red glands near the tip, the inner glabrous or nearly so, slightly expanded at the loose, erose, hyaline-scarious, blue or lavender, somewhat petaloid tip; receptacle flat or merely convex above the shortly columnar involucral portion; flowers mostly 20–40, light blue to purplish or reddish. Achenes ca. 2 (–3) mm long. Fall. Moist prairies, open woods, and borders of fields; La and Miss; WI, Mex, and C Am. *Osmia ivaefolia* (L.) Small—S.

7. E. frustratum B. L. Robinson. Herbs; stems 1–several, freely branching, 2–10 dm tall; herbage hirtellous-puberulent or shortly spreading-hirsute throughout. Leaves opposite, relatively small, with trinerved, lance-ovate to broadly ovate, toothed or subentire blade mostly 1.5–4 × 0.7–2 cm on a definite, slender petiole mostly 4–10 mm long. Heads in small clusters ending the numerous branches, forming a more or less diffuse inflorescence; involucre 5.5–7.5 (8) mm high, its bracts strongly imbricate in several series, at least the upper ones minutely ciliolate or glandular-ciliolate, otherwise glabrous or nearly so, shining and firmly chartaceous, readily deciduous from the shortly columnar to somewhat columnar-obconic bract-bearing portion of the receptacle; floriferous portion of the receptacle flat or merely convex; flowers ca. 20–25 or perhaps sometimes more, blue or lavender. Achenes (3) 3.5–4 mm long. All year. Coastal hammocks; s Fla and the Keys. *Osmia frustrata* (B. L. Robinson) Small—S.

8. E. odoratum L. Coarse herb or half-shrub 6–20 (30) dm tall, freely branched and sometimes scrambling; stem shortly spreading-hairy or glabrate. Leaves opposite, with broadly lanceolate or broadly lance-elliptic to deltoid-ovate, trinerved or triplinerved, toothed or sometimes entire blade mostly (3.5) 5–10 × (1.5) 2–7 cm on a definite slender petiole 0.5–2 cm long; blades shortly spreading-hairy and with scattered reddish sessile glands beneath, sparsely hairy or glabrate above. Heads numerous, involucre cylindric, (8) 8.5–11 mm high, its bracts strongly imbricate in several series, strongly 3 (5)-nerved, firmly scarious, largely stramineous and shining, readily deciduous from the shortly columnar or columnar-obconic bract-bearing portion of the receptacle; floriferous portion of the receptacle merely convex; flowers ca. 17–25, white to rather light blue or lavender. Achenes 4–4.5 mm long. (n = 29, 60, ca. 80) All year. Hammocks and thickets; widespread in tropical Am, n to s Fla and s Tex. *Osmia odorata* (L.) Sch.-Bip.—S.

9. E. villosum Swartz. Freely branching shrub mostly 5–20 dm tall, densely and shortly spreading-hairy throughout, often also atomiferous-glandular, especially on the lower leaf surfaces, the upper leaf surfaces often more sparsely and shortly hairy. Leaves opposite, rather small, the blade narrowly to rather broadly deltoid-ovate (or seldom with subcordate or more rounded base), mostly 2–6.5 × (1) 1.5–4 cm, entire or toothed, somewhat trinerved from the base, on a slender petiole 3–10 mm long. Heads in terminal, corymbiform clusters; involucre 3–4 mm high, hairy like the herbage, its bracts inconspicuously imbricate in 2–3 series, the outer generally at least half as long as the inner; flowers 9–13, white or sometimes (at least in West Indian plants) somewhat anthocyanic. All year. Hammocks and pine woods, often on calcareous soil; WI, n to s Fla.

10. E. incarnatum Walter. Stems weak, 3–12 dm tall, freely branching, each branch ending in a small inflorescence. Leaves opposite, petiolate, the blade deltoid or somewhat cordate at the base, tapering to the acuminate or acute tip, coarsely crenate-serrate, up to 7 × 5.5 cm, trinerved. Involucre 3–5 mm high, glabrous or nearly so, its bracts sharply acute to blunt, the main ones subequal, but some irregularly shorter outer ones also present; receptacle flat; flowers mostly (13) 18–24 per head, mostly pink-purple, seldom pale blue. Achenes 5–8-nerved, glabrous or nearly so. (n = 10) Fall. Woods, ditch banks, and swamps; se Va to ne Fla, w to WVa, Ky, s Mo, Ark, La, Tex, and Mex.

11. E. aromaticum L. Stems solitary or few, 3–8 (10) dm tall, villous-puberulent. Leaves opposite, relatively thick and firm, the blade obtuse or acute, crenate or crenate-serrate, narrowly to broadly deltoid or ovate or even subcordate, trinerved or triplinerved from near the base, usually short-hairy at least along the main veins beneath, mostly 3–7 (10) × 2–5 cm, more than 4 times as long as the short petiole. Inflorescence flat-topped or more rounded; involucre mostly 3.5–5 mm high, slightly to strongly puberulent or villous-puberulent, its principal bracts subequal and sub-biseriate, obtuse or rounded to acute, a few smaller outer bracts sometimes also present; flowers mostly 10–19 per head, white, the corolla lobes commonly villous externally, at least near the tip. Achenes glabrous, or seldom short-hairy near the tip. (n = 17) Fall. Dry woods, especially in sandy soil; Mass to n Fla, w to s Ohio, Ky, Tenn (all), Miss, and e La. *E. latidens* Small—S. Hybridizes throughout its range with *E. rugosum*.

12. E. jucundum Greene. Much like *E. aromaticum*, but commonly more delicately and laxly branched, the branches often somewhat curved or flexuous; leaves more evidently petiolate, the blade up to 6 or 7 cm long, mostly (1.5) 2–4 times as long as the petiole; achenes usually hirtellous-strigose, at least toward the top; corolla lobes more shortly and sparsely hairy, or glabrous. (n = 17) Fall (–winter). In various open or partly shaded, xeric or mesic habitats; peninsular Fla, extending n occasionally to se Ga and w to the Apalachicola region of the Fla panhandle. Intergrades with *E. aromaticum* in the n part of its range.

13. E. rugosum Houttuyn. WHITE SNAKEROOT. Stems 1–3 together, 3–15 dm tall, glabrous below the inflorescence and sometimes glaucous, or shortly and loosely hairy. Leaves opposite, the blade narrowly to broadly ovate or even subcordate, rather thin, glabrous or hairy especially on the main veins beneath, serrate, usually sharply and coarsely so, mostly acuminate, the larger ones mostly 6–18 × 3–12 cm (or smaller in depauperate plants), 1.5–5 times as long as the well-developed petiole. Inflorescence flat-topped or more rounded; involucre mostly 3–5 mm high (to 7 mm in var.

roanense), glabrous or short-hairy, its principal bracts subequal and subbiseriate, acuminate to obtuse, one or two shorter outer bracts sometimes also present; flowers mostly (9) 12–25 per head (up to 34 in var. *roanense*), the corolla bright white, its lobes often short-hairy. Achenes generally glabrous. (n= 17) Fall. Woods, often rich woods; NS to n Fla, w to Sask and Tex, throughout our range except for the Fla peninsula. Poisonous to livestock, the poison transmissible to man in milk, causing the notorious milk sickness of times past. A variable species, but hardly subdivisible, except that plants of the highest s Appalachian Mts., in Tenn and NC, with large, numerously flowered heads as noted in the description, may be segregated as var. *roanense* (Small) Fern. *E. roanense* Small—S; *E. urticaefolium* Reichard—S.

14. E. luciae-brauniae Fern. Resembling *E. rugosum*, but more delicate and slender, 3–6 dm tall, often glabrous throughout; leaves very thin, deltoid or subcordate, about as broad as long, coarsely and irregularly dentate, the petiole nearly or fully as long as the blade; involucral bracts attenuate-acuminate. Late summer. Under overhanging sandstone cliffs; CU of s Ky and n Tenn.

15. E. serotinum Michx. Stems 4–20 dm tall, puberulent especially above. Leaves opposite, with an evident petiole 1 cm long or more, the blade lanceolate to ovate, serrate, mostly acuminate, 5–20 × 1.5–10 cm, 3–5-nerved or -plinerved, commonly less hairy than the stem, the upper surface often subglabrous. Heads numerous; involucre 3–4 mm high, densely villous-puberulent, its bracts broadly rounded to merely obtuse, evidently imbricate in about 3 series, the outer less than half as long as the inner; flowers 9–15 per head, white. (n = 10) Late summer, fall. Mostly in bottomlands and moist woods, sometimes in drier or more open places; s NY to Ill and reputedly Minn, s to Fla, Tex, and n Mex, essentially throughout our range. *E. serotinum* var. *polyneuron* F. J. Hermann is a hybrid with *E. perfoliatum*.

16. E. resinosum Torr. Plants 4–10 dm tall, viscidulous-puberulent throughout, the leaves commonly subtomentosely so beneath. Leaves opposite, sessile, slender, 5–13 × 0.5–1.5 cm, evenly serrate, long-acuminate. Heads more or less numerous in a flat-topped inflorescence; involucre 3.5–5 mm high, its bracts broadly rounded to acutish, imbricate in about 3 series, the inner tending to be somewhat whitish distally, the outer less than half as long as the inner; flowers 9–14 per head, white. (n = 10) Late summer, fall. Pocosins, bogs, and other wet places, often in pine barrens; NJ and LI s to Del and rarely NC, wholly on AC.

17. E. perfoliatum L. BONESET, THOROUGHWORT. Stems 4–15 dm tall, conspicuously crisp-villous with long spreading hairs. Leaves opposite (very rarely ternate), broad-based and strongly connate-perfoliate, tapering gradually to the acuminate tip, crenate-serrate to the base, 7–20 × 1.5–4.5 cm, sparsely pubescent or subglabrous above, more evidently hairy beneath. Inflorescence flat-topped, involucre 4–6 mm high, more or less densely villous-puberulent and commonly also atomiferous-glandular, its bracts evidently imbricate in about 3 series, the outer less than half as long as the inner, mostly obtuse, the inner more acuminate and often pale distally; flowers 9–23 per head, dull white. (n = 10) Late summer, fall. Moist or wet low grounds; NS and Que to n Fla, w to Minn, Neb, Okla, and La. Ours is the common and widespread var. *perfoliatum*. *E. cuneatum* Engelm. is a hybrid with *E. serotinum*, and *E. chapmanii* Small is probably a hybrid with some other species.

18. E. pinnatifidum Elliott. Herbage villous or villous-puberulent and sometimes also atomiferous-glandular; principal leaves up to 13 × 6 cm, irregularly pinnatifid or bipinnatifid, often with fairly broad segments and midstrip; inflorescence terminal, corymbose-paniculiform, not diffuse; involucre often atomiferous-glandular; flowers 5–9 per head; otherwise much like *E. capillifolium* and *E. compositifolium*. Fall. Open pine woods, often in sandy soil, especially in disturbed sites; rare and irregular from SC to s Fla and w to Tex. *E. eugenei* Small—S; *E. pectinatum* Small—S. Plants grouped under this name are hybrids and hybrid segregates with *E. perfoliatum* as one parent and *E. capillifolium* or *E. compositifolium* as the other. The hybrids are frequently produced, but not long-persistent.

19. E. compositifolium Walter. Plants coarse, 5–20 dm tall, forming loose tufts or small colonies, the puberulent stems commonly arising singly from a system of short, coarse rhizomes, freely branched upward, and often with axillary fascicles of leaves or with short, sterile, leafy axillary branches. Leaves very numerous and narrow, opposite or the upper often alternate, glandular-punctate, otherwise puberulent to glabrous, up to ca. 6 cm long, the larger ones trifid or pinnatifid with few segments, the segments or entire leaves mostly 1–2.5 (4) mm wide. Heads very numerous in an elongate panicle, the branches not notably secund or recurved; involucre 3–4 mm high, its bracts more or less imbricate, glabrous or obscurely puberulent and sometimes atomiferous-glandular, greenish or partly anthocyanic with pale, hyaline margins, the inner bracts often mucronulate; flowers 5, white. Achenes 1.3–1.8 mm long, glabrous or nearly so. (n = 10) Fall. Open or lightly shaded places in dry to moist, often sandy soil; NC (mostly CP) and SC (CP and PP) to s Fla, w across CP and PP of Ga to GC of Ala, Miss, La, Ark, and Tex.

20. E. capillifolium (Lam.) Small. DOG-FENNEL. Plants coarse, the stems 5–20 dm tall, clustered on a thick, woody caudex, puberulent, or glabrate below, freely branched upward. Leaves very numerous and narrow, delicate, glandular-punctate, glabrous, the lowest ones opposite, the others alternate, mostly 2–10 cm long, often with axillary fascicles or with short, sterile, leafy axillary branches, the main ones pinnately divided into a few filiform segments mostly ca. 0.5 (–1.0) mm wide, those of the inflorescence mostly simple. Heads very numerous in an elongate panicle that is on the average more diffuse and open than in *E. compositifolium*, but the branches not notably secund or recurved; involucre 2–3.5 mm high, the inner bracts much longer than the outer, usually mucronate or abruptly acuminate; flowers 3–6, white or chloroleucous. Achenes 1–1.3 mm long, glabrous. (n = 10) Fall, or nearly all year southward. Open places, often in old fields and pastures; coastal states from NJ to s Fla and w to Tex and s Ark, chiefly on CP, but inland to PP and occasionally mt. provinces in NC, SC, Tenn, Ga, and Ala.

21. E. leptophyllum DC. Rhizomatous perennial with scattered stems mostly 5–12 dm tall, essentially glabrous throughout. Leaves very numerous and narrow, delicate, glandular-punctate, the lowest ones opposite, the others alternate, mostly 2–10 cm long, often with axillary fascicles or with short, sterile, leafy axillary branches, the main ones pinnately divided into a few filiform segments mostly ca. 0.5 (–1.0) mm wide, those of the inflorescence mostly simple. Heads numerous in a terminal panicle with elongate, recurved-secund, often well-spaced primary branches; involucre 3–4 mm high, the inner bracts much longer than the outer, usually cuspidate or subaristately acuminate; flowers 5, white. Achenes glabrous, 1.3–1.6 mm long; pappus conspicuously barbellate. (n = 10) Fall. Pond margins (often limestone solution ponds) and other wet, low places, sometimes in shallow water; throughout peninsular Fla, and at least a part of the panhandle, extending also into se Ala and n to se NC, wholly on CP. *E. capillifolium* var. *leptophyllum* (DC.) Ahles—R.

22. E. mikanioides Chapman. Perennial from elongate rhizomes, mostly 5–12 dm tall; stem subtomentosely villous-puberulent above. Leaves opposite, with palmately or pinni-palmately veined, crenate, glandular-punctate and somewhat puberulent, narrowly ovate to deltoid-ovate or broadly rhombic-ovate, blunt blade mostly 2–6 × 1–4 cm; petiole (0.5) 1–2.5 cm long, twisted in life so that the blade is vertically oriented. Heads numerous in a terminal, more or less flat-topped corymbiform cyme; involucre 4–6 mm high, its bracts evidently imbricate, closely tomentose-puberulent and often also atomiferous-glandular; flowers 5 in each head, white. Achenes 1.5–1.5 mm long. (n = 10) Midsummer–fall. Moist or wet, low, often saline places; coastal peninsular Fla, n and w only to Wakulla Co.

23. E. album L. Stems mostly solitary from a crown or very short, stout rhizome, 4–10 dm tall, conspicuously spreading-villous at least below (except often in var. *vaseyi*), often merely villous-puberulent above. Leaves opposite, elliptic to elliptic-ovate, lance-elliptic, or elliptic-oblanceolate, sessile or nearly so, 4–13 × 1–4 cm, the larger ones seldom less than 1.5 cm wide except sometimes in var. *subvenosum*, glandular-punctate, evidently hairy to sometimes subglabrous. Inflorescence a dense, terminal, corymbiform, more or less flat-topped cyme; involucre 8–11 mm high, often with dark sessile glands, otherwise generally glabrous or only slightly hairy, its bracts imbricate, conspicuously white-scarious upward (especially the inner), all narrow and long-acuminate, or the inner with broader, more rounded, mucronate tip; flowers 5 per head, the corolla white, 4–5.5 mm long. Achenes 2.5–3.5 mm long. (diploids, triploids, and tetraploids based on x = 10) Midsummer–fall. Dry, open woods, especially in sandy pinelands; coastal states from s NY to c Fla and w to Miss, also inland in the mt. regions to s Ohio and e Ky and Tenn (an outlier on IP), and in Ark (OU). The species consists of a widespread, common diploid phase and two more sporadic or local polyploid phases that reflect hybridization with other species:

Leaves usually evidently pubescent, mostly coarsely serrate, tending to be obtuse or rounded at the tip, the larger ones seldom less than 1.5 cm wide; sexual diploid, common and widespread, with the range of the species: *E. petaloideum* Britton—S; *E. album* var. *glandulosum* (Michx.) DC.—F .... var. *album*.

Leaves sparsely pubescent or subglabrous, serrate to subentire, often acute, sometimes less than 1.5 cm wide; apomictic polyploids, local, not abundant.

- Leaves small, mostly 4–7 × 1–2 cm, acute to obtuse, more or less trinerved or triplinerved, few-toothed, with up to 10 teeth per side, or even entire; s NY to NJ and Del .... var. *subvenosum* A. Gray.
- Leaves larger, mostly 5–11 × 2–4 cm, acute or somewhat acuminate, more pinnately veined, rather closely toothed, with mostly 10–20 teeth per side; very probably derived by hybridization of *E. album* var. *album* and *E. sessilifolium*; DC and Md, s to the mts. of NC, Tenn, nw SC, n Ga, and n Ala; *E. album* var. *monardifolium* Fern.—F .... var. *vaseyi* (T. C. Porter) Cronq.

24. E. leucolepis (DC.) T. & G. Stem 4–10 dm tall from a crown or short, stout rhizome, hirtellous or puberulent to sometimes merely strigose. Leaves opposite, lance-oblong to linear-oblong or oblanceolate, 3–8 cm × 3–10 (13) mm, bluntly few-toothed or entire, sessile, glandular-punctate. Inflorescence a dense, terminal, corymbiform, more or less flat-topped cyme; involucre 5–7 mm high, its bracts imbricate, narrow, tapering to a long-acuminate or mucronately subattenuate point, conspicuously villous-puberulent and often also atomiferous-glandular, their scarious margins mostly inconspicuous; flowers 5 per head; corolla white, 3–4 mm long. Achenes 2–3 mm long. (diploids, triploids, and tetraploids based on x = 10) Midsummer–fall. Pine barrens, wet meadows, and margins of ponds, especially in sandy soil; Mass to c Fla and w to La, almost wholly on CP with us; an outlying station in IP of Tenn.

25. E. sessilifolium L. Plants 6–15 dm tall, single-stemmed from a crown or short caudex, puberulent in the inflorescence, the herbage otherwise glabrous except that the leaves are gland-dotted. Leaves opposite, sessile or subsessile, lanceolate, usually broadly rounded at the base, serrate, acuminate or long-acute, mostly 7–18 × 1.5–5 cm, 2.5–7 times as long as wide, the lower smaller and deciduous. Inflorescence terminal, cymose-corymbiform; involucre 4.5–6.5 mm high, its bracts imbricate, broadly rounded to merely obtuse, villous-puberulent and usually also atomiferous-glandular; flowers 5 (6) per head, white. (diploids and triploids based on x = 10) Summer. Woods; s NH to se Minn, s to Va (all), NC (PP, BR), Ga (PP and mt. provinces), Ala (mt. provinces), and Ark (OU). Rare triploid forms approaching *E. rotundifolium* var. *saundersii* or var. *ovatum* have been called *E. vaseyi* Porter—G, or *E. sessilifolium* var. *vaseyi* (Porter) Fern. & Grisc.—F, R, but the name does not apply. These plants have somewhat hairy, relatively broad leaves, as compared to *E. sessilifolium*, but the venation is strictly pinnate as in *E. sessilifolium*.

26. E. rotundifolium L. Plants 3–15 dm tall, the stems mostly solitary or paired from a short rhizome or crown; herbage pubescent throughout with spreading, usually short and soft hairs; commonly atomiferous-glandular as well. Leaves opposite, or the upper sometimes alternate, sessile or subsessile, lanceolate to ovate or subrotund, toothed or the upper occasionally entire, broad-based, 2–12 × 1–6 cm. Inflorescence terminal, cymose-corymbiform; involucre 4.5–6.5 mm high, its bracts imbricate, sharply acute to obtuse, evidently villous-puberulent and often also atomiferous-glandular; flowers 5 per head (to 7 in var. *ovatum*), white. (x = 10) Summer. Woods; Me to c Fla, w to s Ohio, e Ky, Tenn (e half), Ala, and mainly on CP to Ark, se Okla, La, and Tex. The species as defined here consists of two basic diploids plus a complex superstructure of auto- and allopolyploids involving these two diploids and also several other species. One of the diploids is the core of var. *rotundifolium*, the other of var. *saundersii*. If var. *saundersii* were treated as a distinct species, it would take the name *E. pilosum* Walter. Most of the variants are provided for in the following key.

Leaves ovate (mostly broadly so) to subrotund, mostly evenly toothed except in var. *cordigerum*; upper leaves and main branches of the inflorescence usually all opposite, seldom alternate; dry or seldom wet soil.
- Leaves with broadly rounded, cordate-clasping base, rough-textured and strongly pubescent, regularly to often irregularly toothed or even lobulate; tetraploid probably originating by hybridization of *E. album* with *E. rotundifolium* var. *rotundifolium* and perhaps also var. *saundersii*, possibly not long-persistent, but frequently created in the area where the parental types occur together and have overlapping blooming seasons; Va to SC, w to Ark and c Miss; *E. cordigerum* (Fern.) Fern.—F .... var. *cordigerum* Fern.
- Leaves with cuneate to broadly rounded, but not cordate-clasping base.
  - Principal pair of lateral veins of the leaf diverging from the base of the midrib; leaves blunt, tending to be broadly cuneate or subtruncate at the base, the margins commonly crenate; diploids and chiefly autoploid triploids and tetraploids, common and widespread, nearly throughout the range of the species, except at the north .... var. *rotundifolium*.
  - Principal pair of lateral veins diverging distinctly above the base of the midrib; leaves mostly acute and generally with serrate margins.
    - Leaves distinctly cuneate below, firm, rather harshly short-hairy, up to ca. 5.5 × 3 cm, in life tending to be twisted at the base and vertically oriented; diploids and triploids, originating through hybridization of var. *rotundifolium* and *E. semiserratum*, at least some of the diploids repre-

senting F_1 hybrids; plants perhaps not long-persistent, but frequently created; SC to n Fla, w to Ark, La, and Okla; *E. scabridum* Ell.—S ...... var. *scabridum* (Elliott) A. Gray.

Leaves with broadly cuneate to broadly rounded base, rather smooth-textured, the pubescence tending to be relatively long and soft, and often also rather sparse; leaves not twisted at the base, often larger; triploids and tetraploids reflecting hybridization of var. *rotundifolium* with other species, perhaps mainly *E. sessilifolium*; well established and persistent from Me to Va and e Ky, and s chiefly in the mts. and on PP to n Ga and n Ala, seldom elsewhere, then perhaps sometimes reflecting hybridization with other spp.; *E. pubescens* Muhl.—F .... var. *ovatum* (Bigel.) Torr.

Leaves narrower, mostly lanceolate or lance-ovate to elliptic-ovate, mostly coarsely and unevenly toothed; upper leaves and main branches of the inflorescence tending to be alternate; chiefly in wet soil; mostly triploids, but sometimes diploid or tetraploid, well established and persistent from Mass to e Ky, c Tenn, Va, Fla, and Miss; *E. pilosum* Walt.—F, R; *E. verbenaefolium* Michx.—S .... var. *saundersii* (T. C. Porter) Cronq.

27. E. anomalum Nash. Much like *E. mohrii*, averaging a little more robust, up to ca. 15 dm tall, tending to be more densely short-hairy and sometimes with the rhizomes less conspicuously tuberous-thickened; leaves more often spreading or ascending, wider but not longer, the principal ones mostly 1–2 cm wide; involucre 4–5.5 mm high, averaging more densely hairy; achenes 2–3 mm long. (diploids, triploids, and most often tetraploids based on x = 10) Midsummer–fall. Wet low ground; CP from NC to c Fla. Plants probably reflecting hybridization between *E. rotundifolium* and *E. mohrii*, the diploids probably F_1 hybrids.

28. E. mohrii Greene. Plants with short, conspicuously tuberous-thickened rhizomes, the erect branches therefrom often branching near ground level into two or more aerial stems, these 3–10 (12) dm tall, puberulent or strigose-puberulent. Leaves opposite or the upper alternate, rarely ternate, mostly lance-elliptic, tapering to the sessile or subsessile base, bluntly few-toothed, tending to be recurved or deflexed, small, mostly 1.5–6 cm × 3–10 (12) mm, glandular-punctate, strigose or subglabrous. Heads in a dense, terminal, corymbiform, more or less flat-topped cyme; involucre 3–5 mm high, its bracts imbricate, broadly obtuse to acutish, atomiferous-glandular and slightly to evidently villous-puberulent, the inner inconspicuously scarious-margined and often somewhat pale distally; flowers 5 (6) per head, white. Achenes (1.2) 1.5–2 (2.5) mm long. (diploids, triploids, and seldom tetraploids based on x = 10) Summer, or spring–fall southward. Pond margins, ditches, shores, and moist low ground, often in sandy or peaty soil; se Va to s Fla and w to La, wholly on CP. *E. recurvans* Small—F, G, R, S, the sexual diploid phase. Polyploids, sometimes approaching *E. anomalum*, may perhaps reflect hybridization with *E. rotundifolium*.

29. E. cuneifolium Willd. Stems 3–10 dm tall, more or less clustered on a crown or caudex, commonly branched at or near ground level, copiously provided with short, mostly loosely spreading hairs. Leaves opposite, or the upper alternate, tending to be broadest above the middle, oblanceolate to obovate or nearly elliptic, tapering to the base, few-toothed or entire, usually triplinerved, 2–5 cm × 5–18 mm, glandular-punctate, often less hairy than the stem, the axillary shoots of the middle and lower ones commonly elongating into slender, mostly sterile branches with reduced leaves. Inflorescence terminal or somewhat diffuse; involucre 4–7 mm high, coarsely villous-puberulent and atomiferous-glandular, its bracts imbricate, broadly rounded to acute, the inner inconspicuously scarious-margined; flowers 5 per head; corolla white, 3–5 mm long. Achenes 2–3 mm long. (diploids, triploids, and tetraploids based on x = 10) Midsummer–fall. Pine and oak woods, mostly in dry, sandy soil or in sand hills; se Va to c Fla and w to Miss, mainly on CP. Triploids with oblanceolate leaves ca. 5 mm wide may reflect hybridization with *E. hyssopifolium*. *E. linearifolium* Walt.—F (probably misapplied), and *E. tortifolium* Chapm.—S, may be such triploids.

30. E. semiserratum DC. Stems mostly solitary from a very short, stout rhizome or crown, 5–12 dm tall, densely villous-puberulent, sometimes also atomiferous-glandular, often with loose axillary fascicles of a few reduced leaves. Leaves opposite or the uppermost scattered, firm, twisted at the base to bring the blade into a vertical plane, elliptic or elliptic-oblanceolate, mostly 4–8 cm × 8–30 mm, 2.5–6 times as long as wide, gradually narrowed to the sessile or shortly petiolar base, usually serrate or crenate-serrate, especially above the middle, finely and densely puberulent on both sides (and often also atomiferous-glandular), sometimes more shortly so above than beneath, triplinerved, the principal pair of lateral veins arising as branches from the midrib. Inflorescence terminal, flat-topped or somewhat elongate; involucre 2.5–4 mm high, softly short-hairy like the peduncles, its bracts imbricate, broadly rounded or obtuse to sometimes submucronately acute, the inner obscurely scarious-margined; flowers 5 per head, the corolla white, 2.5–3.5 mm long. Achenes 1.5–2.5 mm long. (n = 10) Late summer, fall. Low woods, clearings, and swampy places; Va to s Fla, w to Tex, and n to Ark, se Mo, and s Tenn, chiefly on CP and ME, but extending into IP in Tenn. *E. cuneifolium* var. *semiserratum* (DC.) Fern. & Grisc.—F.

31. E. lancifolium (T. & G.) Small. Similar in most respects to *E. semiserratum*, but less pubescent; stem often glabrous below the middle; leaves thick and firm, sometimes more coarsely and divergently toothed, of a duller, more blue-green color in life, subglabrous or the upper surface often wholly glabrous, the hairs on the lower surface often chiefly confined to the main veins; principal pair of lateral veins sometimes arising from the midrib, as in *E. semiserratum*, sometimes distinct to the base, as in *E. altissimum*; involucre averaging a little larger, up to 4.5 mm high. (n = 10) Summer. Prairies and open woods, usually in drier soil than *E. semiserratum*; s Ark (OU, GC), n La, and e Tex.

32. E. altissimum L. Plants stout, 8–20 dm tall, the stems arising singly or in pairs from a system of stout, branching rhizomes, softly villous-puberulent, or glabrate below. Leaves numerous, opposite, hairy like the stem, strongly trinerved to the base, lance-elliptic, gradually and narrowly acute or acuminate, and gradually narrowed to a sessile or shortly petioliform base, serrate above the middle or subentire, mostly 5–12 cm × 8–30 mm. Inflorescence large, terminal; involucre 4.5–6.5 mm high, short-hairy like the herbage, its bracts imbricate, broadly rounded or obtuse; flowers 5 per head, the corolla white, 4–5 mm long. Achenes 2–3 mm long. (diploids, triploids, and tetraploids based on x = 10) Late summer, fall. Woods, thickets, and clearings; Pa to s Minn and e Nebr, s to Va (all), NC (PP, BR), SC (BR), Tenn (all), n Ala (CU), the black belt on GC of Miss, Ark (all), and Tex. *E. saltuense* Fern.—F, G, from se Va, may reflect hybridization between *E. altissimum* and some other species such as *E. album*, or possibly between *E. hyssopifolium* and *E. album*.

33. E. hyssopifolium L. Stems solitary (seldom several) from a short crown, strigose or scabrous-puberulent, especially above. Leaves verticillate in 3's or 4's, or sometimes merely opposite, or even alternate above, narrow, mostly 6–40 times as long as wide, sessile or tapering to a slender base, spreading or ascending, glandular-punctate, otherwise glabrous or slightly hairy chiefly on the main veins beneath, the principal ones subtending conspicuous axillary fascicles of reduced leaves. Inflorescence terminal, corymbiform, commonly flat-topped; involucre 4–7 mm high, its bracts imbricate, broadly rounded to acute, the inner narrowly scarious-margined; flowers 5 per head, the corolla white, 3.5–4 mm long. (diploids, triploids, and tetraploids based on x = 10) Late summer, fall. Fields and other open places, especially in dry, sandy soil; coastal states from Mass to n Fla and w to La, and sometimes inland to Ohio, Ky, and Tenn. Two well-marked varieties:

Plants up to 1 m tall; leaves usually quaternate but sometimes ternate, opposite, or even alternate above, entire or inconspicuously and irregularly few-toothed, linear or nearly so, the main ones 2–7 cm × 1–5 mm and 10–40 times as long as wide. (diploids and polyploids) Mass to Ga and w to La, and sometimes inland as far as Tenn (all). *E. lechaeafolium* Greene—S; *E. hyssopifolium* var. *calcararum* Fern & Schub.—F. .... var. *hyssopifolium*

Plants more robust, up to 1.5 m tall; leaves sometimes quaternate, but more often ternate or merely opposite (or alternate above), conspicuously and divergently toothed, up to 10 cm long, the main ones mostly 5–10 (12) mm wide and 6–15 times as long as wide. (all polyploids, mainly tetraploid) s NY to Ga and n Fla and w to La, inland occasionally as far as s Ohio, c Ky (IP), and e Tenn (CU). *E. torreyanum* Short—S. This taxon probably originated through hybridization between *E. hyssopifolium* and some other species, but now is a stable entity .... var. *laciniatum* A. Gray

92. MIKANIA Willd., nom. conserv. CLIMBING HEMPWEED

Perennial twining vines, or rarely (some extralimital spp.) erect shrubs or herbs. Leaves opposite, simple, mostly petiolate, often cordate or hastate. Heads small, numerous, borne in open to dense corymbiform clusters, or sometimes (in extralimital spp.) in other sorts of inflorescences, strictly discoid, the flowers 4, all tubular and perfect; involucre narrow, of 4 essentially equal bracts, ours and many other spp. with a looser, greener, shorter bract of the inflorescence closely subtending each head; receptacle small, naked; corolla white to pink or ochroleucous; anthers with a small, hyaline, apical appendage, minutely rounded-auriculate at the base; style branches papillate, elongate, linear, acutish, with short ventromarginal stigmatic lines near the base. Achenes 5-angled, in our spp. atomiferous-glandular; pappus a single series of numerous capillary bristles. Holmes, W. C. 1975. A revision of *Mikania scandens* and relatives (Compositae). Ph.D. thesis. Mississippi State Univ.

1 Heads smaller, the involucre mostly (3.5) 4–5.5 (6) mm high, the achenes mostly 1.5–2.5 (2.7) mm long; herbage and involucres inconspicuously short-hairy or subglabrous .... 1. *M. scandens*.

1 Heads larger, the involucre mostly 6.5–8 mm high, the achenes mostly 3.5–4.5 mm long; herbage and involucres evidently and loosely spreading-hairy ..... 2. *M. cordifolia*.

1. M. scandens (L.) Willd. Twining and climbing or sprawling herbaceous vine to 5 m long, puberulent and somewhat atomiferous to subglabrous; stem weakly hexagonal; roots fleshy, fascicled. Leaves petiolate, the blade deeply cordate, usually obviously longer than wide, evidently acuminate, entire or sinuately few-toothed, palmately veined, 2.5–14 × 1.5–8.5 cm. Inflorescences small and numerous, on axillary peduncles; heads with white or often pinkish flowers, relatively small, the involucre mostly (3.5) 4–5.5 (6) mm high, its bracts ca. 1 mm wide or less, the corolla limb mostly 1.5–2 mm long, its lobes mostly 0.5–1 mm, the anthers mostly 1.2–1.5 mm. Achenes mostly 1.5–2.5 (2.7) mm long. (n = 19) Midsummer–fall, or all year southward.

Climbing on bushes in moist places; coastal states (especially CP) from s Me to s Fla, Tex, and adj. Mex, also up the ME to s Ill, and at scattered more inland stations, as in Ind and Mich; Bahama Isl; Cuba. *M. batatifolia* DC.—S.

2. M. cordifolia (L.f.) Willd. Habitally similar to *M. scandens*, but more evidently hairy, with relatively broader, blunt to merely acute leaves often as wide as or wider than long, and with larger heads; stem more distinctly hexagonal; involucre mostly 6.5–8 mm high, its bracts mostly ca. 1.5 (2) mm wide, the corolla limb mostly 2–3 mm long, its lobes 1.5–2 mm, the anthers mostly 1.7–2.5 mm long. Achenes mostly 3.5–4.5 mm long. (n = 18, 19) All year, but especially late fall–winter. Widespread in tropical Am, in our range principally in hammocks in s Fla, but also at scattered stations n to the Fla panhandle and w to s La.

93. AGERATUM L. AGERATUM

Annual (all ours) or perennial herbs, or shrubs. Leaves opposite, simple, in our spp. petiolate, evidently toothed, and tending to have 3 main veins. Heads rather small, in small, corymbiform clusters, often forming a flat-topped inflorescence, strictly discoid, the flowers all tubular and perfect; involucre campanulate to hemispheric or turbinate, its bracts narrow, firm but largely green or greenish, mostly 2 (3)-seriate, but not much imbricate; receptacle conic, generally (including our spp.) naked; flowers mostly blue-lavender, seldom white or ochroleucous; anthers with a small, hyaline, apical appendage, minutely rounded-auriculate at the base; style branches elongate, linear-clavate, papillate, with ventromarginal stigmatic lines toward the base. Achenes prismatic, 5-angled and -veined; pappus of 5–6 distinct, often shortly setaceous-tipped scales, or these more or less united into an irregularly toothed crown, or sometimes reduced to a mere annular crown, or obsolete. (x = 10) Johnson, M. F. 1971. A monograph of the genus *Ageratum* L. (Compositae-Eupatorieae). Ann. Mo. Bot. Gard. 58: 6–88.

1 Herbage essentially glabrous; leaves somewhat succulent .... 1. *A. littorale*.
1 Herbage evidently hairy or glandular or both; leaves not succulent.
  2 Involucre stipitate-glandular as well as long-hairy, or the long hairs themselves obviously viscid; principal leaves mostly with truncate to somewhat cordate base .... 2. *A. houstonianum*.
  2 Involucre glabrous or with some long, nonviscid hairs, not glandular; principal leaves with obtuse to sometimes truncate base .... 3. *A. conyzoides*.

1. A. littorale A. Gray. SEASHORE AGERATUM. Essentially glabrous, freely branching annual, mostly 1.5–5 dm tall, often decumbent and sometimes rooting below, or with more or less erect branches from prostrate main stems. Leaves somewhat succulent, but drying thin, with rhombic-ovate to broadly triangular-ovate blade mostly 1.5–5 × 0.7–4 cm, on a well-developed petiole sometimes fully as long. Heads several in a compact, corymbiform, naked-pedunculate terminal cyme; involucre 3–4 mm high, its innermost bracts sometimes minutely scabrous distally. Achenes glabrous; pappus a minute, annular, virtually obsolete crown, or seldom of several small, unequal, irregularly toothed or cleft, distinct or partly connate scales. All year. Open, wet or dry, often sandy or coral soil, in hammocks and along the shore; s Fla to Cuba, Bahama Isl, and Brit Hond.

2. A. houstonianum Miller. Erect or basally decumbent, freely branching, malodorous annual, mostly 3–10 dm tall, viscidly pilose to merely stipitate-glandular throughout. Larger leaf blades mostly 3–9 × 3–6.5 cm, triangular-ovate to broadly cordate-ovate, truncate to often more or less cordate (seldom merely obtuse) at the base, on a petiole sometimes more than half as long. Heads generally more or less numerous in a branching, often flat-topped, terminal cyme; involucre 4–5 mm high, viscid-pilose and commonly also stipitate-glandular, its bracts narrow, with slender, firm, subcaudately acuminate tip. Achenes sparsely scaberulous along the angles; pappus of 5 well-developed, distinct, shortly setose-tipped scales, or seldom the scales reduced and not setose-tipped. (n = 10) Summer, or all year southward. Native to Mex, C Am, and WI, widely cult., and sometimes escaped in our range, especially from Fla to NC.

3. A. conyzoides L. Much like no. 2, and not always easy to distinguish; leaf blades broadly obtuse to sometimes truncate at the base, but not at all cordate; involucre evidently hairy to largely glabrous, but scarcely viscid or glandular, its bracts tending to be broader and more shortly acuminate. (n = 10, 20). Summer, or all year southward. Native to tropical Am, now a pantropical weed, and occasionally found in our range as a temporary escape from cult.

94. SCLEROLEPIS Cass.

1. S. uniflora (Walter) B.S.P. Slender, rhizomatous, aquatic perennial, the stems emergent for 1–3 dm, simple or nearly so, atomiferous-glandular or subglabrous. Leaves verticillate in 4's to 6's, linear, entire, 7–23 mm long, up to 2 mm wide, punctate. Heads terminal, mostly solitary, naked-pedunculate, discoid, the rather numerous flowers all tubular and perfect; disk 4–12 mm wide; involucre campanulate or hemispheric, 3–4 mm high, its bracts subequal, subherbaceous or distally more scarious and anthocyanic; receptacle hemispheric or conic, naked; corollas pink-lavender; anthers with a short, broad hyaline terminal appendage, rounded-truncate at the base; style branches with short, ventromarginal stigmatic lines and an elongate, papillate, linear-clavate appendage. Achenes 5-angled and -nerved, atomiferous-glandular; pappus of several short, broad scales. Spring–summer. In still, shallow water; in the coastal states from NH to c Fla and w to sw Ala, wholly CP with us.

95. HARTWRIGHTIA A. Gray ex S. Watson

1. H. floridana A. Gray ex S. Watson. Fibrous-rooted perennial from a thickened, short, praemorse rhizome, 6–12 dm tall, atomiferous-glandular (or in part impressed-glandular) throughout, even to the corollas and achenes. Leaves alternate, simple, entire, basally disposed, the lowest ones with elliptic-oblanceolate to elliptic blade 5–25 × 1–8 cm tapering to a petiole nearly as long, or longer; cauline leaves progressively reduced, the upper linear and distant. Heads more or less numerous in an open, corymbiform or more or less flat-topped, branching cyme, pink, discoid; flowers mostly 7–10, all tubular and perfect; involucre 3–4 mm high, of a few blunt, linear, obscurely biseriate subequal bracts, or some of the outer shorter; receptacle convex, with a few bracts near its margin resembling the inner involucral ones; corollas white or pink, surpassing the involucre; anthers with a small, elliptic-ovate, deeply retuse, hyaline, apical appendage, truncate at the base; style branches exserted, linear or slightly clavate, rounded above, papillate externally, with short ventromarginal stigmatic lines at the base. Achenes ca. 3–3.5 mm long, obpyramidal, acutely 5-angled and -veined, slightly contracted at the callous-thickened tip, which is minutely 5-lobed in alignment with the angles of the body. Pappus of several fragile, deciduous awns, or more often wanting. Late summer. Wet, low ground, sometimes in pine woods; c to ne Fla.

96. GARBERIA A. Gray

1. **G. fruticosa** (Nutt.) A. Gray. Freely branched, aromatic, more or less evergreen shrubs 1–2.5 m tall; young twigs mealy-puberulent and sometimes atomiferous-glandular. Leaves numerous, alternate (the lower reputedly opposite), simple, entire, not veiny, spatulate to spatulate-obovate or rotund-obovate, often slightly retuse, tapering to the shortly petiolar or subsessile base, mostly 1.5–3.5 cm long and 0.7–2 cm wide, viscid and minutely glandular-punctate, mealy when young. Heads in small, corymbiform clusters terminating the branches, discoid, the 5 flowers all tubular and perfect; involucre narrow, 9–12 mm high, mealy, its bracts lance-triangular to linear-oblong, acute or acuminate, the outer subherbaceous, the inner more chartaceous and striate, arranged in 5 vertical ranks of 3 (less often 4) each; receptacle small, naked; corolla pink-purple, 8–10 mm long, narrowly goblet-shaped, the limb about equaling the slender tube, the lobes as long as or longer than the throat; anthers with a small, hyaline-scarious terminal appendage, minutely rounded-auriculate at the base; style branches elongate, linear, blunt, papillate, with short ventromarginal stigmatic lines near the base. Achenes slender, ca. 10-ribbed, 7–8 mm long, conspicuously short-hairy; pappus of numerous brownish barbellate capillary bristles in about 2 series, the outer tending to be shorter and more slender than the inner. Nearly all year, but mainly late summer and fall. Dry, sandy, pine or pine-oak scrub and prairies; c and n peninsular Fla. If the names in Bartram's Travels are considered to be validly published, our plant must take the name *G. heterophylla* (Bartram) Merrill & F. Harper.

97. BRICKELLIA Elliott, nom. conserv.

Perennial herbs (our spp.) or sometimes shrubs or annual herbs. Leaves chiefly or wholly cauline, opposite or alternate (rarely whorled), simple, in ours petiolate and with deltoid-ovate to subcordate or cordate, acuminate or narrowly acute, evidently trinerved blade. Heads discoid, the flowers all tubular and perfect, the corollas rather slender, with short teeth, white or ochroleucous or chloroleucous to pink or purple; involucral bracts imbricate in several series, the principal ones striate and scarcely herbaceous; receptacle naked; anthers with a small, hyaline, apical appendage, minutely rounded-auriculate at the base; style branches papillate, elongate, linear-clavate, rounded-obtuse, with short ventromarginal stigmatic lines near the base. Achenes mostly 10-ribbed; pappus of 10–80 barbellate or nearly smooth (rarely subplumose) bristles in a single series. (x = 9) Robinson, B. L. 1917. A monograph of the genus *Brickellia*. Mem. Gray Herb. 1: 1–147. *Coleosanthus* Cass.

1 Corollas anthocyanic; pappus bristles purplish; leaves all or nearly all opposite .... 1. *B. cordifolia.*
1 Corollas ochroleucous or chloroleucous; pappus bristles white; middle and upper leaves, or many of them, generally alternate .... 2. *B. grandiflora.*

1. B. cordifolia Elliott. Perennial, apparently fibrous-rooted, 5–12 dm tall, softly velvety-puberulent, especially on the lower surfaces of the leaves, sometimes also atomiferous-glandular; stems mostly simple up to the inflorescence. Leaves all or nearly all opposite, the blade mostly 5–10 × 3–6.5 cm, crenate or sometimes crenate-serrate. Heads several in a terminal, corymbiform inflorescence; flowers fairly numerous (ca 45), the corolla purple to dark rose or lavender, 6–9 mm long; involucre mostly 8–11 mm high, the very outermost bracts abruptly differentiated from the others, more herbaceous, slender, elongate, linear-caudeate, scarcely striate. Achenes ca. 5 mm long, slightly hairy distally; pappus of numerous (ca. 40) purplish, barbellate, capillary bristles. Late summer, fall. In woods; CP of Ga and e Ala, s to c Fla. *Coleosanthus cordifolius* (Ell.) Kuntze—S.

2. B. grandiflora (Hook.) Nutt. Perennial from long, thickened roots, the stems clustered, 2.5–7 dm tall, often freely branched; herbage minutely (and sometimes sparsely) puberulent or hirtellous throughout, often also with sessile, semiimpressed glands especially on the lower surface of the leaves. Leaves opposite below, otherwise generally many or all of them alternate, the blade 2–11 × 1–6 cm, crenate to serrate. Heads several or numerous in a short, corymbiform or subumbelliform inflorescence; involucre 7–12 mm high, regularly imbricate, the outer bracts tipped with a well developed, slender appendage; flowers mostly 20–40, sometimes more, the corolla ochroleucous or chloroleucous, 6–8 mm long. Achenes 4–5 mm long, short-hairy at least distally; pappus of ca. 20–25 white, barbellate, capillary bristles. (n = 9) Late summer, fall. Rocky woods and bluffs; widespread in w US, and extending e to the Ozark region of Mo and Ark.

98. KUHNIA L. FALSE BONESET

1. K. eupatorioides L. Puberulent to subglabrous perennial from a stout, cylindrical taproot; stems several, 3–13 dm tall. Leaves all cauline, numerous, opposite to more often subopposite or scattered, gland-dotted beneath, variable in form and size, up to 10 × 4 cm, tapering to a sessile or often shortly petiolar base. Heads mostly in small corymbiform clusters terminating the branches, strictly discoid, the flowers all tubular and perfect, with cylindrical corolla and short lobes, ochroleucous, in our varieties mostly 7–15 per head; involucre 7–14 mm high, its bracts in several series, the inner greenish but scarcely herbaceous, striate, only slightly unequal inter se, the outer several series evidently differentiated from the inner, more herbaceous, narrower, and attenuate-tipped, the involucre thus somewhat calyculate; receptacle flat or nearly so, naked; anthers with a small, hyaline, apical appendage, minutely rounded-auriculate at the base; style branches papillate, elongate, linear-clavate, rounded-obtuse, with inconspicuous, short, ventromarginal stigmatic lines near the base. Achenes mostly 10-nerved, cylindric, 3.5–5.5 mm long, short-hairy; pappus of 20 plumose bristles in a single series. (n = 9) Late summer, fall, or all year southward. Dry, open places, especially in sandy soil; NJ to s Fla, w to Mont, Tex, and Ariz. Three varieties in our range:

Outer involucral bracts with much prolonged setaceous tip, only slightly or scarcely shorter than the inner; vegetatively like var. *eupatorioides*, Ozarkian var., in Ark, Mo, and s Ill .... var. *ozarkana* Shinners.
Outer involucral bracts not so strongly setaceous-prolonged, seldom much more than half (three-fourths) as long as the inner; not Ozarkian.
  Leaves very narrow, mostly linear and entire, commonly 1–3 mm wide, seldom some of the lower ones up to 5 mm wide; s CP, mostly in Fla, but n, often in forms transitional to the next var., to Ala, Ga, and SC. *K. mosieri* Small—S .... var. *gracilis* T. & G.
  Leaves wider, from broadly linear or narrowly lanceolate to lance-ovate or broadly rhombic-lanceolate, at least the better developed ones mostly 5–40 mm wide, often some of them toothed; widespread in our range, but not Ozarkian and not extending far into peninsular Fla. *K. glutinosa* Ell.—S .... var. *eupatorioides*.

99. CARPHEPHORUS Cass.

Fibrous-rooted perennial herbs from a stout, simple or branched caudex, crown, or short rhizome. Leaves alternate, simple, entire or seldom toothed, basally disposed, the basal (and sometimes lower cauline) ones evidently the largest and mostly tapering to a petiolar base (in one species the leaves all linear, without an expanded blade), the others obviously smaller and generally sessile. Inflorescence corymbiform or sometimes thyrsoid-paniculate; heads discoid, the flowers all tubular and perfect; involucral bracts more or less herbaceous or partly hyaline-scarious, sometimes anthocyanic in part, from few and subequal to numerous and imbricate; receptacle chaffy (often only toward the margin) or naked; corolla pink-purple; anthers with a small, hyaline, apical appendage, minutely rounded-auriculate at the base; style branches papillate, elongate, linear or linear-clavate, with short ventromarginal stigmatic lines near the base. Achenes 10-ribbed, columnar-clavate, short-hairy; pappus of 20–55 coarse, unequal, barbellate-bristles in a single series. (n = 10) *Trilisa*, *Litrisa*. The genus consists of only the following 7 well-defined species. Correa, M. D. and R. L. Wilbur, 1969. A revision of the genus *Carphephorus*. J. Elisha Mitchell Soc. 85: 79–91.

1 Heads relatively large, the involucre mostly (6) 7–12 mm high, composed of mostly 15–30 or more bracts, containing mostly (10) 15–35 flowers; receptacle chaffy, at least toward the margin, with several or many pales.
  2 Principal leaves oblanceolate or broader, mostly 5–25 mm wide.
    3 Stem glabrous or subglabrous below the inflorescence; cauline leaves relatively few and remote, not clothing the stem; involucral bracts broadly rounded distally, glabrous on the back .... 1. *C. bellidifolius*.
    3 Stem evidently spreading-hairy below the middle or throughout; cauline leaves relatively numerous, erect and more or less clothing the stem.
      4 Involucral bracts glabrous on the back, broadly rounded and often erose; inflorescence compact and dense .... 2. *C. corymbosus*.
      4 Involucral bracts strongly viscid-hairy, acute or acutish, often callous-pointed, inflorescence open .... 3. *C. tomentosus*.
  2 Principal leaves narrowly linear, mostly 1–2 (3) mm wide; stem and involucre much as in *C. tomentosus* .... 4. *C. pseudo-liatris*.
1 Heads relatively small, the involucre 3.5–6 mm high, composed of 5–12 bracts, containing mostly 4–10 (15) flowers; receptacle naked or sometimes with 1 or 2 (3) pales.
  5 Stem densely spreading-hairy throughout; involucre often hairy as well as glandular.
    6 Inflorescence short, broad, open-corymbiform; involucral bracts evidently imbricate in several series .... 5. *C. carnosus*.
    6 Inflorescence thyrsoid-paniculate, generally more or less elongate and narrow; involucral bracts subequal .... 6. *C. paniculatus*.
  5 Stem glabrous; involucral bracts somewhat glandular but not otherwise hairy, subequal or the outer shorter; inflorescence open-corymbiform .... 7. *C. odoratissimus*.

1. C. bellidifolius (Michx.) T. & G. Stems generally several or numerous, 2–6 dm tall, puberulent in the inflorescence, otherwise glabrous or subglabrous. Leaves punctate and atomiferous-glandular, otherwise glabrous, the basal and lower cauline ones ascending or erect, oblanceolate, acute or obtuse, mostly 4–20 × 0.5–2.5 cm; cauline leaves relatively few and remote, not clothing the stem. Heads several or numerous in an open-corymbiform, often flat-topped inflorescence, involucre 7–12 (15) mm high, composed of ca. 15–30 bracts, these well imbricate in several series, minutely ciliolate, otherwise glabrous, broadly rounded distally, sometimes very narrowly hyaline-margined and anthocyanic; receptacle chaffy at least toward the margin, with several or many pales; flowers ca. 15–30; corolla mostly 7–10 mm long, its lobes 1.5–2.5 mm long. Achenes 4–6 mm long. (n = 10) Late summer, fall. Sandy, mostly rather dry soil, often in pine barrens or pine-oak woods or in sand hills; AC from se Va to Ga.

2. C. corymbosus (Nutt.) T. & G. Stem solitary, 3–12 dm tall, shortly spreading-hirsute throughout. Leaves punctate and atomiferous-glandular, otherwise glabrous, obtuse or broadly-rounded, the basal and lowermost cauline ones ascending or erect, oblanceolate, mostly 6–20 × 1–2 (2.5) cm, the others abruptly reduced, numerous, and more or less clothing the stem, erect or ascending, mostly sessile and oblong to oblong-oblanceolate or oblong-obovate. Heads several to fairly numerous in a short, broad, dense, often flat-topped, corymbiform inflorescence; involucre (6) 7–10 mm high, composed of ca. 15–20 bracts, these well imbricate in several series, broad, blunt, at least the inner with hyaline-scarious, often erose or ciliate, often anthocyanic margins; receptacle chaffy at least toward the margin, with several or fairly numerous pales; flowers mostly (10) 12–20; corolla mostly 7–9 mm long, its lobes ca. 1.5 mm long or less. Achenes 2–4 mm long. (n = 10) Midsummer–fall. Mostly in dry, sandy pine woods; peninsular Fla and s Ga.

3. C. tomentosus (Michx.) T. & G. Stems 1–several, 2–8 dm tall, evidently spreading-hirsute below, the hairs commonly shorter and more appressed upward. Leaves glabrous or usually more or less hairy, atomiferous-glandular but not punctate, the basal and lower cauline ones ascending or erect, oblanceolate or a little broader, acute or somewhat obtuse, mostly 3.5–15 × 0.5–2 cm; cauline leaves quickly reduced and becoming sessile, erect, more or less numerous and clothing the stem. Heads 2–40 in an open-corymbiform, often flat-topped inflorescence; involucre 7–11 mm high, conspicuously viscid-hairy, composed of ca. 20–40 bracts, these well imbricate in several series, firm, with rather lax or spreading, thick, often callous-pointed, acute or acutish tip; receptacle chaffy toward the margin, commonly with 4–5 pales; flowers ca. 15–30; corolla mostly 6–9 mm long, its lobes ca. 2 mm long. Achenes 3–4 mm long. (n = 10) Late summer, fall. In wet to fairly dry, often sandy or peaty soil, especially in pine barrens, sometimes in pine-oak woodland or on sand hills; AC from se Va to Ga.

4. C. pseudo-liatris Cass. Stem solitary, mostly 3–10 dm tall, rather shortly spreading-villous-hirsute throughout, surrounded at the base by the elongate, fibrous remnants of the leaves of previous years. Leaves glabrous, inconspicuously punctate, or the cauline ones ciliate-margined, all very narrow, the lowest ones elongate and needlelike, thick and firm, mostly 10–40 cm long and only 1–2 (3) mm wide; cauline leaves progressively reduced, numerous, erect, more or less clothing the stem. Heads mostly 3–15 in a compact, corymbiform cluster only 2–6 cm wide; involucre 6–9 mm high, strongly villous-hirsute, composed of ca. 15–25 bracts, these thick and firm, narrowly acute, mostly callous-tipped; receptacle chaffy nearly throughout, the pales about as many or more numerous than the involucral bracts; flowers mostly 20–35; corolla ca. 5–8 mm long, its lobes 1–2 mm. Achenes 2–3 mm long. (n = 10) Fall. Moist to dry, sandy pine woods and savannas; sw Ga and the Fla panhandle, w on GC to La.

5. C. carnosus (Small) C. W. James. Stems solitary, 2–5 (9) dm tall, densely and rather shortly viscid-villous throughout, beset with scattered, narrow, erect or ascending bracts 1–3 (5) cm long, not appearing leafy. Foliage leaves all in a flat basal rosette, thick and succulent, punctate and sometimes inconspicuously atomiferous-glandular, commonly ciliate, otherwise glabrous, more or less oblanceolate, acute, mostly 3–7 (9) cm × 4–13 mm. Heads in a short and broad, terminal, fairly compact, often flat-topped, corymbiform cyme; involucre 4.5–6 mm high, composed of mostly 5–10 bracts, these densely viscid- or glandular-hairy and more or less ciliate, evidently imbricate in several series, relatively broad, the middle and outer ones acute or abruptly acuminate or somewhat mucronate, the inner more obtuse or rounded; receptacle naked or sometimes with 1 or 2 pales; flowers mostly 5–10, conspicuously surpassing the involucre; corolla mostly 4–5.5 mm long, the lobes mostly less than 1 mm long. Achenes ca. 2 mm long. (n = 10) Summer. Moist low ground, in flat woods or wet prairies or margins of swamps; c Fla. *Litrisa carnosa* Small—S.

6. C. paniculatus (J. F. Gmel.) Hebert. DEER'S TONGUE. Stem apparently solitary, (3) 5–12 (18) dm tall, densely and rather coarsely spreading-villous-hirsute throughout. Basal and lowermost cauline leaves coarse and firm, more or less ascending, oblanceolate to rather narrowly elliptic, mostly 5–35 × 1–4 cm, inconspicuously punctate, essentially glabrous; cauline leaves much reduced, numerous, erect and clothing the stem, often somewhat hairy. Heads numerous in a more or less elongate, thyrsoid-paniculiform to almost racemiform inflorescence; involucre (3.5) 4–6 mm high, composed of 6–9 (11) bracts, these subequal, acute to rounded, glandular or glandular-hairy, often purplish; receptacle naked or seldom with 1 or 2 pales; flowers mostly 4–10; corolla ca. 4–5 mm long, its lobes ca. 1–1.5 mm long. Achenes 2–3 mm long. (n = 10) Late summer, fall. Moist or wet, low ground, in flatwoods or wet prairies or savannas or margins of swamps; CP of NC, s to c Fla, and w to se Ala and the Fla panhandle. *Trilisa paniculata* (Walt.) Cass.—S.

7. C. odoratissimus (J. F. Gmel.) Hebert. DEER'S TONGUE, VANILLA PLANT. Plants fragrant, bearing coumarin; stem apparently solitary, mostly 5–14 (18) dm tall, glabrous. Leaves somewhat succulent, punctate, minutely atomiferous, otherwise glabrous, glaucous beneath, the lowest ones ascending, oblanceolate to almost obovate, mostly 9–50 × 2–10 cm, those on the lower half or third of the stem (but above the base) smaller but still relatively well developed, mostly sessile, commonly broad-based and clasping, often coarsely few-toothed, those on the upper half of the stem progressively reduced and distant, becoming bractlike. Heads more or less numerous in a short, broad, often flat-topped, open corymbiform cyme; involucre 3.5–5 mm high, its bracts 6–12, subequal or the outer shorter, somewhat glandular but not otherwise hairy; receptacle naked or often with 1 or 2 pales; flowers 5–10 (15); corolla mostly 4–5 mm long, its lobes ca. 1 mm long or less. Achenes 2–3 mm long. (n = 10) Aug–Nov, or all year southward. Moist or wet, low ground, in flat woods or wet prairies or savannas or margins of swamps; throughout Fla, n on AC to NC, and w on GC to La. *Trilisa odoratissima* (Walt.) Cass.—S.

100. LIATRIS Schreb., nom. conserv. BLAZING STAR, GAY FEATHER, BUTTON SNAKEROOT

Perennial herbs with a thickened, usually cormlike rootstock. Leaves alternate, simple, entire, evidently to obscurely punctate, narrow and sessile, or with an evident blade tapering to the petiole, the basal or lower cauline ones usually the largest. Heads small to fairly large, in a mostly spiciform or racemiform (seldom corymbiform or evidently branched) inflorescence, strictly discoid, the flowers all tubular and perfect; involucral bracts imbricate, often punctate, green or greenish at least in part, but often with scarious, anthocyanic or pale margins or tip, the innermost bracts occasionally encroaching onto the margin of the otherwise naked receptacle; flowers pink or pink-purple, or casually white, anthers with a small, hyaline, apical appendage, minutely rounded-auriculate at the base; style branches elongate-clavate, obtuse, papillate, with inconspicuous ventromarginal stigmatic lines near the base. Achenes ca. 10-ribbed, hairy; pappus of 1 or 2 series of rather stout, barbellate or plumose bristles. (x = 10) *Lacinaria*—S. Gaiser, L. O. 1946. The genus *Liatris*. Rhodora 48: 165 et seq.

Many of the species are not sharply limited. There are numerous hybrids, some of them between species that are not closely related.

The morphological nature of the rootstock is not clear. In most species it is more or less globose and produces a cluster of slender, fibrous roots at the base. Although it lacks any external suggestion of leaves, this globose body has often been called a corm. In other species what is evidently the same morphological structure looks more like a tuberously thickened, sometimes horizontally segmented and sometimes branched taproot, or even like a fleshy-thickened, scaleless rhizome or a cluster of fingerlike tuberous roots.

In the following descriptions, measurements of the length of the leaves include the petiole (if any) as well as the blade.

1 Pappus more or less strongly barbellate to sometimes subplumose, the barbels mostly 0.1–0.3 (0.4) mm long.
 2 Heads relatively broad, with mostly 14–many flowers; larger leaves, except in *L. ohlingerae*, mostly 1–4 cm wide.
  3 Leaves all very narrow, only 1–2.5 mm wide; heads relatively very large, the involucre mostly 17–23 mm high .... 1. *L. ohlingerae*.
  3 Larger (lower) leaves mostly (7) 10–40 mm wide; heads not so large, the involucre mostly 7–17 mm high.
   4 Middle involucral bracts herbaceous (or somewhat coriaceous) throughout, or with narrow, entire or slightly erose, but scarcely lacerate scarious margins.
    5 Flowers mostly 25–80 per head; heads seldom more than 20 .... 2. *L. scariosa*.
    5 Flowers fewer, mostly 14–24 per head; heads often more than 20.
     6 Middle and outer involucral bracts generally loose or squarrose above the middle; heads tending to be turned away from the axis; involucral bracts glabrous or often evidently short-hairy on the back ..... 3. *L. squarrulosa*.
     6 Middle and outer involucral bracts, like the inner, appressed; heads closely ascending, not turned away from the axis; involucral bracts glabrous .... 5. *L. turgida*.
   4 Middle involucral bracts with wide, uneven, irregularly lacerate scarious margins; flowers mostly 16–35 per head .... 4. *L. aspera*.
 2 Heads relatively narrow, with mostly 3–14 flowers; leaves very often all less than 1 cm wide, though sometimes wider in some species.
  7 Corolla not much elongate, mostly 6–11 mm long, including the 1.5–3 (3.5) mm lobes.
   8 Corolla tube evidently hairy toward the base within.
    9 Pappus notably short, not reaching the sinuses of the corolla .... 6. *L. helleri*.

9 Pappus of ordinary length, reaching or surpassing the sinuses of the corolla.
10 Involucral bracts mostly obtuse or rounded (in *L regimontis* often acutish and commonly abruptly callous-mucronulate); flowers often more than 4.
11 Involucral bracts firm, coriaceous, strongly glandular-punctate, tending to be keeled or cupped above and often acutish, the midrib generally excurrent as a short, broad, blunt mucro ...... 7. *L. regimontis*.
11 Involucral bracts thinner and more chartaceous, puntate or not, but only slightly or not at all keeled, cupped, or mucronate, instead mostly with flat, broadly rounded (seldom acutish) tip.
12 Flowers mostly 7–14 per head; inflorescence tending to be dense and spikelike, the heads more or less erect on closely ascending peduncles, or nearly sessile but still suberect; involucre 7–12 mm high.
13 Larger leaves mostly (7) 10–20 mm wide; leaves and involucral bracts only weakly or scarcely punctate .... 5. *L. turgida*.
13 Larger leaves rarely more than 7 (10) mm wide; leaves and involucral bracts evidently to scarcely punctate .... 8. *L. graminifolia*.
12 Flowers mostly 3–6 (7) per head, rarely more; inflorescence tending to be loose and racemelike, the heads borne on divergent or arcuate peduncles, or nearly sessile but still spreading away from the axis; involucre (4) 5–7 (8) mm high .... 9. *L. gracilis*.
10 Involucral bracts all strongly acute or acuminate; flowers 3–4 .... 10. *L. provincialis*.
8 Corolla glabrous (or very nearly so) within.
14 Plants with a thickened, cormlike base and slender, fibrous roots.
15 Involucral bracts all appressed or merely a little loose at the tip; axis of the inflorescence glabrous or nearly so.
16 Involucral bracts, or at least the inner ones, obtuse or broadly rounded (but commonly abruptly mucronulate in *L. tenuifolia*).
17 Heads evidently pedunculate; flowers 3–5 (6) per head.
18 Involucral bracts all broadly rounded or obtuse; pappus short, not reaching the sinuses of the corolla, cauline leaves fairly gradually reduced upward .... 13. *L. microcephala*.
18 Middle and outer involucral bracts sharply acute or acuminate, the innermost ones more broadly rounded and usually mucronulate; pappus of ordinary length, generally reaching or surpassing the sinuses of the corolla; cauline leaves tending to be abruptly reduced and setaceous .... 14. *L. tenuifolia*.
17 Heads strictly sessile; flowers (5) 6–10 (14) per head .... 15. *L. spicata*.
16 Involucral bracts all sharply acute or acuminate .... 16. *L. acidota*.
15 Involucral bracts with abruptly spreading or recurved, acuminate tip; axis of the inflorescence usually evidently spreading-hairy, seldom glabrous .... 17. *L. pycnostachya*.
14 Plants with a cluster of tuberous-thickened, fingerlike roots, and without a well-defined cormlike base .... 18. *L. garberi*.
7 Corolla much elongate, mostly 11–16 mm long, including the 3–5 mm lobes.
19 Heads evidently (though sometimes shortly) pedunculate, tending to be turned away from the axis, forming a racemiform, usually strongly secund inflorescence .... 11. *L. pauciflora*.
19 Heads essentially sessile, erect or closely ascending, forming a spiciform, not at all secund inflorescence .... 12. *L. chapmanii*.
1 Pappus evidently plumose, the barbels mostly 0.5–1 mm long.
20 Flowers 10–60 per head; corolla lobes coarsely hairy within.
21 Involucral bracts appressed or merely a little loose, the tip mostly broadly rounded and shortly mucronate .... 19. *L. cylindracea*.
21 Involucral bracts with loosely spreading or squarrose, acuminate tip ...... 20. *L. squarrosa*.
20 Flowers 4–6 per head; corolla lobes glabrous.
22 Involucral bracts firm, mostly sharply mucronate-acuminate, not at all petaloid; corolla tube evidently hairy toward the base within .... 21. *L. punctata*.
22 Inner involucral bracts with prolonged, slightly expanded, scarious, petaloid, pink or white, loosely spreading tip; corolla tube glabrous within .... 22. *L. elegans*.

1. L. ohlingerae (Blake) B. L. Robinson. Plants with a tuberous-thickened, cylindric, segmented primary root; stems 3–10 dm tall, villous-puberulent. Leaves numerous, glabrous or nearly so, all linear and only 1–2.5 mm wide, the lowermost ones up to 15 cm long but commonly deciduous, the others mostly 3–8 cm long, gradually reduced upward. Heads mostly 3–30 in an open-corymbiform to very openly racemiform inflorescence, on arcuate or ascending-spreading peduncles mostly 1.5–7 cm long, relatively very large, the involucre mostly 17–23 mm high, its bracts loosely erect, obtuse or broadly rounded, subherbaceous with scarious and ciliolate margins, sometimes partly anthocyanic; flowers ca. 20–30, the corolla ca. 2 cm long including the 5–8 mm lobes, glabrous within. Achenes 7–10 mm long; pappus strongly barbellate, inconspicuously biseriate, the outer series somewhat the shorter. Late summer. Sand hills in Polk and Highland Cos., c Fla. *Ammopursus ohlingerae* (Blake) Small—S.

2. L. scariosa (L.) Willd. Plants glabrous or sometimes hairy, 3–8 (10) dm tall. Lowermost leaves long-petiolate, with elliptic to linear-elliptic or broadly oblanceolate blade, mostly 10–35 cm long and (ours) 2–5 cm wide. Heads few to fairly numerous, seldom more than 20 (35), subsessile or more often ascending on arcuate to sometimes spreading peduncles up to 5 cm long; involucre more or less hemispheric, 9–17 mm high, its bracts appressed to more often loose or sometimes distally squarrose, glabrous or atomiferous to evidently puberulent on the back, broadly rounded, often anthocyanic distally, the middle ones only narrowly or not at all scarious-margined, often ciliolate but not erose-lacerate, the innermost ones sometimes more obviously scarious and erose; flowers (21) 25–80 per head; corolla tube hairy toward the base within. Pappus strongly barbellate. (n = 10) Late summer, fall. Prairies, open woods, and other dry, open places; Me to Pa and in the mts. to n Ga and se Tenn, w to Mich, Mo, and Ark. The species consists of 3 geographic vars., two in our range, the third (var. *novae-angliae* Lunell) in New England and adjacent NY. Var. *scariosa* has relatively few and well-spaced cauline leaves, mostly 8–20 (25) below the inflorescence, and seldom has more than ca. 40 flowers per head; it occurs in the s Appalachian region, often on shale barrens. *Lacinaria scariosa* (L.) Hill—S. Var. *nieuwlandii* Lunell has more numerous and crowded cauline leaves, commonly 20–60 below the inflorescence, and has up to 80 flowers per head; it occurs in inland ne and c US, from NY, Pa, and n WVa to n Mich, Mo, and Ark (OZ). *L. novae-angliae* var. *nieuwlandii* (Lunell) Shinners—G; *L. borealis* Nutt.—F.

3. L. squarrulosa Michx. Similar to *L. scariosa* and evidently hybridizing with it, but averaging more robust, up to 15 dm tall, and more often evidently hairy; heads 6–60, often more than 20, commonly in a spiciform inflorescence with the heads turned away from the axis, or the peduncles sometimes more elongate and ascending-spreading, up to ca. 5 cm long; involucre glabrous or often evidently short-hairy, the middle and outer bracts commonly notably loose or squarrose above the middle; flowers (11) 14–24 (28) per head. Midsummer, fall. Dry woods and open places, especially in rocky or sandy soil; s WVa and s Ohio and w Va to coastal SC and n Fla, w to s Mo, Ark, and La. *L. earlei* (Greene) K. Schum.—F, R; *L. scabra* (Greene) K. Schum.—F, G, the more hairy phase; *Lacinaria ruthii* Alexander—S; *Lacinaria shortii* Alexander—S; *Lacinaria tracyi* Alexander—S, a narrow-leaved, perhaps varietally separable form of CP.

4. L. aspera Michx. Plants short-hairy, or glabrous throughout, 4–15 dm tall. Lowest leaves long-petiolate, 5–40 × 1–4.5 cm; cauline leaves 25–90 below the inflorescence. Heads (10–) more or less numerous in an elongate-spiciform inflorescence, or the peduncles occasionally more elongate and up to 5 cm long; involucre 8–15 mm high, campanulate or subhemispheric, glabrous, its bracts loosely spreading or squarrose, tending to be bullate, often purplish upward, with conspicuous, lacerate, often crisped, scarious margins that are sometimes folded under; flowers 16–35 per head; corolla hairy toward the base within. Pappus strongly barbellate. (n = 10) Late summer, fall. Dry, open places and thin woods, especially in sandy soil; chiefly OZ and prairie sp., from Minn and e ND to c Neb, c Okla, and e Tex, e to Mich, Ark, La, and Miss, and occasionally to Ohio, w Va, NC (BR), SC (PP), sw Ga, and nc Fla. *Lacinaria aspera* (Michx.) Greene—S.

5. L. turgida Gaiser. Relatively small plants, mostly 2–8 dm tall, glabrous or somewhat hirsute. Leaves relatively few, weakly or scarcely punctate, the lowest ones oblanceolate, 7–30 cm × 7–20 mm, usually irregularly ciliate-margined toward the base. Heads mostly 5–40, closely ascending in a more or less spiciform inflorescence, subsessile or on peduncles up to 1 (3) cm long; involucre turbinate, 7–12 mm high, its bracts appressed, broadly rounded, scarious-margined, ciliate, not strongly punctate, generally purplish above; flowers mostly 8–18 (20), often ca. 13; corolla hairy toward the base within. Pappus strongly barbellate. (n = 10) Late summer. Dry, rocky woods; mts. of Va, WVa, and NC to n Ga and ne Ala. *Lacinaria pilosa* (Ait.) Heller—S, probably misapplied.

6. L. helleri T. C. Porter. Small, glabrous plants, 1–5 dm tall. Leaves rather numerous and crowded, scarcely punctate, all narrow, the lowest ones mostly 2–20 cm × 3–7 (10) mm, not ciliate. Heads turbinate, 3–20 (30), closely ascending in a more or less elongate, spiciform inflorescence; involucre 7–10 mm high, its bracts glabrous or minutely ciliolate, scarcely punctate, the outer commonly lance-triangular and wholly herbaceous, the others more oblong (or linear) and distally rounded, thinner, often with narrow hyaline-scarious margins and often partly anthocyanic; flowers mostly 7–10 per head, the corolla tube short-hairy toward the base within. Pappus notably short, not reaching the sinuses of the corolla. (n = 10) Late summer. Open, rocky outcrops at high elev.; BR of NC, notably Blowing Rock, Roan Mt., and Grandfather Mt. *Lacinaria helleri* Porter—S.

7. L. regimontis (Small) K. Schum. Stem glabrous, 4–10 dm tall. Leaves markedly punctate, linear or nearly so, 6–30 cm × 2–7 (10) mm, the margins irregularly ciliate toward the base, the surfaces sometimes also sparsely hairy. Heads sessile or on stout, often bracteate peduncles seldom approaching 1 cm in length, closely ascending or somewhat divergent, seldom fewer than 20, forming an elongate, spike-like (sometimes branched), sometimes secund inflorescence; involucre cylindric or obconic, 7–11 mm high, its bracts firm, coriaceous-herbaceous, strongly punctate, obscurely scarious-margined and slightly or scarcely ciliate, with keeled or cupped, slightly spreading, often acutish (or sometimes distinctly acute) tip, the midvein commonly projecting as a short, broad, blunt mucro; flowers 5–12 per head; corolla mostly 6–8 mm long, hairy toward the base within. Pappus strongly barbellate. (n = 10) Fall. Pine or oak woods, especially in sandy soil, mainly on PP and AC, seldom encroaching into the mts.; common in SC, extending into Ga and NC and to Bedford Co., Va. Hybridizes with or passes into *L. turgida*, *L. graminifolia*, and *L. gracilis*. *Lacinaria regimontis* Small—S; *Lacinaria smallii* Britton—S; *Liatris graminifolia* var. *smallii* (Britton) Fern. & Grisc.—F.

8. L. graminifolia Willd. Plants 2–12 dm tall, hairy or subglabrous. Leaves numerous, strongly to scarcely punctate, all linear or nearly so, the lowest ones 6–30 cm × 2–7 mm (or a little wider in very robust plants), the margins irregularly ciliate toward the base. Heads 10–many in an elongate, spiciform (seldom branched) inflorescence, suberect on short, closely ascending peduncles, or subsessile, only seldom turned away from the axis; involucre turbinate or subcylindric, 7–12 mm high, its bracts relatively thin, scarcely to sometimes evidently punctate, more or less ciliolate on the scarious margins, mostly broadly rounded and often purplish upward; flowers mostly (6) 7–14 per head, the corolla hairy toward the base within. Pappus strongly barbellate. (n = 10) Late summer, fall. Dry, open woods, especially in sandy soil among pines; CP and PP from NJ to Ga, nw Fla, and Ala, and occasionally extending into the mt. provinces. *Lacinaria graminifolia* (Walt.) Kuntze—S. In SC much of the material appears to be genetically contaminated with *L. regimontis*.

9. L. gracilis Pursh. Plants 2–10 dm tall, hairy or subglabrous, the lower leaves generally with at least a few long, scattered marginal cilia toward the base, the stem often densely spreading-hairy. Leaves mostly numerous, strongly to scarcely punctate, generally all linear or nearly so, the lowest ones 6–30 cm × 2–10 (20) mm. Heads usually more or less numerous in an elongate, racemiform (seldom branched) inflorescence, borne on slender, divaricate or arcuate-ascending peduncles up to several cm long, or subsessile but tending to be turned away from the axis; involucre turbinate to cylindric, (4) 5–7 (8) mm high, its bracts strongly ciliolate and sometimes also short-hairy on the back, scarcely to strongly punctate, sometimes thin, appressed, and broadly rounded distally, as in *L. graminifolia*, sometimes with a tendency toward the thicker and looser, abruptly mucronulate form of *L. regimontis*; flowers mostly 3–6 (7) per head, the corolla 6–8 mm long, hairy toward the base within. Pappus strongly barbellate. (n = 10) Fall, or all year southward. Sandy pinelands; throughout peninsular Fla, n to se SC and sw Ala, almost wholly on CP; apparently isolated in Laurens Co., SC. *Lacinaria gracilis* (Pursh) Kuntze—S.

Some plants from peninsular Fla, notably in Dade Co., far beyond the geographic range of *L. graminifolia*, appear to constitute an ill-defined phase of *L. gracilis*, tending to differ from the more ordinary forms in their more compact inflorescence, with closely ascending, 4- to 9-flowered heads. Most of these plants are short, less than 5 dm tall, and have relatively few leaves. The name *L. laxa* (Small) K. Schum. is available for such plants. Other plants that combine the features of *L. gracilis* and *L. graminifolia* in various ways occur where the ranges of the two species overlap.

10. L. provincialis Godfrey. Plants 4–8 dm tall; stem densely and shortly spreading-hairy. Leaves glabrous, numerous, all linear or nearly so, the larger (lower) ones up to 15 cm × 7 mm, the others progressively reduced. Heads numerous in an elongate, virgate, spiciform or racemiform inflorescence, sessile or on short peduncles up to 5 mm long, evidently spreading away from the axis; involucre narrow, 8–11 mm high, its bracts firm, evidently punctate, green or often anthocyanic, narrow, strongly acute or acuminate, appressed or the inner with somewhat spreading tip, the outer inconspicuously ciliolate, at least the inner narrowly scarious-margined; flowers 3–4 per head; corolla 8–10.5 mm long including the 2.5–3.5 mm lobes, the tube evidently hairy toward the base within. Pappus evidently barbellate. Fall. Sand dunes and sandy pine-oak woods near the coast; irregularly distributed but locally abundant from Wakulla Co. to Bay Co. in the Fla panhandle.

11. L. pauciflora Pursh. Plants 2–8 dm tall. Leaves numerous, all linear or nearly so, the lowest ones mostly 6–20 cm × 2–10 mm, the others evidently reduced.

Heads in well-developed plants more or less numerous, evidently (though sometimes shortly) pedunculate, tending to be turned away from the axis, forming an elongate, racemiform, strongly secund inflorescence; bracts of the inflorescence mostly inconspicuous and shorter than the involucres; involucre (8) 10–15 mm high, its bracts greenish or partly anthocyanic, only narrowly or scarcely scarious-margined, appressed or a little loose at the tip, more or less strongly acute to acuminate, sometimes callous-pointed; flowers 3–6 per head; corolla 11–16 mm long, including the 3–5-mm lobes, only sparsely and inconspicuously (or scarcely) hairy within. Pappus evidently barbellate. (n = 10) Late summer. Sand hills and open, sandy pine or pine-oak woods on CP; c Fla, n to s NC and w to sw Ala and reputedly La. *Lacinaria pauciflora* (Pursh) Kuntze—S. Typical *L. pauciflora* is essentially glabrous and is restricted mainly to peninsular Fla, rarely extending as far n as Tatnall Co., Ga. Plants with the stem densely and minutely spreading-hirtellous, and sometimes also with the leaves hairy, are chiefly more northern, seldom extending into peninsular Fla, but the two phases sometimes occur together and may prove to be Mendelian variants. The hairy-stemmed plants have been segregated as *Liatris secunda* Elliott. *Lacinaria secunda* (Ell.) Small—S.

12. L. chapmanii T. & G. Much like *L. pauciflora*, sometimes 1.5 m tall; stem consistently spreading-hirtellous, the leaves shortly spreading-hairy or glabrous; heads essentially sessile, erect or closely ascending, crowded, forming a dense, spiciform, not at all secund inflorescence; bracts of the inflorescence tending to be prominent, often equalling the involucres. (n = 10) Midsummer, fall, or all year southward. Dry, sandy soil with pines or scrub oak; throughout peninsular Fla, w to the Apalachicola region, and occasionally n into s Ga and w into se Ala. *Lacinaria chapmanii* (T. & G.) Kuntze—S.

13. L. microcephala (Small) K. Schum. Plants slender, mostly 3–8 dm tall, essentially glabrous throughout. Leaves numerous, linear, the lower ones 3–16 × 1–5 mm, the others fairly gradually reduced upward. Inflorescence racemiform and sometimes secund, the more or less numerous heads borne on slender, erect peduncles mostly 3–20 mm long (or a few of the uppermost ones subsessile), often turned away from the axis; involucre narrow, subcylindric, 6–8 mm high, its green or greenish bracts narrowly scarious-margined, blunt, commonly broadly rounded above; flowers 4–5 (6) per head; corolla mostly 6–8 mm long, including the 1.5–2-mm lobes, glabrous within. Pappus strongly barbellate, relatively short, seldom reaching the sinuses of the corolla. (n = 10) Late summer, fall. Exposed, rocky places, glades, open woods, and sandy shores; mt. provinces and PP of Ga and Ala (where barely extending onto GC), and mt. provinces and IP of Tenn, n and e occasionally to Ky (CU), NC (BR), and SC (PP). *Lacinaria microcephala* Small—S.

14. L. tenuifolia Nutt. Plants rather coarse, mostly 4–20 dm tall, glabrous or nearly so, appearing subscapose. Leaves firm, numerous and narrow, the basal and lowermost cauline ones linear to setaceous, up to 40 cm × 6 (9) mm, the others more or less abruptly reduced and setaceous-bracteate, especially in var. *laevigata*. Inflorescence elongate and narrow, virgate, sometimes secund, the more or less numerous heads erect or sometimes turned a bit aside, borne on short, slender, erect or closely ascending (seldom branched) peduncles rarely more than 1 cm long; involucre mostly 6–9 mm high, often anthocyanic in part, its middle and outer bracts strongly acute or acuminate, the inner ones broadly rounded (or even retuse) above and usually shortly mucronulate; flowers (3) 4–5 (6) per head, the corolla mostly 6–8 mm long including the 2–3-mm lobes, glabrous within. Pappus strongly barbellate, generally reaching or surpassing the sinuses of the corolla. Fall, or all year southward. Sandy or gravelly soil in open oak or pine woods; CP of SC to s Fla, and w to s Ala. The species consists of two well-marked but wholly confluent geographic vars.:

Occurring from SC to n (rarely c) Fla and w to s Ala, has setaceous lower leaves up to 2 (2.5) mm wide, sparsely and irregularly long-ciliate toward the base, and the cauline leaves are not always very abruptly reduced; its rootstock is relatively small, seldom more than 2 (2.5) cm thick. *Lacinaria tenuifolia* (Nutt.) Kuntze—S. .... var. *tenuifolia*.

Occurring throughout Fla, has linear lower leaves mostly (1) 2–6 (9) mm wide; the leaves are strictly glabrous, and the cauline ones are very abruptly reduced and setaceous; the rootstock averages a little larger, sometimes as much as 3.5 or 4 cm thick .... var. *quadriflora* Chapman.

15. L. spicata (L.) Willd. Plants glabrous or seldom somewhat hirsute, 6–20 dm tall, often with fibrous remains of old leaf bases persistent on the rootstock. Leaves numerous, linear or nearly so, the lowest ones 10–40 cm × (2) 5–20 mm, the others reduced upward. Heads sessile, crowded into an elongate, densely spiciform inflorescence; involucre subcylindric or narrowly turbinate-campanulate, 7–11 mm high, its bracts narrowly scarious-margined, often anthocyanic, at least the inner ones mostly rounded or broadly obtuse, the outer sometimes acutish; flowers (5) 6–10 (14), the corolla glabrous within, 7–11 mm long including the 2–3-mm lobes. Pappus strongly barbellate. (n = 10) Midsummer, fall. Wet meadows and other moist, open places; NY to c Fla, w to Mich, se Mo, and se La, and occasionally across the plains to Wyo and NM. *Lacinaria spicata* (L.) Kuntze—S; *Liatris spicata* var. *resinosa* (Nutt.) Gaiser—F.

16. L. acidota Engelm. & A. Gray. Plants slender, mostly 3–10 dm tall, essentially glabrous. Leaves numerous, firm, the basal and lower cauline ones elongate-linear, 10–40 cm × (1) 2–5 mm, the others gradually reduced; fibrous remains of old leaf bases persistent on the rootstock. Heads sessile or on short peduncles up to ca. 3 mm long, crowded and ascending or suberect in an elongate, spiciform inflorescence; involucre mostly 8–10 (11) mm high, its bracts firm, often somewhat anthocyanic but not notably scarious, appressed or a little loose at the tip, strongly acuminate, tending to be shortly spinulose-tipped; flowers 3–4 (5), the corolla mostly 7–10 mm long including the 2–3-mm lobes, glabrous within. Pappus strongly barbellate or subplumose, the barbels sometimes up to ca. 0.4 mm long. (n = 10) Summer. Moist, low ground; GC of La and Tex.

17. L. pycnostachya Michx. Plants strict, 5–15 dm tall, hirsute (seldom glabrous) in the inflorescence or sometimes throughout; fibrous remains of old leaf bases commonly persistent on the rootstock. Leaves numerous, linear or nearly so, the lowest ones 10–50 cm × 3–13 mm, the others reduced upward. Heads sessile, crowded in an elongate, densely spiciform inflorescence; involucre subcylindric or narrowly turbinate, 8–11 mm high, its bracts tapering to an acuminate, conspicuously squarrose tip, or the inner ones sometimes blunter and loosely erect; flowers 5–7 (12) per head, the corolla 7–9 mm long including the 2–2.5-mm lobes, glabrous or very nearly so within. Pappus strongly barbellate. (n = 10, 20) Summer. Moist or dry prairies and open woods; Ind to Miss and La, w to Minn, SD, and Tex. *Lacinaria pycnostachya* (Michx.) Kuntze—S.

18. L. garberi A. Gray. Plants 2–8 dm tall from a cluster of tuberous-thickened, fingerlike roots; stem spreading-hirsute, or the upper part merely hirsute-puberulent.

Leaves numerous and narrow, the lower ones linear or linear-oblanceolate, 7–25 cm × 2–5 mm, the petiole commonly somewhat ciliate-margined; middle and upper leaves strongly reduced, setaceous. Heads more or less numerous in a dense, spiciform inflorescence, sessile or on short, stout peduncles seldom more than 5 mm long; involucre 8–11 mm high, its bracts firm, appressed or a little loose at the tip, acuminate and shortly spinulose-tipped to sometimes obtuse and merely mucronate, sometimes hirsute on the back; flowers 6–10, the corolla 8–10 mm long including the 2–3-mm lobes, glabrous within. (n = 10) Summer, or all year southward. Low pinelands and moist prairies; c and s peninsular Fla.; Bahama Isl. *Lacinaria garberi* (A. Gray) Kuntze —S; *Lacinaria chlorolepis* Small—S.

19. L. cylindracea Michx. Plants glabrous or rarely short-hairy, 2–6 dm tall from a cormose-thickened rootstock with slender, fibrous roots. Leaves more or less numerous, firm, linear or nearly so, the lowest small and subsheathing, the next longer, 10–25 cm × 2–12 mm, the rest reduced upward. Heads few or even solitary, stiffly pedunculate or sessile; involucre 11–20 mm high, broadly cylindric or cylindric-campanulate, its bracts firm, appressed or a little loose, generally broadly rounded and shortly mucronate, occasionally more tapering or without the mucro; flowers 10–35 per head, the corolla lobes coarsely hairy within. Pappus evidently plumose. Summer. Dry, open places; w NY and s Ont to s Ohio, n Ind, Mich, Minn, and thence s to s Mo and reputedly n Ark; isolated stations in c Tenn (IP) and c Ala (GC). Occasionally difficult to distinguish from *L. squarrosa*.

20. L. squarrosa (L.) Michx. Plants mostly 3–8 dm tall from a cormose-thickened rootstock with slender, fibrous roots. Leaves linear or a little broader, firm, those near the base 6–25 cm × 4–13 mm, often partly sheathing, the very lowest often smaller and deciduous. Heads mostly few or even solitary, on stiff, erect peduncles or sessile; involucre 12–25 (30) mm high, its bracts firm, with loose or squarrose, acuminate tip; flowers 20–45 per head, or up to 60 in the terminal head; inner surface of the corolla lobes coarsely hairy. Pappus evidently plumose. (n = 10) Summer. Dry, open places; Del to n Fla, w to SD, Colo, and Tex. Three vars. in our range:

Plants hairy, at least in the inflorescence.
- Involucral bracts mostly tapering to a long, loosely spreading tip; pubescence mostly short and curly or appressed; widespread eastward and southward, but not in Ark and n La; var. *gracilenta* Gaiser—F .... var. *squarrosa*.
- Involucral bracts abruptly contracted to the short, squarrose tip; pubescence tending to be longer and more spreading; chiefly from s Iowa to n La, w to e Neb, e Kans, e Okla, and se Tex, but with scattered outlying stations to NC and SC .... var. *hirsuta* (Rudb.) Gaiser.

Plants essentially glabrous throughout (except the corolla lobes); involucre similar to that of var. *squarrosa*; OU region of Ark .... var. *compacta* T. & G.

21. L. punctata Hook. Rootstock more or less elongate and pointed at the base, like a short, fleshy-thickened taproot, or seldom horizontal and resembling a thickened rhizome, producing scattered aerial stems; stems mostly 4–8 dm tall; herbage essentially glabrous, or some of the leaf margins inconspicuously short-ciliate. Leaves numerous, all linear, only 1–3 mm wide, the lowest small and often soon deciduous, those next above the largest, the others reduced upward. Heads several or many in a spiciform inflorescence, sessile or nearly so; involucre subcylindric, mostly 13–18 mm high, its bracts punctate, mostly sharply mucronate-acuminate; flowers mostly 4–6 per head, the corolla lobes glabrous, the tube hairy toward the base within. Pappus evidently plumose. (n = 10, 20) Late summer, fall. Dry, open places, often in sandy soil; sw Mich to Man, Alta, nw Ark, Tex, NM, and n Mex. Our plants, as here described, belong to var. *nebraskana* Gaiser. The smaller, somewhat wider-leaved and more ciliate var. *punctata* occurs chiefly on the high plains. The closely related *L. mucronata* DC., differing notably in its globose, cormlike rootstock, occurs chiefly in Tex and adj. Okla and Mex, but has been reported and may be sought in w Ark. It is possible that *L. mucronata* should be treated as a var. of *L. punctata*.

22. L. elegans (Walter) Michx. Plants with a globose or ellipsoid corm and slender, fibrous roots; stem 3–12 dm tall, hirsute-puberulent. Leaves numerous, glabrous, strongly punctate, narrow, the lower sometimes elongate and up to 30 cm × 13 mm, but often deciduous, the others gradually or sometimes more abruptly reduced upward, the upper ones commonly deflexed. Heads numerous in an elongate, slender and virgate or seldom broader and branching inflorescence, sessile or often shortly bracteate-pedunculate; involucre (12) 15–25 mm high, its bracts rough-hirsute, all slender, the outer ones tapering gradually to an often softly spinulose point, the inner ones with prolonged, slightly expanded, scarious, petaloid, pink or white, loose or spreading tip; flowers 4–5 per head, the corolla glabrous within, often white. Pappus evidently plumose. (n = 9?) Late summer, fall. Open woods, sand hills, and prairies; CP from SC to n Fla and w to Tex, and encroaching into OU in Ark and Okla. *Lacinaria elegans* (Walt.) Kuntze—S; *Lacinaria flabellata* Small—S; *L. boykinii* T. & G.—S, appears to be a hybrid with *L. tenuifolia*.

101. VERNONIA Schreber, nom. conserv. IRONWEED*

Our spp. perennial herbs with one introduced annual; leaves alternate, pinnately veined, usually cauline, or sometimes mostly basal; inflorescence of corymbiform cymes; heads, numerous discoid, our spp. with 9–many flowers; involucre cylindric to broadly hemispheric or campanulate; phyllaries loosely or closely imbricate in several series, the inner phyllaries progressively longer; receptacle flat to subconvex, naked; pappus in 2 series, the inner of numerous long, slender bristles, the outer of short, sometimes flattened bristles or scales; corollas tubular, regular, 5-lobed, in our spp. deep reddish purple (rarely white); anthers sagittate at the base; style branches slender, gradually tapering, hispidulous throughout with stigmatic lines near the base; achenes ribbed or ribless in one sp., commonly resin-dotted between the ribs. Hybrid swarms very common.

1 Achenes terete, without ribs, or faintly ribbed; annual pantropical weed; s Fla .... 1. *V. cinerea*.
1 Achenes ribbed or furrowed; perennials.
  2 Phyllaries densely grey-tomentose on the back; stems densely grey-tomentose .... 3. V. lindheimeri.
  2 Phyllaries glabrous, ciliate, or very thinly pubescent on the back; stems glabrous to pubescent but not grey-tomentose.
    3 Tips of middle and inner phyllaries long-acuminate to filiform.
      4 Heads 1 cm or more wide, 60–120 flowered; Ozark Mts. .... 5. *V. arkansana*.
      4 Heads less than 1 cm wide, usually less than 60-flowered.
        5 Pappus whitish or straw-colored.

*The treatment of *Vernonia, Elephantopus,* and *Stokesia* is by Samuel B. Jones, University of Georgia, Athens.

6 Leaves mostly basal, cauline leaves greatly reduced .... 11. *V. acaulis.*
6 Leaves cauline .... 10. *V. glauca.*
5 Pappus tawny or brownish-purple.
7 Leaves auriculate at the base, crisped; veins pubescent beneath with conspicuous brown hairs; se Ga & sw SC .... 14. *V. pulchella.*
7 Leaves attenuate at the base, not crisped; veins sometimes pubescent with inconspicuous hairs.
8 Leaves linear to broadly linear, less than 1 cm wide, 4–11 cm long; heads usually with less than 30 flowers .... 12. *V. angustifolia.*
8 Leaves lanceolate, 1.3–5.5 cm wide, 12–28 cm long; heads with 30–65 flowers .... 9. *V. noveboracensis.*
3 Tips of inner phyllaries rounded, acute, obtuse, mucronate or short-acuminate.
9 Leaves conspicuously pitted beneath, the pits appearing as black dots under low magnification .... 2. *V. texana.*
9 Leaves not conspicuously pitted beneath.
10 Middle cauline leaves linear, usually less than 1 cm wide.
11 Leaves scattered on stem; pappus pale yellow; s Fla .... 13. *V. blodgettii.*
11 Leaves crowded on stems; pappus tawny to purple.
12 Stems 2–5 dm high; chert rocks and gravel bars, Ouachita River drainage in Ark, and Okla .... 8. *V. lettermanii.*
12 Stems 4–11 dm high; sandy soil, NC into Fla and Miss .... 12. *V. angustifolia.*
10 Middle cauline leaves elliptic, lanceolate, or oblanceolate, usually more than 1 cm wide.
13 Stems glabrous, glaucous; pappus pale straw-colored; lower Appalachian region, nw Ga, ne Ala and sc Tenn .... 15. *V. flaccidifolia.*
13 Stems sparsely to densely pubescent especially on the branches of the inflorescence, not glaucous; pappus brown to purple.
14 Heads usually with 30–65 flowers.
15 Pappus whitish or straw-colored; PP and mts., Ala to NJ ...... 10. *V. glauca.*
15 Pappus brown to purple.
16 Leaves pubescent to tomentose beneath; ME of sc US .... 6. *V. missurica.*
16 Leaves scabrous to tomentulose beneath; Mass sw into Ala and Fla .... 9. *V. noveboracensis.*
14 Heads usually with less than 30 flowers.
17 Middle phyllary tips usually recurved or spreading, acute to acuminate, stems pubescent; OZ, OU, and Great Plains ...... 4. *V. baldwinii.*
17 Middle phyllary tips not recurved or spreading; acute to obtuse or mucronate; stems puberulent to glabrate; throughout e US into e Neb, Kans, Okla, and Tex .... 7. *V. gigantea.*

1. V. cinerea (L.) Less. Annual. Stems tomentulose with T-shaped hairs, sometimes becoming glabrate in age below, 3–6 dm high. Leaves scattered along stem; blades of middle stem leaves 1.5–2.5 cm wide, 2–5 cm long, lanceolate, pubescent above, pubescent with T-shaped hairs, and punctate beneath, apically acute, basally attenuate, margins remotely toothed; petioles margined, ca. 1.5–2.5 cm long, pubescent; inflorescence loose, open and often spreading. Heads 12–16 flowered. Involucre campanulate, 6–7 mm high, 5–6 mm wide; phyllaries loosely and irregularly imbricate; inner phyllaries linear-oblong, 5–5.5 mm long, 0.6–0.8 mm wide, with acuminate to subulate purplish tips 0.5–1 mm long. Pappus whitish, deciduous; inner bristles ca. 4 mm long; outer bristles ca. 0.2 mm long. Corollas purplish-lavender, 6–7 mm long. Achenes rounded, nearly ribless, ca. 1.5 mm long. (n = 9) All year. An intro. pantropical weed of disturbed areas; s Fla. *Seneciodes cinerea* (L.) Kuntze—S.

2. V. texana (A. Gray) Small. Stems puberulent to glabrate below and above, 4–10 dm high. Leaves scattered, cauline; blades of middle stem leaves 0.2–1.4 cm wide, 6.5–13.5 cm long, linear to linear-lanceolate or sometimes lanceolate, scabrous above, scabrous to puberulent beneath, apically acute, basally attenuate, margins revolute, remotely serrate, or sometimes prominently toothed; petioles 1–4 mm long, puberulent. Inflorescence loose and open. Heads 15–25 flowered. Involucre broadly campanulate, 4–6 mm high, 5–7 mm wide; phyllaries imbricate, slightly appressed, greenish-purple, ciliate; inner phyllaries oblong, oblong-lanceolate, 3.9–6 mm long, 1–1.7 mm wide, with acute to short-acuminate tips ca. 0.1–0.2 mm long. Pappus brownish to straw-colored, often tinged with purple; inner bristles 5.5–7 mm long; outer scales ca. 0.6 mm long. Corollas 9–11 mm long. Achenes 1.5–3 mm long. (n = 17) Summer. Well-drained soil of pinelands; GC and OU of sw Ark, w La, and sw Miss [Tex, Okla]. This sp. hybridizes with *V. baldwinii, V. gigantea* and *V. missurica*.

3. V. lindheimeri A. Gray & Engelm. Stems usually 1–3 per plant, thinly tomentose below and tomentose above, 2–6 dm high. Leaves numerous, crowded, cauline; blades of middle stem leaves 1–4 mm wide, 4–8 cm long, linear, punctate and pitted with a few hairs above, tomentose with dense longhorn hairs beneath, apically acute, basally attenuate and sessile, margins entire, extremely revolute. Inflorescence compact. Heads 12–29 flowered. Involucre narrowly campanulate, 6–10 mm high, 5–8 mm wide; phyllaries appressed, greenish with purple margins, tomentose; inner phyllaries oblong, 5–9 mm long, 1–1.7 mm wide, with acute tips. Pappus purple to tawny; inner bristles ca. 7 mm long; outer scales 0.5–1 mm long. Corollas 9–10 mm long. Achenes deeply ribbed, densely glandular, purple to tawny, ca. 4 mm long. (n = 17) Summer. Alkaline soil of black prairies, chalky limestone; one collection from sw Ark, probably not now extant in Ark [Tex]. This sp. hybridizes with *V. baldwinii*.

4. V. baldwinii Torr. Stems pubescent below and pubescent above, 8–15 dm high. Leaves numerous, cauline; blades of middle stem leaves 2–6 cm wide, 7–17 cm long, lanceolate to ovate-lanceolate, puberulent to glabrate above, puberulent to scabrous beneath, apically acute to acuminate, basally cuneate to attenuate, margins serrate; petioles absent to ca. 3 mm long, puberulent; inflorescence loose to irregular. Heads 17–34 flowered. Involucre cylindric to campanulate 4.7–8 mm high, 4–6 mm wide; phyllaries imbricate, appressed, greenish-brown to purplish, ciliate and sometimes puberulent, inner phyllaries oblong to oblong-lanceolate, 4.8–8.1 mm long, 1.2–1.5 mm wide, with acute to acuminate tips 0.1–0.2 mm long. Pappus light brown to brown with a purple tinge; inner bristles 5–6.5 mm long; outer scales 0.5–0.8 mm long. Corollas 8–10 mm long. Achenes puberulent to glabrous on the ribs, 2.5–4 mm long. (n = 17) Summer.

Heads 23–34 flowered; involucres usually 4.7–6.7 mm high; phyllary tips recurved, mostly cuneate to acuminate; recurved tips of inner phyllaries pubescent on inner surfaces .... 4a. ssp. *baldwinii.*

Heads 17–27 flowered; involucres usually 5.6–8 mm high; phyllary tips loosely spreading, mostly acute; loosely spreading inner bracts usually not pubescent on inner surfaces .... 4b. ssp. *interior.*

4a. ssp. **baldwinii**. Stems pubescent to tomentose above; heads 23–34 flowered; involucres 4.7–6.7 mm high; phyllary tips recurved. Well-drained soil of upland pastures, thin woods, roadsides, waste places; OZ, OU, Ark [Okla, Mo, Ill]. This ssp. hybridizes with *V. arkansana, V. gigantea, V. lettermannii, V. missurica* and *V. texana*.

4b. ssp. **interior** (Small) Faust. Stems puberulent to pubescent above; heads 17–27 flowered; involucres 5.6–8 mm high; phyllary tips loosely spreading, not recurved. Well-drained soil of overgrazed land, pastures, roadsides, often weedy; OZ, OU, Ark [Tex, Okla, Mo]. This ssp. hybridizes with ssp. *baldwinii, V. missurica* and *V. texana*.

5. V. arkansana DC. Stems glabrate below and puberulent above, 0.7–1.2 (2) m high. Leaves numerous, cauline; blades of middle stem leaves 0.7–2.5 cm wide, 9–20 cm long, narrowly lanceolate, punctate above, punctate beneath, apically acuminate, basally attenuate, margins slightly revolute with callous teeth; petioles 2–5 mm long, glabrate. Inflorescence compact. Heads 60–120 flowered. Involucre broadly hemispheric, 9–15 mm high, 10–15 mm wide; phyllaries imbricate with long spreading tips, greenish, ciliate, puberulent; inner phyllaries oblong-lanceolate, 6–8 mm long, ca. 1.5 mm wide, with long-filiform, flexuous, spreading tips 2–4 mm long. Pappus brownish-purple; inner bristles 6–7 mm long; outer scales 0.7–1 mm long. Corollas 10–11 mm long. Achenes strongly ribbed, resinous in the furrows, 4–5 mm long. (n = 17) Summer. Rocky intermittent stream beds, low pastures, along streams; OZ, Ark [Okla, Mo, Ill]. It hybridizes with *V. baldwinii*, *V. gigantea* and *V. missurica*. *V. crinita* Raf.—G, F.

6. V. missurica Raf. Stems pubescent to tomentose becoming puberulent with age below, pubescent to tomentose above, 1–2 m high. Leaves numerous, cauline, blades of middle stem leaves 2–6 cm wide, 8–20 cm long, lanceolate to ovate-lanceolate, puberulent above, pubescent to tomentose beneath, apically acute to acuminate, basally attenuate to cuneate, margins serrate; petioles 2–3 mm long, usually tomentose. Inflorescence loose to irregular. Heads 32–58 flowered. Involucre campanulate to cylindric, 7–10.3 mm high, 5.5–8 mm wide; phyllaries imbricate, appressed, greenish to brownish, ciliate and sometimes pubescent on outer surface, inner phyllaries oblong, 5.5–10.5 mm long, 1.5–2.5 mm wide, with obtuse to subacute tips ca. 0.1–0.2 mm long. Pappus tawny to brown; inner bristles 6–8 mm long; outer scales 0.6–1 mm long. Corollas 9–11 mm long. Achenes puberulent to nearly glabrous on the ribs, 4–6 mm long. (n = 17) Summer. Moist prairies, bottomland pastures, low roadsides; ME, GC of Ark, Ky, La, Miss, Tenn [Tex, Okla, Mo, Ill, Ind]. This sp. hybridizes with *V. arkansana*, *V. baldwinii*, *V. fasciculata*, *V. gigantea* and *V. texana*.

7. V. gigantea (Walter) Trelease ex Branner & Coville. Stems puberulent to glabrate below and puberulent above, 1–2 (3.5) m high. Leaves numerous, cauline; blades of middle stem leaves 1–7.5 cm wide, 6–30 cm long, lanceolate to linear-lanceolate or oblanceolate, essentially glabrous to slightly scabrous along the margins above, puberulent and often pubescent on the veins beneath, apically acuminate, basally attenuate, margins serrate; petioles 0.3–2 cm long, essentially glabrous. Inflorescence usually large, loose to irregular. Heads 9–30 flowered. Involucre cylindric to cylindric-campanulate, 2.3–7 mm high, 2–5.5 mm wide; phyllaries imbricate, appressed, usually purplish, ciliate and pubescent along the outer surfaces; inner phyllaries oblong-lanceolate, oblong-obovate, 3.5–5.3 mm long, 1–2 mm wide, with acute to obtuse or mucronate tips 0.1–0.3 mm long. Pappus tan to brown, sometimes with a purple tinge; inner bristles ca. 6 mm long; outer scales 0.2–0.8 mm long. Corollas 9–11 mm long. Achenes pubescent on the ribs, ca. 3.5 mm long. (n = 17) Summer.

Heads 13–30 flowered; leaf blades linear-lanceolate, 10–30 cm long, 1.2–7.5 cm wide; involucre width 2.3–5.5 mm .... 7a. ssp. *gigantea*.
Heads 9–20 flowered; leaf blades elliptic to oblanceolate, 6–20 cm long, 1.2–5 cm wide; involucre width 2–4 mm .... 7b. ssp. *ovalifolia*.

7a. ssp. **gigantea**. Blades of middle stem leaves 1.2–7.5 cm wide, up to 30 cm long, lanceolate to linear-lanceolate; heads 13–30 flowered; involucre 2.3–5.5 mm wide; inner phyllaries 1.2–3.1 mm wide. Weedy in pastures, roadsides, thin woods, along streams; throughout except Del [Tex, Okla, Mo, Ill, Ind, Ohio, Pa]. This ssp. commonly hybridizes with *V. angustifolia, V. baldwinii, V. flaccidifolia, V. missurica* and *V. noveboracensis. V. altissima* Nutt.—S, G, F, R: *V. altissima* var. *taeniotricha* Blake—F.

7b. ssp. **ovalifolia** (T. & G.) Urbatsch. Blades of middle stem leaves 1.2–5 cm wide, 6–20 cm long, lanceolate to oblanceolate; heads 9–20 flowered; involucre 2–4 mm wide; inner phyllaries 0.9–1.8 mm wide. Low roadsides and pastures, around the margins of streams and lakes; sw Ga, n Fla. This ssp. hybridizes with *V. angustifolia;* the hybrids are called *V. concinna* Gl.—S. *V. ovalifolia* T. & G.—S.

8. V. lettermanii Engelm. ex A. Gray. Stems glabrous, 2–5 dm high; leaves numerous, crowded, cauline; blades of middle stem leaves 0.8–2 mm wide, 5–12 cm long, linear, punctate above, punctate beneath, apically acute, basally attenuate and sessile, margins revolute and entire; inflorescence compact; sometimes highly branched; heads 9–15 flowered; involucre narrowly cylindric, 7.5–10 mm high, 3.5–5.6 mm wide; phyllaries closely imbricated and appressed, purple, ciliate, inner phyllaries oblong-lanceolate, 6.5–7.5 mm long, 1.2–1.6 mm wide, with acuminate tips ca. 0.3 mm long; pappus purple or tawny-purple; inner bristles 6.5–7.5 mm long; outer scales ca. 0.5 mm long; corollas ca. 11 mm long; achenes glandular, ca. 3–3.5 mm long. (n = 17) Summer. Gravel bars, the cracks of chert rocks; along the drainage of the Ouachita River, OU, Ark [Okla]. It hybridizes with *V. baldwinii.*

9. V. noveboracensis (L.) Michx. Stems puberulent becoming glabrate with age below and puberulent above, 1–2 m high. Leaves numerous, cauline; blades of middle stem leaves 2–6 cm wide, 12–28 cm long, narrowly lanceolate, glabrous to scabrous above, scabrous to tomentulose beneath, apically acuminate, basally attenuate, margins denticulate; petioles 3–12 mm long, puberulent. Inflorescence loose and spreading. Heads 30–50 (65) flowered. Involucre campanulate, 7–17 mm high, 6–10 mm wide; phyllaries imbricate, greenish-purple, ciliate; inner phyllaries triangular-ovate to oblong, 8–12 mm long, 1.5–2.3 mm wide, with acuminate to long-acuminate tips (1) 2–8 (10) mm long. Pappus brownish-purple; inner bristles 6–7 mm long; outer scales 0.4–0.8 mm long. Corollas 8–11 mm long. Achenes puberulent, ribbed, 4–4.5 mm long. (n = 17) Summer. Moist soil of low meadows, roadsides, pastures, along streams; BR, RV, PP, AC from Md s into Ala and Fla [Pa, NJ]. This sp. hybridizes with *V. acaulis, V. angustifolia, V. gigantea* and *V. glauca. V. harperi* Gl.—S; *V. noveboracensis* var. *tomentosa* (Walt.) Britt.—F.

10. V. glauca (L.) Willd. Stems glabrate below and puberulent above, 6–10 dm high. Leaves numerous, cauline; blades of middle stem leaves 3.4–7.4 cm wide, 12–19 cm long, lanceolate to obovate to oblong, almost glabrous above, glabrous to scabridulous, sometimes punctate and light green beneath, apically acute to acuminate, basally attenuate, blades often abruptly narrowed, margins irregularly dentate; petioles 0.3–2 cm long, glabrate. Inflorescence compact; heads 32–48 flowered. Involucre broadly campanulate, 5–8.8 mm high, 6.1–8.5 mm wide; phyllaries tightly imbricate, purplish to greenish, ciliate; inner phyllaries 4.5–7 mm long, 1–1.5 mm wide, oblong to narrowly ovate, tips acute to long acuminate, 0.5–3 mm long. Pappus straw-colored; inner bristles 6–7 mm long; outer scales 0.4–0.8 mm long. Corollas ca. 10 mm long. Achenes nearly glabrous, ribbed, 2.7–3.3 mm long. (n = 17) Summer. Well-drained soil of upland deciduous woods, semidisturbed roadsides; PP, Md s to Ala [Pa, NJ]. This sp. hybridizes with *V. angustifolia, V. flaccidifolia,* and *V. noveboracensis.* Hybrids with *V. angustifolia* in e Ala and w Ga have been called *V. dissimilis* Gl.—S. *V. glauca* forma *longiaristata* Fern.—F.

11. V. acaulis (Walt.) Gleason. Stems puberulent to glabrate below and puberulent above, 6–10 dm high; leaves mostly basal, cauline leaves bractlike. Blades of basal leaves 4–10 cm wide, 12–30 cm long, oblong to obovate, scabrous above, glabrate to sometimes punctate or pubescent beneath, apically acute, basally attenuate, margins coarsely and irregularly serrate; petioles absent or up to 2 cm long, glabrate to pubescent. Inflorescence loose and open. Heads 31–50 (59) flowered. Involucre broadly campanulate, 5–8.5 mm high, 5.7–8.5 mm wide; phyllaries loosely imbricate, greenish-purple, ciliate; inner phyllaries oblong-lanceolate, 5–7 mm long, 1–1.5 mm wide, with acute to long acuminate tips 0.5–1.5 mm long. Pappus straw-colored; inner bristles 5.5–9 mm long; outer scales 0.5–1 mm long. Corollas ca. 10–11 mm long. Achenes sparsely pubescent, strongly ribbed, resinous, 2.7–3.2 mm long. (n = 17) Summer. Well-drained sandy clay soil of thin woods or semidisturbed roadsides; PP, AC of NC and SC, se Ga. This sp. hybridizes with *V. angustifolia* and *V. noveboracensis*. The hybrid with the first has been called *V. georgiana* Bart.—S. *V.* × *georgiana* Bart—R.

12. V. angustifolia Michx. Stems puberulent to glabrate below and puberulent above, 4–11 dm high. Leaves crowded, cauline; blades of middle stem leaves 0.1–0.6 cm wide, 4–11 cm long, linear to linear-lanceolate, glabrous to scabrous above, short sparse hairs beneath, apically acute to acuminate, basally attenuate, sessile, margins entire to serrate, revolute. Inflorescence compact to loose and open. Heads 8–30 flowered. Involucre campanulate, 4–10 mm high, 4–10 mm wide; phyllaries imbricate, loosely appressed, greenish to purple, often puberulent and ciliate; inner phyllaries 2–8.5 mm long, 1.4–2 mm wide, with variable (depending on subspecies) acute to long acuminate tips 0.1–4.8 mm long. Pappus tawny to purplish; inner bristles 5.5–7 mm long; outer scales 0.4–0.8 mm long. Corollas 11–13 mm long. Achenes puberulent on the ribs, 2.5–3.2 mm long. (n = 17) Summer.

Tips of the inner phyllaries acute to acuminate, 0.1–1 mm long.
  Heads 16–19 flowered; phyllary tips acuminate .... 12a. ssp. *angustifolia*.
  Heads 8–15 flowered; phyllary tips acute .... 12b. ssp. *mohrii*.
Tips of the inner phyllaries long-acuminate, 1.4–4.8 mm long .... 12c. ssp. *scaberrima*.

12a. ssp. **angustifolia**. Heads 16–19 flowered; involucre 5–7 mm high, 5–7.5 mm wide; inner phyllaries 3.5–7 mm long, with acuminate tips 0.1–1.0 mm long; tips of the outer phyllaries 0.3–2.5 mm long. Sandy, well-drained soil of pinelands and sand ridges; AC, Ga, NC, SC. This ssp. hybridizes with *V. gigantea* and *V. acaulis*. *V. angustifolia* var. *angustifolia*—R.

12b. ssp. **mohrii** S. B. Jones & Faust. Heads 8–15 flowered; involucre 4–6 mm high, 4–6.9 mm wide; inner phyllaries 2.5–6.5 mm long, with acute tips 0.1–1.0 mm long; tips of the outer phyllaries 0.1–1.2 mm long. Sandy, well-drained soil of pinelands and sand ridges; GC, PP, s Ala, nw Fla, sw Ga, se Miss. This ssp. hybridizes with *V. gigantea* and *V. glauca*.

12c. ssp. **scaberrima** (Nutt.) S. B. Jones & Faust. Heads 20–30 flowered; involucre 6.5–10 mm high, 6–10 mm wide; inner phyllaries 5–8.5 mm long, with long-acuminate tips 1.4–4.8 mm long; tips of the outer phyllaries 1.5–5.3 mm long. Sandy, well-drained soil of pinelands and sand ridges; AC, se Ga, sw SC. Ssp. *scaberrima* hybridizes with *V. pulchella*. *V. scaberrima* Nutt.—S. *V. angustifolia* var. *scaberrima* (Nutt.) Gray—R.

13. V. blodgettii Small. Stems puberulent below and glabrate to puberulent above, 2–5 dm high. Leaves mostly basal; blades of middle stem leaves 0.1–1 cm wide, 1.8–6.9 cm long, linear to linear-lanceolate, faintly punctate and glabrous above, sparsely punctate and with scattered short hairs beneath, apically acute to rounded, basally attenuate; margins entire and revolute; petioles absent or up to 3 mm long, glabrous. Inflorescence loose, irregular, with few heads. Heads ca. 21 flowered. Involucre campanulate, 5–8.5 mm high, 5–10.5 mm wide; phyllaries loosely and irregularly imbricate, purple, ciliate; inner phyllaries oblong, oblong-lanceolate, obovate, 3.9–6.7 mm long. 1–1.7 mm wide, with acute to short acuminate tips up to 0.2 mm long. Pappus straw-colored; inner bristles 5.5–7.8 mm long; outer scales 0.5–0.8 mm long. Corollas 10–11 mm long. Achenes puberulent, ribbed, 2.5–3.2 mm long. (n = 17) All year. Low pinelands; s Fla [Bahama Isl]. Plants of this sp. from the Bahamas have been called *V. insularis* Gl.

14. V. pulchella Small. Stems pilose below and puberulent above, 4–7 dm high. Leaves abundant, cauline; blades of middle stem leaves 0.5–1.8 cm wide, 2.7–6 cm long, lanceolate, oblanceolate, oblong-lanceolate, scabrous to puberulent above, minutely punctate, scabrous to puberulent, with conspicuous long, brownish pubescence on the veins beneath, apically acute, basally auriculate, sessile, margins revolute. Inflorescence open, loosely branched. Heads 20–36 flowered. Involucre campanulate, 6–10.5 mm high, 5–9.1 mm wide; phyllaries imbricate, appressed, greenish to purplish, often ciliate, puberulent on back; inner phyllaries oblong, 6.5–10.5 mm long, 1–1.5 mm wide, with long-acuminate to filiform tips that are conspicuously recurved, 1–6 mm long. Pappus tawny to straw-colored; inner bristles 5.5–9 mm long; outer scales 0.6–1 mm long. Corollas 9–13.5 mm long. Achenes with pubescent ribs, 2.7–3.1 mm long. (n = 17) Summer. Sandy scrub, pinelands; AC, se Ga, sw SC. This sp. has hybridized with *V. angustifolia* ssp. *scaberrima*; the hybrids are known as *V. recurva* Gl.—S.

15. V. flaccidifolia Small. Stems glaucous, glabrous below and glabrous above, 1–2 (2.5) m high. Leaves numerous, cauline; blades of middle stem leaves 2–7 (8) cm wide, 10–25 (35) cm long, narrowly to broadly lanceolate, glabrate above, puberulent beneath, apically acuminate, basally attenuate, margins dentate; petioles 0.5–1.5 (2) cm long, glabrate. Inflorescence loose. Heads 16–26 flowered. Involucre cyanthiform, 3.5–5 mm high, 3.5–5.5 mm wide; phyllaries tightly appressed, sometimes entirely purple or usually green with purple margins or tip, glabrous to ciliate; inner phyllaries ovate to oblong, 3.2–5 mm long, 1.5–2 mm wide, with obtuse to acute tips. Pappus straw-colored; inner bristles ca. 6 mm long: outer scales irregular, 0.2–0.8 mm long. Corollas 11–12 mm long; achenes pubescent on ribs, ca. 3.5 mm long. (n = 17) Summer. Upland deciduous woods, pastures, semidisturbed roadsides; sw CU, RV, PP, of Ala, Ga, Tenn. This sp. hybridizes with *V. gigantea* and *V. glauca*.

102. ELEPHANTOPUS L. ELEPHANT'S FOOT*

Perennial herbs with 1–5 erect, branched stems from a sturdy rootstock. Leaves simple, alternate, cauline or basal, sessile, pinnately veined. Inflorescence corymbo-paniculate or spicate (in one intro. species). Glomerules terminal or lateral, subtended by 3, foliaceous bracts. Involucre of 4 phyllaries. Flowers discoid, perfect; corolla tube

*Adapted from the treatments of H. A. Gleason, N. Amer. Fl. 33: 106–109. 1922, and J. A. Clonts, 1972. A revision of the genus *Elephantopus*, including *Orthopappus* and *Pseudoelephantopus* (Compositae). Unpublished Dissertation. Mississippi State University.

slender, the limb unequally 5-cleft; anthers 5, sagittate; style slender, terminated by 2 stigmatic branches reflexed at anthesis. Pappus of 5–8 rigid, flattened scales prolonged into terminal bristles.

1 Glomerules mostly lateral; bracts linear; pappus with 2 bristles plicate at the tip and with several shorter straight bristles .... 1. *E. spicatus.*
1 Glomerules terminal; bracts cordate to ovate; pappus of straight bristles only.
  2 Leaves distinctly cauline .... 2. *E. carolinanus.*
  2 Leaves mostly basal, cauline leaves greatly reduced.
    3 Longest phyllaries 10–13 mm long; pappus 5–8 mm long; leaves pubescent on the midvein beneath with spreading or reflexed hairs .... 3. *E. tomentosus.*
    3 Longest phyllaries 6–9 mm long; pappus 3–5 mm long; leaves pubescent on the midvein beneath with spreading or appressed hairs.
      4 Pappus bristles abruptly dilated below into broadly triangular bases; phyllaries glandular or resinous and sparsely pubescent with short hairs .... 4. *E. nudatus.*
      4 Pappus bristles gradually dilated below into narrowly triangular bases; phyllaries densely villous with long white hairs, occasionally resinous .... 5. *E. elatus.*

1. E. spicatus Juss. ex Aubl. Stems glabrate to pilose-hispid below and pilose-hispid above, 4–10 dm high. Leaves cauline; blades 1–5 cm wide, 3–16 cm long, lanceolate, oblanceolate, obovate, or occasionally elliptic, sparsely pilose-hispid above, sparsely pilose-hispid and glandular beneath, apically acute or obtuse, basally attenuate, margins remotely serrate. Inflorescence spicate to racemo-spicate, glomerules lateral and terminal, mostly lateral; bracts 1 or 2, lanceolate, 0.2–0.3 cm wide, 1.5–2.3 cm long. Heads 4 flowered; involucre cylindrical, ca. 10 mm high, ca. 2.5 mm wide. Phyllaries greenish, keeled, membranous along margin, sparsely pubescent, lanceolate, 9–12 mm long, 2–3 mm wide, with sharply acuminate tips. Pappus greenish-white of 4–10 bristles in 2 series, 2 plicate, 5–7 mm long, the others straight and shorter, gradually dilated at base. Corollas white to lavender, ca. 10 mm long. Achenes ribbed, pubescent, 4–7 mm long. (n = 11) Flowering throughout the year. Weed of cultivated fields and waste places; subtropical Fla. *Pseudoelephantopus spicatus* (Juss.) Rohr of authors.

2. E. carolinianus Raeuschel. Stems hirsute below and pilose or almost strigose above, 4–10 dm high. Leaves cauline; blades 5.5–9.5 cm wide, 9–25 cm long, ovate to broadly lanceolate, sparsely pilose above, pilose beneath, apically acute or short acuminate, basally long attenuate, clasping, margins crenate or occasionally serrate. Inflorescence corymbo-paniculate; glomerules terminal, to 2.5 cm wide; bracts 3, cordate, 0.7–1.8 cm wide, 1.1–3 cm long. Heads 4 flowered; involucre cylindric, ca. 9–10 mm high, ca. 2 mm wide. Phyllaries membranous along margins, resin-dotted, pubescent, lanceolate, 6–10 mm long, 1.5–2 mm wide, with abruptly acute tips. Pappus of 5 bristles in 1 series, 4–5 mm long. Corollas white to violet, 7–8 mm long. Achenes ribbed, pubescent, 3–4 mm long. (n = 11) Midsummer–fall. Moist habitats along margins of streams and swamps, deciduous woodlands, moist hillsides; throughout [Tex, Okla, Mo, Ill, Ind, Ohio, Pa, NJ].

3. E. tomentosus L. Stems pilose-hispid, 4–6 dm high. Leaves mostly basal; blades 5.5–10.5 cm wide, 9–25 (35) cm long, oblanceolate to obovate, occasionally broadly elliptic, sparsely pilose above, densely tomentose beneath, apically acute, obtuse or rounded, basally long attenuate to cuneate, margins crenate becoming serrate at the base. Inflorescence corymbo-paniculate; glomerules terminal, to 2.5 cm wide, bracts 3, resin-dotted, cordate, 0.7–1.8 cm wide, 0.9–2.1 cm long. Heads 4 flowered; involucre cylindric, 10–13 mm high, ca. 2 mm wide. Phyllaries membranous along margins, resinous, pubescent, lanceolate, 8–13 mm long, 1.5–2.5 mm wide, with sharply acuminate tips. Pappus of 5 bristles in 1 series, 5–8 mm long, gradually dilated at the base. Corollas white to purple, 7–8 mm long. Achenes ribbed, pubescent, ca. 4–5 mm long. (n = 11) Midsummer–fall. Bottomlands, thickets, sandhills, roadsides, dry pinelands, dry deciduous woodlands; throughout except WVa [Tex, Okla].

As C. W. James, *Rhodora* 61 (1959): 309–311, noted, introgressants or hybrid forms of this species may be found where it is sympatric with either *E. elatus* and/or *E. nudatus*. The hybrids are intermediate with respect to their key characters. The complex of *E. tomentosus*, *E. elatus*, and *E. nudatus* is in need of a thorough biosystematic-chemosystematic study. Generally speaking, the differences between these three taxa have been emphasized and their similarities overlooked.

4. E. nudatus A. Gray. Stems hirsute below and pilose-hispid above, 4–8 dm high. Leaves mostly basal, reduced above; blades 1.5–6 cm wide, 8–25 (30) cm long, oblanceolate, occasionally elliptic or spatulate, sparsely pilose and resinous beneath, apically acute, obtuse, or rounded, basally long-attenuate to cuneate, margins remotely serrate. Inflorescence corymbo-paniculate; glomerules terminal; bracts 3, cordate, pilose, resin-dotted, 0.5–0.7 mm wide, 0.7–1.2 mm long. Heads 4 flowered; involucre cylindrical, ca. 8 mm high, ca. 1.5 mm wide. Phyllaries membranous along margins, purple-tinged, resin-dotted, sparsely pubescent with short hairs, lanceolate, 5–9 mm long, 1–2 mm wide. Pappus of 5 bristles in 1 series, abruptly dilated at the base, 4 mm long. Corollas lavender to purple, ca. 9 mm long. Achenes ribbed, pubescent, 2–3 mm long. (n = 11) Midsummer–fall. Pine flatwoods, bogs, swamp borders, and low deciduous woodlands; AC and GC Md to La. Occasional PP, OU, OZ [Tex].

5. E. elatus Bertol. Stems hirsute below and pilose above, 4–10 (12) dm high. Leaves mostly basal; blades 1.5–7.5 cm wide, 7.5–25 (30) cm long, oblanceolate, occasionally obovate, sparsely pilose above, densely pilose beneath, apically acute, obtuse or rounded, basally long attenuate, margins crenate to serrate. Inflorescence corymbo-paniculate, glomerules terminal, to 2.5 cm wide or wider, bracts 3, cordate to ovate, 0.4–1.2 cm wide, 0.5–2 cm long. Heads 4 flowered; involucre cylindrical, ca. 10 mm high, ca. 2 mm wide. Phyllaries membranous along the margin, densely villous with long, white hairs, lanceolate, 5–9 mm long, 1–2 mm wide, with acute or sharply acuminate tips. Pappus of 5 bristles in 1 series, 3–4 mm long, gradually dilated at the base. Corollas pink to pale purple, ca. 7–8 mm long. Achenes ribbed, pubescent, 3–4 mm long. (n = 11) Midsummer–fall. Pine flatwoods, deciduous woodlands, borders of swamps, sandhills; AC, GC, SC into Fla and La.

103. STOKESIA L'Her.

1. S. laevis (Hill) Greene. STOKES' ASTER. Herbaceous perennial. Stems pubescent becoming glabrate below and woolly above, 2.5–5 dm high. Leaves alternate, clustered toward base, the upper sessile and clasping, pinnately veined, blades of middle cauline leaves 0.3–2.5 cm wide, 7–12 cm long, variable, linear-lanceolate to lanceolate, punctate above and beneath, apically blunt, basally long attenuate and clasping, margins entire and slightly revolute, basal leaves 1–5 cm wide, 10–30 cm long; petioles variable and winged, glabrous. Inflorescences variable. Heads large, showy, ca. 60–70 flowered, solitary or a few in a corymb on terminal peduncles. Involucre campanulate, 2–3 cm high, 2–3 cm wide; phyllaries foliaceous, bristly, green, punctate, imbricate in several series, inner phyllaries narrower, 1.5–2 cm long, 2–3 mm wide, with acuminate-bristle tips 1.5–2 mm long; outer phyllaries broader. Receptacle flat, naked. Pappus whitish of 4–5 awnlike chaffy scales, 8–9 mm long. Corollas blue, rarely whitish (when fresh), outer ligulate, 20–30 mm long, inner tubular, 15–20 mm long. Anthers sagittate, rounded at base. Style branches long and slender, stigmatic surface flat and glabrous, back rounded and pubescent. Achenes short, plump, 3- or 4-sided, shiny, ca. 7–8 mm long. (n = 7) Summer. Moist soil of low pinelands, pitcher plant bogs; AC, GC, Ala, Fla, Ga, La, Miss, SC. The monotypic *Stokesia laevis* is of minor importance as an ornamental in the e US.

104. ARCTIUM L. BURDOCK

Coarse biennial herbs. Leaves alternate, large, entire or toothed to rarely laciniate, mostly cordate or cordate-based. Heads discoid, the flowers all tubular and perfect; involucre subglobose, its bracts multiseriate, narrow, firm but somewhat greenish, appressed at the base, with a spreading, subulate, inwardly hooked tip; receptacle flat, densely bristly; corolla pink or purplish, with a slender tube, short throat, and narrow lobes; anthers very shortly awn-pointed at the tip, evidently tailed at the base; style with an abrupt change of texture below the minutely papillate branches. Achenes oblong, slightly compressed, few-angled, many-nerved, truncate, glabrous; pappus of numerous short, subpaleaceous, separately deciduous bristles.

1 Inflorescence mostly of racemiform or thyrsoid branches, with mostly short-peduncled or subsessile heads, not corymbiform; involucre mostly 1.5–2.5 cm thick .... 1. *A. minus*.
1 Inflorescence more or less corymbiform, the heads mostly long-pedunculate.
 2 Involucre evidently tomentose, mostly 2–3 cm thick .... 2. *A. tomentosum*.
 2 Involucre not tomentose, mostly 2.5–4 cm thick .... 3. *A. lappa*.

1. A. minus Schkuhr. COMMON BURDOCK. Plants up to 1.5 m tall, rarely more. Leaves petiolate, the lower petioles mostly hollow, the blade narrowly to very broadly ovate, up to 5 × 4 dm, thinly tomentose or eventually glabrate beneath, subglabrous above. Branches of the inflorescence ascending to widely spreading, racemiform or subthyrsoid, the heads mostly short-pedunculate or subsessile, 1.5–2.5 cm thick, glabrous or slightly glandular to sometimes evidently arachnoid-tomentose, usually a little shorter than the flowers, greenish-stramineous or a little purplish, the inner bracts often more flattened than the others and scarcely hooked. (n = 16, 18). Midsummer–fall. Native of Eurasia, now established as a common weed of roadsides and waste places throughout most of the US and s Can. Plants with somewhat larger, more pedunculate heads, called *A. nemorosum* Lej. & Courtois, may reflect present or past hybridization with *A. lappa*.

2. A. tomentosum Miller. Resembling *A. minus* and *A. lappa*; plants seldom more than 1.5 m tall; lower petioles mostly hollow; inflorescence corymbiform, the heads mostly long-pedunculate; involucre 2–3 cm thick, more or less strongly arachnoid-tomentose, the bracts only weakly or scarcely hooked. (n = 18) Summer–fall. Native of Eurasia, very sparingly established in disturbed sites in e US. *A. nemorosum* Lej. & Courtois—R, misapplied.

3. A. lappa L. GREAT BURDOCK. Plants up to 1.5 or even 3 m tall. Leaves petiolate, the petioles mostly solid, progressively shorter upward, the blade ovate or broader, cordate, up to 5 × 3 dm, thinly tomentose beneath, subglabrous above. Inflorescence corymbiform, with long, glandular or glandular-hairy peduncles; heads large, the involucre 2.5–4 cm wide, generally equaling or surpassing the flowers, glabrous or slightly glandular, and often with a few long, cobwebby hairs. (n = 16, 18) Midsummer–fall. Native of Eurasia, sparingly established as a weed in disturbed sites over most of n US and s Can, s at least to DC and sometimes NC.

105. CARDUUS L. THISTLE

Much like *Cirsium*, differing chiefly in its pappus of merely capillary, not at all plumose bristles; stem commonly spiny-winged by the decurrent leaf-bases; achenes quadrangular or somewhat flattened, with 5–10 or more nerves, or nerveless.

1 Involucral bracts mostly 2 mm wide or more; heads mostly nodding .... 1. *C. nutans*.
1 Involucral bracts rarely as much as 2 mm wide; heads erect.
  2 Plants very strongly spiny; stem tough .... 2. *C. acanthoides*.
  2 Plants weakly spiny; stem brittle .... 3. *C. crispus*.

1. C. nutans L. MUSK THISTLE, NODDING THISTLE. Biennial, 3–20 dm tall. Leaves glabrous, or long-villous chiefly along the main veins beneath, deeply lobed, up to 25 × 10 cm. Heads mostly solitary and nodding at the ends of the branches, usually large, the disk (1.5) 4–8 cm wide (pressed); middle and outer involucral bracts conspicuously broad (2–8 mm), with long, flat, spreading or reflexed, spine-pointed tip; inner bracts narrower and softer. Achenes 3–5 mm long. (n = 8) Summer–fall. Roadsides and waste places; native of Eurasia, now widely established in the US, in our range s to Ga, Miss, and La.

2. C. acanthoides L. Biennial, 3–10 dm tall, very strongly spiny; stem tough. Leaves deeply lobed or pinnatifid, up to 25 × 8 cm, loosely villous beneath, chiefly along the main veins, with long multicellular hairs, or glabrous, the upper surface glabrous or similarly hairy. Heads clustered or solitary at the ends of the branches, erect, small, the disk 1.5–2.5 cm wide when pressed; involucral bracts narrow, rarely as much as 2 mm wide, erect or loosely spreading, the middle and outer ones spine-tipped, the inner softer and flatter. Achenes 2–3 mm long, with ca. 10 (sometimes more) embedded nerves or striae. (n = 11) Midsummer–fall. Roadsides, pastures, and waste places; native of Eurasia, now widely but sparingly established in the US, in our range s to NC.

3. C. crispus L. Biennial, 6–20 dm tall, rather weakly spiny, the stem brittle. Leaves broader or less deeply cleft than in no. 2, cottony-tomentose beneath, at least when young. Heads clustered at the ends of the often short branches, similar to those of no. 2, but less pungent. (n = 8) Summer. Roadsides and waste places; native of Eurasia, very sparingly intro. in ne US, s to WVa.

106. CIRSIUM Miller THISTLE

Spiny, biennial or perennial herbs. Leaves alternate, toothed to more often pinnatifid, with weakly to strongly spiny margins. Heads medium-sized to large, discoid, the flowers all tubular and perfect, or seldom the heads functionally unisexual and the plants partly or wholly dioecious; involucral bracts imbricate in several or many series, firm, in most species some or all of them spine-tipped, often also with a glutinous dorsal ridge; receptacle flat to subconic, densely bristly; corollas slender and elongate, with relatively long, narrow lobes, purple, lavender, pink, or red to sometimes white, seldom rather light yellow, filaments usually papillose-hairy; anthers with a firm, narrow, apical appendage, evidently tailed at the base; style with a thickened, hairy ring and an abrupt change of texture below the more or less connate, papillate branches. Achenes basifixed or nearly so, glabrous, firm, thick-compressed, often curved, nerveless, commonly with a cartilaginous, pale yellow collar, which may be conspicuous before maturity as an expanded ring at the summit of the young achene; pappus of numerous plumose bristles, deciduous in a ring. Hybrids are frequent.

1 Plant perennial, strongly colonial by deep-seated creeping roots, polygamo-dioecious, the heads nearly unisexual; involucre 1–2 (2.5) cm high .... 12. *C. arvense*.
1 Plants biennial or less often perennial, but not becoming colonial; heads all alike, with perfect flowers; involucre often but not always more than 2.5 cm high.
 2 Stem conspicuously winged by the spiny, decurrent leaf bases, the wings nearly or fully as long as the internodes; leaves scabrous-hispid above; involucral bracts all spine-tipped, and without any glutinous dorsal ridge .. 1. *C. vulgare*.
 2 Stem inconspicuously or not at all winged, the leaf bases not decurrent, or (*C. lecontei*) only shortly so; involucral bracts often with a glutinous dorsal ridge, the innermost ones generally innocuous.
  3 Heads conspicuously and closely invested by a series of narrow and rather small, strongly spiny-toothed leaves that form a sort of secondary or false involucre; innermost involucral bracts with attenuate, narrow tip .... 2. *C. horridulum*.
  3 Heads otherwise, though sometimes subtended by 1 or 2 reduced leaves; involucral bracts various.
   4 Involucral bracts wholly innocuous, or with a vestigial spinule up to ca. 0.5 mm long .... 7. *C. muticum*.
   4 Middle and outer involucral bracts tipped with an evident spine that is generally at least 1 mm long.
    5 Lower surface of the leaves arachnoid-villous or only thinly and loosely tomentose, often eventually glabrate.
     6 Heads relatively small, the involucre 1.5–2.5 cm high; plants 5–35 dm tall, branched and many-headed when well developed .... 6. *C. nuttallii*.
     6 Heads larger, the involucre 2.5–5 cm high; plants 2–10 dm tall, simple or sparingly branched, with few or solitary heads.
      7 Stem and leaves loosely arachnoid-villous or crisp-hirsute, or seldom arachnoid-tomentose when young; heads solitary or terminating short branches, scarcely to sometimes more or less strongly pedunculate.
       8 Inner involucral bracts with expanded, chartaceous, crisped and erose tip; plants generally with well-developed and persistent basal leaves; cauline leaves not especially crowded, except sometimes near the base .... 3. *C. pumilum*.
       8 Inner involucral bracts very gradually tapering to the narrowly pointed tip; plants without well-developed basal leaves, at least at flowering time; cauline leaves crowded, the internodes commonly less than 1 cm long .... 4. *C. repandum*.
      7 Stem and lower surfaces of the leaves thinly and rather loosely white-tomentose when young, later often glabrate; head solitary and strongly pedunculate, or the stem with a very few long branches, each ending in a pedunculate head; involucral bracts as in *C. repandum* .... 5. *C. lecontei*.

5 Leaves densely, closely, and persistently white-tomentose beneath.
9 Heads relatively large, the involucre mostly 2.5–3.5 cm high; plants robust, 10–30 (40) dm tall, the larger leaves generally more than 2.5 cm wide when lobeless, and more than 5 cm wide when lobed.
10 Leaves deeply pinnatifid .... 8. *C. discolor*.
10 Leaves merely toothed or shallowly lobed .... 9. *C. altissimum*.
9 Heads smaller; the involucre mostly 1.5–2.5 cm high; plants more slender, 5–15 dm tall, the larger leaves up to ca. 2.5 cm wide when lobeless, and up to ca. 5 cm wide when lobed.
11 Cauline leaves numerous, mostly 30–70; plants flowering in Aug and Sept .... 10. *C. virginianum*.
11 Cauline leaves relatively few, mostly 10–25; plants flowering in Apr, May, and June .... 11. *C. carolinianum*.

1. C. vulgare (Savi) Tenore. BULL THISTLE. Biennial weed, 5–15 dm tall; stem conspicuously spiny-winged by the decurrent leaf bases, copiously spreading-hirsute to sometimes arachnoid. Leaves pinnatifid, the larger ones with the lobes again toothed or lobed, scabrous-hispid above, thinly white-tomentulose to sometimes green and merely hirsute beneath. Heads several, purple; involucre 2.5–4 cm high, its bracts all spine-tipped, without any well-developed glutinous dorsal ridge. Achenes 3–4 mm long. (n = 28–30, 34) Midsummer–fall. Pastures, fields, roadsides, and waste places; native of Eurasia, now widely established in N Am, and found in much of our range, s to c Ga, c Miss, La, and Ark. *C. lanceolatum* (L.) Hill—S, misapplied.

2. C. horridulum Michx. Stout biennial 2–15 dm tall, simple or with short, stout, ascending, pedunclelike branches, some of the roots commonly fleshy-thickened. Herbage thinly tomentose and eventually generally glabrate, or the leaves merely arachnoid and glabrate; leaves strongly spiny, variously broad and pinnatifid (in either var.) to narrow and merely spiny-toothed (especially in var. *vittatum*), the largest ones at or near the base. Heads light yellow or white to lavender or purple, several or solitary, each subtended by a number of narrow, erect, strongly spiny, reduced leaves; involucre (2.5) 3–5 cm high, the outer bracts with erect spine tip, the inner merely attenuate, all with more or less modified margin but without any glutinous dorsal ridge. Achenes 4–6 mm long. (n = 16, 17) Spring–summer, or all year southward. Open places, especially in sandy soil along salt or fresh marshes, less commonly in uplands; coastal states (and Pa) from Me to Fla and Tex, avoiding the mt. provinces; disjunct in Mex; intro. in RV of se Tenn. The species consists of 2 vars. Var. *horridulum* has the modified margin of the involucral bracts only shortly scabrous or scabrous-ciliolate, the barbs commonly ca. 0.1 mm long, only the outer involucral bracts sometimes with longer ones; it has nearly the range of the species, but is uncommon s of n Fla. *Carduus spinosissimus* Walter—R. Var. *vittatum* (Small) R. W. Long has the modified margin of the involucral bracts, or at least the middle and outer ones, evidently setulose-ciliate or lacerate-ciliolate, many of the barbs or bristles in the range of 0.3–0.5 mm, and the plants average smaller and more slender and less pubescent than var. *horridulum*; it is the characteristic phase of s Fla, extending n less commonly to Wakulla and Gadsden Cos., and even to se NC. *C. vittatum* Small—S; *C. smallii* Britton—S; *Carduus smallii* (Britton) Ahles—R.

3. C. pumilum (Nutt.) Sprengel. PASTURE THISTLE. Stout biennial 3–8 dm tall, with 1–several coarse, slightly thickened roots. Herbage green and coarsely arachnoid-villous or crisp-hirsute; leaves lobed or pinnatifid, beset with numerous short marginal spines in addition to the scattered longer ones; basal leaves well developed and generally persistent, these and the lower cauline ones 12–25 × 2–7 cm; cauline leaves not especially crowded except sometimes near the base. Heads few or solitary, large, the involucre 3.5–5 cm high, usually some of its bracts with a narrow, glutinous dorsal ridge; middle and outer involucral bracts tipped by a short, more or less erect spine; inner bracts elongate, with expanded, chartaceous, crisped and erose tip; flowers sweet-scented, purple or sometimes white. Achenes 3–4 mm long. (n = 15) Summer. Pastures, old fields, and open woods; Me to w NY, s to Del, to PP of NC, and to the mts. of Va and WVa. *C. odoratum* (Muhl.) Petrak—S; *Carduus pumilus* Nuttall—R.

4. C. repandum Michx. Deep-rooted perennial, 2–6 dm tall, leafy throughout, loosely and rather copiously arachnoid-villous (or seldom arachnoid-tomentose) when young, subglabrate or merely hirsute in age. Leaves crowded (the internodes commonly less than 1 cm long), all cauline, narrow, sessile, 6–15 × 1–2.5 cm, somewhat ruffled along the margins, beset by numerous small spines in addition to the scattered larger ones, coarsely toothed or shallowly lobed, and often finely toothed as well. Heads solitary, or more often several and terminating short branches from near the summit, only shortly or scarcely pedunculate, purple; involucre 2.5–4 cm high, its middle and outer bracts with short, mostly erect spine-tip, the glutinous dorsal ridge poorly developed or wanting; inner bracts looser and merely attenuate. Achenes 3.5–4 mm long. (n = 15) Summer. Dry, sandy soil, especially in pine barrens and sand hills; se Va to Ga. *Carduus repandus* (Michx.) Persoon—R.

5. C. lecontei T. & G. Perennial with 1 or several stout roots, 4–10 dm tall, simple or sometimes with 1–3 long branches, the main axis and each such branch terminating in a solitary, evidently pedunculate head; stem and lower surface of the leaves thinly and rather loosely white-tomentose, at least when young, often eventually more or less glabrate, the upper leaf surfaces green and glabrous or with a few long, coarse hairs. Leaves narrow, less densely spiny than in *C. repandum*, pinnately lobed, often appearing scalloped, varying to merely spiny-toothed, tending to be shortly decurrent on the stem, the largest ones at or near the base, mostly 10–30 × 0.8–4 cm, the others gradually reduced upward. Heads purple; involucre 2.5–4 cm high, its bracts with a narrow glutinous dorsal ridge, the innermost ones very slender and long-tapering, innocuous or nearly so, the others tapering to a short, erect spine tip. Achenes ca. 5 mm long. (n = 16) Summer. Moist or wet pinelands and savannas; CP from NC to n Fla, and w to La. *Carduus lecontei* (T. & G.) Pollard—R.

6. C. nuttallii DC. Coarse biennial 15–35 dm tall, branching and many-headed when well developed, the heads commonly solitary at the ends of long, slender, subnaked branches; stem glabrous or with crisp spreading hairs. Leaves arachnoid-tomentose beneath when young, generally eventually glabrate, smooth or somewhat crisp-hairy on the upper surface, thin, deeply pinnatifid, the lobes generally again toothed or cleft; larger (lower) leaves up to 60 × 15 cm, but often deciduous. Involucre 1.5–2.5 cm high, the middle and outer bracts with a glutinous dorsal ridge and tipped with a weak, abruptly spreading spine mostly 1–2 (3) mm long, the inner bracts innocuous and merely attenuate, often crisped but not expanded; flowers pink or lavender (often very pale) to white. Achenes 3–4 mm long. (n = 12, 14) Summer, or all year southward. Wet or dry, usually sandy soil, often in thickets; peninsular Fla to La and SC, and (apparently) disjunct in se Va, largely on CP. *Carduus nuttallii* (DC.) Pollard—R.

7. C. muticum Michx. SWAMP THISTLE. Coarse biennial with several stout roots, 5–20 dm tall, branching and many-headed when well developed, villous or arachnoid when young, later glabrescent. Leaves deeply pinnatifid into lanceolate or oblong, entire, lobed, or dentate segments, only weakly spiny, often very large, up to 55 × 20 cm, sometimes thinly tomentose beneath, otherwise arachnoid-hirsute to subglabrous. Involucre 2–3.5 cm high, multiseriate, the bracts more or less tomentose, especially marginally, generally with a glutinous dorsal ridge, innocuous or with a vestigial spinule up to ca. 0.5 mm long, the inner commonly with loose, crisped tip, but not expanded; flowers purple, pink, or deep lavender, rarely white. Achenes ca. 5 mm long. (n = 10; 2n = 30—Fla) Late summer–fall. Swamps and bogs, wet meadows, and moist woods and thickets; Nf to Sask, s to Del and Mo and in the mts. to NC and Tenn; also at widely scattered stations s to n Fla, La, and e Tex. *Carduus muticus* (Michx.) Persoon—R.

8. C. **discolor** (Muhl.) Sprengel. Much like *C. altissimum*, but with the leaves deeply pinnatifid, generally firmer and spinier; larger leaves generally more than 6 cm wide, sometimes as much as 50 cm long and 25 cm wide; plants averaging smaller than in *C. altissimum*, mostly 10–20 dm tall, the peduncles tending to be a little leafier, the heads often a little broader-based. (n = 10) Midsummer–fall. Fields, open woods, river bottoms, and waste places; sw Que and s Ont to Man, s to NC, Tenn, Miss, La, Mo, and Kans. Hybridization with the closely related *C. altissimum* is somewhat restricted by the cytological difference. *Carduus discolor* (Muhl. ex Willd.) Nuttall—R.

9. C. altissimum (L.) Sprengel. Robust, fibrous-rooted perennial, 10–30 (40) dm tall, openly branched when well developed; stem crisply spreading-hirsute to subglabrate, sometimes slightly tomentose in the inflorescence. Leaves large, the lower ones up to 50 cm long and 20 cm wide, broadly oblanceolate to obovate or elliptic, densely white-tomentose beneath, scabrous-hirsute to subglabrous above, merely spiny-toothed or coarsely toothed to sometimes lobed (seldom more than halfway to the midrib), the reduced ones of the inflorescence sometimes more evidently lobed than the others. Heads several or numerous on the more or less leafy peduncles; involucre (2) 2.5–3.5 (4) cm high; middle and outer bracts tipped with a spine 2–5 mm long; inner bracts merely attenuate or often with a scarious, slightly dilated and erose tip; flowers mostly pink-purple. Achenes 4.5–6 mm long. (n = 9) Late summer–fall. Fields, waste places, river bottoms, and open woods; Mass to n Fla, w to SD and Tex. *Carduus altissimus* L.—R.

10. C. virginianum (L.) Michx. Slender biennial 5–10 (15) dm tall from a cluster of fleshy-fibrous roots; stem glabrous or arachnoid, sometimes more evidently tomentose when young. Leaves closely white-tomentose beneath, glabrous or hirsute on the upper surface, the basal ones soon deciduous, the others numerous, mostly 30–70, 5–20 cm long, sometimes merely spinose-ciliate and up to ca. 1 cm wide, sometimes evidently pinnatifid and up to 5 cm wide; peduncles with a few reduced leaves or bracts. Heads several to rather numerous, terminating the branches, or sometimes solitary; involucre 1.5–2.5 cm high, the middle and outer bracts with a glutinous dorsal ridge and a slender, suberect or spreading spine tip (0.7) 1–2.5 mm long, the inner ones merely attenuate and commonly crisped; flowers purple. Achenes 3–4 mm long. (n = 14) Late summer. Savannas, bogs, and wet pinelands; AC from s NJ to n Fla. *C. revolutum* Small—S; *Carduus virginianus* L.—R.

11. C. carolinianum (Walter) Fern. & Schubert. Slender, fibrous-rooted biennial 5–15 (18) dm tall; stem glabrous or arachnoid, sometimes more evidently tomentose when young. Leaves closely white-tomentose beneath, glabrous or hirsute on the upper surface, from merely spinose-ciliate and up to ca. 2.5 cm wide to evidently pinnatifid and up to 5 cm wide, the basal ones up to 3 dm long, the cauline ones relatively few, mostly 10–25, 8–15 cm long and (except when lobed) seldom more than 1.5 cm wide, narrow-based, reduced upward. Heads several or occasionally solitary, on long, naked peduncles terminating the branches; involucre 1.5–2 cm high, the middle and outer bracts with a glutinous dorsal ridge and a slender, suberect or spreading spine mostly 1.5–4 mm long, the inner ones merely attenuate and commonly crisped; flowers pink-purple. Achenes 3–4 mm long. (n = 10, 11) Spring. Open woods and dry, sandy soil; s Ohio to the mts. of NC, Ga, and Ala, w to Mo, Miss, Ark, (OZ, OU), La, and Tex. *Carduus carolinianus* Walter—R; *Cirsium flaccidum* Small—S.

12. C. arvense (L.) Scop. CANADA THISTLE. Perennial weed 3–15 (20) dm tall from deep-seated creeping roots, subglabrous, or the leaves more or less white-tomentose beneath. Heads more or less numerous in an often flat-topped inflorescence, unisexual or nearly so, the plants polygamo-dioecious; involucre 1–2 cm high, its bracts all innocuous, or the outer with a weak spine tip about 1 mm long; flowers pink-purple or occasionally white. Achenes 2.5–4 mm long; pappus of the pistillate heads surpassing the corollas, that of the staminate heads surpassed by the corollas. (n = 17) Summer. A noxious weed of fields and waste places; native of Eurasia, now widely intro. in n US and s Can, and found along the n margin of our range, s as far as NC and Tenn. *Carduus arvensis* (L.) Robson—R. The var. *arvense* has merely toothed or shallowly lobed, weakly spiny leaves. (var. *mite* Wimm. & Grab.—F; var. *integrifolium* Wimm. & Grab.—F; var. *vestitum* Wimm. & Grab.—F). The var. *horridum* Wimmer & Grabowski, more common in the US, has more spiny, deeply pinnatifid leaves. (treated by F as typical *C. arvense*)

107. SILYBUM Gaertn., nom. conserv. MILK THISTLE

1. S. marianum (L.) Gaertn. Glabrous or slightly tomentose thistle, winter annual or biennial, 6–15 (30) dm tall, the stems simple or sparingly branched, not winged. Leaves alternate, pinnately lobed (less so upward), up to 80 cm long and nearly half as wide, petiolate below, becoming sessile and strongly auriculate-clasping above, spiny-margined, more or less marked with white along the main veins. Heads terminating the branches, large, 3–6 cm wide, discoid, the flowers all tubular and perfect; involucral bracts imbricate in several series, the middle and outer ones with a broad, firm, spinulose-ciliate base and a coarse, subfoliose, divaricately spreading, spine-tipped and basally spine-margined appendage; receptacle flat, densely setulose; corollas purple, with slender tube, short throat, and narrow lobes; filaments glabrous, connate below; anthers with a firm, slender, terminal appendage, shortly tailed at the base; style with an abrupt change of texture below the connate, papillate branches. Achenes basifixed, somewhat compressed, glabrous, 6–7 mm long; pappus of numerous slender, unequal, subpaleaceous bristles, deciduous in a ring. (n = 17) Spring–summer. A weed of waste places, native to the Mediterranean region, rare and casual with us. *Mariana Mariana* (L.) Hill—S.

108. ONOPORDUM L. SCOTCH THISTLE

1. O. acanthium L. Coarse, branching, white-tomentose, very strongly spiny, biennial thistle up to 2 m tall; stem broadly spiny-winged. Leaves alternate, toothed or slightly lobed, sessile and decurrent, or the lower petiolate, the blade up to 6 × 3 dm, spiny-margined. Heads large, 2.5–5 cm wide, purple, discoid, the flowers all tubular and perfect; involucral bracts imbricate in several series, all slender and spine-tipped; receptacle flat, fleshy, honey-combed, commonly with short bristles on the partitions, but not densely bristly; corollas elongate, with long, slender tube, short throat, and slender lobes; anthers with a firm, narrow, terminal appendage, evidently tailed at the base; style with an abrupt change of texture below the papillate, connate branches. Achenes basifixed, 4–5 mm long, slightly compressed, glabrous, transversely rugulose; pappus of numerous unequal, capillary or somewhat chaffy-flattened, merely barbellate bristles. (n = 17) Midsummer–fall. A Eurasian weed, sparingly naturalized in disturbed sites over much of the US.

109. ECHINOPS L. GLOBE THISTLE

1. **E. sphaerocephalus** L. Coarse, branching thistle up to 2.5 m tall; stem spreading-hairy, also tomentose above. Leaves alternate, white-tomentose beneath, green and merely scabrous or hirsute above, sessile and clasping (at least the middle and upper ones) but not decurrent, up to 35 × 20 cm, pinnatifid and spiny-margined. Heads 1-flowered, aggregated into globose secondary heads, these 3.5–6 cm thick, pale bluish, naked-pedunculate at the branch tips; proper involucres narrow, 1.5–2 cm long, each subtended by a tuft of flattened bristles seldom half as long; involucral bracts firm and dry, somewhat imbricate, shortly spine-tipped, some of them coarsely ciliate; corolla blue, with slender tube, short throat, and long narrow lobes; anthers with a firm, narrow, terminal appendage, shortly tailed at the base, the tails fringed-hairy; style with a ring of hairs and an abrupt change of texture below the rather short, papillate branches. Achenes strongly hairy; pappus a well-developed crown of concrescent scales. (n = 15, 16) Summer. Native of Eurasia, occasionally cult. for ornament, and casually established here and there in the US and s Can.

110. CENTAUREA L. STAR THISTLE, KNAPWEED

Herbs. Leaves alternate or all basal, entire to pinnatifid, scarcely or not at all prickly. Heads discoid, the flowers all tubular and perfect, or more often the marginal flowers neutral, with enlarged, irregular, falsely radiate corolla; involucral bracts imbricate in several series, dry, either spine-tipped or more often some of them with an enlarged, scarious or hyaline, erose to lacerate or pectinate appendage, or at least with a pectinate or lacerate fringe near the tip; receptacle nearly flat, densely bristly; corollas pink-purple or blue to yellow or white, with slender tube, short throat, and long, narrow lobes; anthers with a narrow, firm, apical appendage, evidently (though often shortly) tailed at the base; style with a thickened, often hairy ring and an abrupt change of texture at the base of the papillate branches. Achenes obliquely or laterally attached to the receptacle, seldom evidently nerved; pappus of several series of graduated bristles or narrow scales, often much reduced, or wanting. We have only one native species, the others all introduced from the Old World.

1 Involucral bracts not at all spiny, commonly more or less lacerate or fringed.
 2 Pappus well developed, mostly 6–12 mm long; heads large, the involucre mostly 2–4 cm high; plants annual .... 1. *C. americana*.
 2 Pappus much reduced, not more than about 3 mm long, or wanting; heads smaller, the involucre mostly 1–2 cm high; plants annual or perennial.
  3 Leaves entire or toothed, or some of the larger ones few-lobed.
   4 Plants perennial; some of the lower leaves, at least, more than 1 cm wide.
    5 Scarious tips of the involucral bracts conspicuously blackish at least in part, those of the middle and outer ones regularly pectinate, seldom any of them obviously bifid.
     6 Scarious tips of the involucral bracts small, 1–3 mm long; heads relatively narrow, the pressed involucre generally higher than broad; marginal flowers enlarged and raylike .... 2. *C. dubia*.
     6 Scarious tips of the involucral bracts large, the larger ones (3) 4–6 mm long; heads relatively broad, the involucre broader than high; marginal flowers not enlarged, except in a frequent hybrid with *C. jacea* .... 3. *C. nigra*.
    5 Scarious tips of the involucral bracts tan to dark brown, those of the middle and outer ones irregularly lacerate, those of the inner ones expanded and often strongly bifid; marginal flowers generally enlarged and raylike .... 4. *C. jacea*.
   4 Plants annual or winter annual; leaves linear or nearly so, less than 1 cm wide (sometimes some of them with a few linear lobes, and then more than 1 cm wide) .... 5. *C. cyanus*.
  3 Leaves all (except the reduced ones of the inflorescence) pinnatifid, with narrow lobes .... 6. *C. maculosa*.
1 Involucral bracts, or some of them, evidently spine-tipped.
 7 Stem merely angled; not winged; flowers purple; pappus none .... 7. *C. calcitrapa*.
 7 Stem evidently winged by the decurrent leaf bases; flowers yellow; pappus of the central flowers well developed .... 8. *C. solstitialis*.

1. C. americana Nutt. AMERICAN KNAPWEED. Simple or sparingly branched annual, 2–10 dm tall. Leaves numerous, sparsely to densely scabrous or scabrous-puberulent, and more or less evidently glandular-punctate, entire or slightly toothed, mostly sessile and narrowly to broadly lanceolate or lance-oblong, or the deciduous lower ones more oblanceolate and subpetiolate, the middle cauline ones mostly 3.5–10 cm × 7–35 mm. Heads few or solitary, large; involucre mostly 2–4 cm high, each bract with a long, narrow, loosely pectinate terminal appendage; flowers purple, the marginal ones enlarged and sterile, sometimes making the head 10 cm wide. Longer pappus bristles 6–12 mm long. (n = 13) Summer. Prairies, fields, and other open places; Mo to La, Ariz, and Mex.

2. C. dubia Suter. SHORT-FRINGED KNAPWEED. Habitally like the next 2 spp.; involucre 11–18 mm high, generally higher than broad, even as pressed; appendages of the involucral bracts conspicuously blackish at least in part, 1–3 mm long, the middle and outer ones deeply and rather regularly pectinate; marginal flowers enlarged and raylike. Pappus mostly wanting. (n = 11, 22) Midsummer–fall. Fields, roadsides, and waste places; native of Europe, now widely established in se Can and ne US, s to Va and WVa. *C. nigrescens* Willd.—F; *C. vochinensis* Bernh.—F.

3. C. nigra L. BLACK KNAPWEED. Perennial, 2–8 dm tall, rough-puberulent, and sometimes arachnoid when young. Leaves entire or toothed, the basal ones broadly oblanceolate or elliptic, entire or toothed to sometimes few-lobed, mostly (1) 1.5–4 (6) cm wide, petiolate, the cauline ones reduced upward and becoming sessile. Heads terminating the often numerous branches; involucre 12–19 mm high, broader than high, appendages of the involucral bracts well developed, conspicuously blackish at least in part, the middle and outer deeply and fairly regularly pectinate, the larger ones mostly (3) 4–6 mm long, seldom any of them markedly bifid; flowers pink-purple, the marginal ones typically not enlarged. Pappus ca. 1 mm long or less. (n = 11, 22) Midsummer–fall. Fields, roadsides, and waste places; native of Europe, now widely established in s Can and n US, s to Va and WVa. Hybridizes with no. 4, producing segregating or stabilized intermediates called *C.* X *pratensis* Thuill., these often approaching *C. nigra* as to the involucre, but subradiate as in *C. jacea*. *C. nigra* var. *radiata* DC.—F.

4. **C. jacea** L. BROWN KNAPWEED. Habitally similar to *C. nigra*. Involucre 12–18 mm high, from a little narrower to a little broader than high; appendages of the involucral bracts well developed, broad, tan to dark brown, the middle and outer ones rather irregularly lacerate, the inner ones less so and often deeply bifid; marginal flowers almost always enlarged. Pappus none. (n = 11, 22) Summer. Fields, roadsides, and waste places; native of Europe, now widely established in s Can and n US, s to WVa.

5. **C. cyanus** L. CORNFLOWER, BACHELOR'S BUTTON. Annual or winter annual, 2–12 dm tall, usually loosely white-tomentose when young, the lower leaf surfaces often linear, entire, or the lower ones a little toothed or with a few narrow lobes, up to 13 cm × 8 mm (excluding the lobes). Heads terminating the branches; involucre 11–16 mm high, its bracts more or less striate, with a relatively narrow, often darkened, pectinate or lacerate fringe near the tip; flowers mostly blue, sometimes pink, purple, or white, the marginal ones enlarged. Pappus 2–3 mm long. (n = 12) Spring. Fields, roadsides, and waste places; native to the Mediterranean region, widely cult. and now a cosmopolitan weed.

6. **C. maculosa** Lam. SPOTTED KNAPWEED. Biennial or short-lived perennial, 3–12 dm tall, sparsely scabrous-puberulent, and with a thin and loose, evanescent tomentum. Leaves pinnatifid with narrow lobes, or the reduced ones of the inflorescence entire. Heads terminating the numerous branches, constricted upward in life; involucre 10–13 mm high, its bracts striate, the middle and outer ones with short, dark, pectinate tip; flowers pink-purple, the marginal ones enlarged. Pappus up to 2 mm long, or rarely none. (n = 18) Summer–fall. Fields, roadsides, and waste places; native of Europe, now commonly established in ne US, s to NC and occasionally SC, and to Tenn and Mo, and to be expected in Ark.

7. **C. calcitrapa** L. PURPLE STAR THISTLE. Branching biennial 1–8 dm tall, arachnoid-villous or glabrate; stem angled but not winged. Leaves small, pinnatifid with narrow lobes, or the upper entire. Heads numerous, narrow; involucre 10–18 mm high, its bracts weakly spinose-ciliate and stoutly spine-tipped, the larger spines 10–30 mm long; flowers few, purple, the outer ones not much if at all enlarged. Pappus none. (n = 10) Summer. Native of Eurasia, now widespread in the US as a roadside weed, but only occasional in our range.

8. **C. solstitialis** L. BARNABY'S THISTLE, YELLOW STAR THISTLE. Annual or biennial, 2–8 dm tall, thinly but persistently tomentose, the stem evidently winged by the decurrent leaf bases. Basal leaves lyrate or pinnatifid, up to 20 × 5 cm, the middle and upper smaller, becoming linear and entire. Involucre 10–15 mm high, broad-based; middle and outer bracts spine-tipped, the larger central spines 11–22 mm long; inner bracts with a small hyaline appendage; flowers yellow, the marginal ones not enlarged; pappus of the marginal flowers wanting, that of the others 3–5 mm long. (n = 8) Summer. A weed in fields and waste places, native to the Mediterranean region, now widely established in the US, but only occasional in our range.

C. melitensis L., the Maltese star thistle, is sparingly intro. in the US and has been collected in SC and Ga. It has shorter spines than C. *solstitialis*, the larger ones mostly 5–9 mm long, and the flowers all have a pappus 1.5–3 mm long.

111. CNICUS L., nom. conserv. BLESSED THISTLE

1. **C. benedictus** L. Branching annual thistle 1.5–8 dm tall; stem spreading-villous. Leaves alternate, toothed or pinnatifid, weakly spiny-margined, scarcely or not at all decurrent, up to 20 × 5 cm, the lower petiolate, the others sessile. Heads yellow, terminating the branches, closely subtended by ovate or lance-ovate foliage leaves crowded beneath the proper involucre and generally surpassing it, essentially discoid, the flowers all tubular and perfect except for a few inconspicuous marginal neutral ones with very slender, 2–3-lobed corolla; involucre 3–4 cm high, of several series of broad, spine-tipped bracts, the spines of the inner ones pinnatisect; receptacle flat, densely bristly; anthers with a firm, narrow, terminal appendage, shortly sagittate-tailed at the base; style with a ring of hairs and an abrupt change of texture at the base of the very short, scarcely divergent, papillate branches. Achenes obliquely attached to the receptacle, ca. 8 mm long, subterete, strongly 20-ribbed, glabrous, with a firm 10-toothed crown; pappus biseriate, the outer of 10 firm smooth awns about as long as the achene, alternating with as many much shorter, minutely hairy and sparsely pectinate inner ones. (n = 11) Spring–summer. A Mediterranean weed, now sparingly established in fields and waste places in the US and s Can. *Centaurea benedicta* (L.) L.—R.

112. CHAPTALIA Vent., nom. conserv. SUNBONNETS

Fibrous-rooted perennial herbs with a rosette of basal leaves, and with one or more monocephalous, naked or merely bracteate, more or less tomentose flowering stems. Leaves simple, entire or toothed to lyrately lobed, more or less tomentose at least beneath. Heads turbinate or campanulate, often nodding; involucral bracts slender, acute, more or less tomentose, well imbricate in several series, semiherbaceous or subscarious, sometimes partly anthocyanic; receptacle flat, naked; flowers anthocyanic to white, trimorphic, the outermost one to several rows pistillate, with strap-shaped, more or less tridentate ligule and with or without 1 or 2 minute teeth on the inner side, the next inner row of flowers also pistillate, but with a slender, tubular corolla shorter than the style (this set wanting in some extralimital spp.), the central flowers perfect or functionally staminate, with bilabiate corolla, the outer lip 3-toothed, the inner with 1 or 2 teeth; anthers tailed at the base, the tails entire; style branches of the perfect flowers short, rounded, externally papillate-hairy, with inconspicuous ventromarginal stigmatic lines, sometimes tardily or scarcely separating. Achenes columnar to fusiform, usually with a more or less prominent beak, 5- to several-nerved, glabrous to scabrous, villous, or glandular; pappus of numerous capillary bristles.

1 Achenes with a slender beak half to twice as long as the body; flowers all fertile.
  2 Leaves firm, narrow, mostly 0.5–2 cm wide, densely white-tomentose beneath; involucre 5–10 mm high at anthesis, elongating to 11–22 mm in fruit; heads rarely nodding .... 1. *C. dentata*.
  2 Leaves thin and soft, wider, mostly (1.5) 2–7 cm wide, thinly gray-tomentulose

beneath; involucre 11–15 mm high at anthesis, elongating to 15–25 mm in fruit; fruiting heads mostly nodding .... 2. *C. nutans*.
1 Achenes with a short, stout beak less than half as long as the body; central flowers functionally staminate, their achenes empty .... 3. *C. tomentosa*.

1. C. dentata (L.) Cass. Leaves firm, oblanceolate to narrowly elliptic-obovate, mostly 3–15 (25) × 0.5–2 cm, inconspicuously dentate to sometimes shallowly lobulate, densely white-tomentose beneath, green and glabrous above or with a little deciduous tomentum. Scapes usually 2–several, naked, (0.5) 1–3 (4) dm tall. Head erect or seldom very loosely nodding at maturity; involucre 5–10 mm high at anthesis, elongating to 11–22 mm in fruit; ligules ca. (8–) 13 (–21), white, shortly surpassing the involucre and pappus. Achenes 6–12 mm long, glabrous or commonly scabrous-hispidulous in part, the slender, elongate beak from a little more than half as long as the body to about twice as long. (n = 16, 24) All year. Pinelands; s Fla, and in parts of WI and Mex.

2. C. nutans (L.) Polakowsky. Leaves thin, elliptic-oblanceolate to elliptic-obovate, tending to be lobulate or somewhat lyrate as well as callous-denticulate, mostly 5–20 × (1.5) 2–7 cm, thinly gray-tomentulose beneath, green and glabrate above. Scapes usually 2–several, naked, 1–4 (–8) dm tall. Heads usually nodding in fruit; involucre mostly 11–15 mm high at anthesis, elongating to 15–25 mm in fruit; ligules mostly ca. 21 to ca. 34, commonly pink or lavender on the back and white on top, short, often scarcely surpassing the involucre. Achenes 7–15 mm long, glabrous to usually glandular-scabrous or scabrous-hispidulous at least in part, the slender, elongate beak from half to fully twice as long as the body. (n = 24) Spring. Dry, open or wooded, often rocky places; tropical Am, n to s Tex and s La.

3. C. tomentosa Vent. Leaves oblanceolate to elliptic or almost obovate, mostly 3–15 × 1–3.5 cm, minutely callous-denticulate, densely white-tomentose beneath, glossy green and glabrous or nearly so above. Scapes 1–several, naked, 1–4 dm tall. Heads tending to be more or less erect at anthesis, but distinctly and abruptly nodding both before and after; involucre 9–14 mm high at anthesis, scarcely elongating in fruit; ligules mostly from ca. 13 to ca. 21, white on top, pink or lavender on the back, surpassing the involucre by 3–6 mm; disk flowers ochroleucous. Achenes 5–6 mm long, glabrous, with a short, stout beak less than half as long as the body; central achenes fully as long as the outer ones, but empty. (n = 24) Spring, or winter southward. Low pinelands, savannas, and bogs; CP from NC to c (and barely in s) Fla, w to e Tex.

113. LYGODESMIA D. Don, RUSH PINK, SKELETON WEED, ROSE RUSH, RUSHWEED

Perennial herbs with milky juice, ours glabrous and with the stems arising singly from a deep-seated system of creeping roots; stems rushlike, green and photosynthetic. Leaves alternate, mostly linear or subulate, sometimes most or all of them reduced to mere scales. Heads borne singly at the branch tips; flowers all ligulate and perfect, pink to lavender or purple, rarely white; involucre cylindric, of 4–8 principal bracts and a few much reduced outer ones; receptacle naked. Achenes glabrous, obscurely angled or subterete, linear-columnar, generally narrowed toward the summit, sometimes also below; pappus of numerous capillary bristles.

1 Heads relatively large, the ligules commonly surpassing the involucre by 1.5–2 cm, the achenes 10–15 mm long .... 1. *L. aphylla.*
1 Heads smaller, the ligules commonly surpassing the involucre by 0.5–1 cm, the achenes 5–7 mm long .... 2. *L juncea.*

1. L. aphylla (Nutt.) T. & G. Stems 4–8 dm tall, rather sparingly branched or simple. Basal leaves 1–several, elongate-linear, mostly 10–30 cm × 1–3 mm, sometimes deciduous before anthesis; cauline leaves reduced to inconspicuous scales, or sometimes a single linear leaf present near the base. Heads relatively large and showy; involucre (13) 15–20 (24) mm high, with ca. 8 principal bracts; flowers ca. 10, the ligule commonly surpassing the involucre by 1.5–2 cm, mostly 5–7 mm wide. Achenes mostly 10–15 mm long. Spring, or all year southward. Commonly in sandy soil of open pine woods or oak scrub; throughout peninsular Fla, w onto the panhandle and n to s Ga.

2. L. juncea (Pursh) D. Don. Stems 2–6 dm tall, much branched and rigid. Leaves linear, up to 4 cm × 3 mm, the upper (or nearly all) reduced to subulate scales; elongate basal leaves wanting. Heads relatively small; involucre 9–16 mm high, with ca. 5 (very rarely 8) principal bracts; flowers ca. 5 (very rarely up to 10), the ligule surpassing the involucre by ca. 0.5–1 cm. Achenes mostly 5–7 mm long. Summer. Prairies and other dry, open places; Minn to Ark (OU, where rare and perhaps disjunct), w to s BC, e Wash, and Ariz, mainly on the Great Plains.

114. PRENANTHES L. WHITE LETTUCE, RATTLESNAKE ROOT

Perennial herbs with milky juice and slightly to strongly tuberous-thickened roots. Leaves alternate, well developed (the larger ones at least 1 cm wide), variously simple and entire or merely toothed to few-lobed or deeply cleft, or the lower ones sometimes several-foliolate. Heads more or less numerous in an elongate and racemiform or thyrsoid to open-paniculiform inflorescence, erect or more often nodding; flowers all ligulate and perfect, 5–35 in number, white to ochroleucous, chloroleucous, grayish, pink, or pale lavender; involucre cylindric or seldom campanulate, of 4–15 principal bracts and some much reduced outer ones, or the outer occasionally better developed and passing into the inner; receptacle small, naked. Achenes elongate, cylindric or nearly so, glabrous, mostly reddish-brown, in ours more or less ribbed-striate; pappus of numerous deciduous capillary bristles. *Nabalus* Cass.—S. Milstead, W. L. 1964. A revision of the North American species of *Prenanthes*. Ph.D. thesis. Purdue University.

The taxonomy of *Prenanthes* is complicated by frequent hybridization and by extreme instability of leaf form in several spp.

1 Involucre evidently pubescent with long, coarse hairs (these sometimes very few in *P. serpentaria*).
  2 Principal involucral bracts ca. 13 (12–15); flowers mostly 20–35 per head ..... 1. *P. crepidinea.*
  2 Principal involucral bracts ca. 8 (6–10); flowers from 5 to ca. 13 (–19) per head.
    3 Inflorescence cylindric, more or less thyrsoid, the branches all very short.
      4 Heads loosely ascending to suberect; leaves toothed or entire, the lower ones somewhat obovate, tapering to the petiole, but often deciduous, the others sessile or nearly so and often clasping .... 2. *P. aspera.*
      4 Heads nodding; lower and middle leaves petiolate, with broad-based, deltoid to mostly sagittate or hastate, toothed or sometimes deeply lobed or cleft blade .... 3. *P. roanensis.*

3 Inflorescence generally broader, corymbiform or paniculiform, at least some of the branches elongate.
5 Involucre rather copiously setose; leaves merely toothed, or seldom shallowly lobed .... 4. *P. barbata.*
5 Involucre only rather sparsely setose; principal leaves usually evidently lobed, but sometimes merely toothed .... 5. *P. serpentaria.*
1 Involucre glabrous, or sometimes with a few short cilia or inconspicuous tufts of fine, short, soft hairs at the tip.
6 Principal involucral bracts 7–10, often 8; flowers 8–15 per head.
7 Inflorescence relatively open, corymbiform or paniculiform, some of the branches usually elongate; flowers variously white to ochroleucous or chloroleucous or sometimes pink or lavender.
8 Pappus stramineous or light brown .... 6. *P. trifoliolata.*
8 Pappus cinnamon-brown .... 7. *P. alba.*
7 Inflorescences very narrow and elongate; flowers mostly pink or pale lavender .... 8. *P. autumnalis.*
6 Principal involucral bracts (4) 5 (6); flowers 5 or 6 per head .... 9. *P. altissima.*

1. P. crepidinea Michx. Stem 9–25 dm tall, glabrous below the somewhat glandular-puberulent or villous-puberulent inflorescence. Leaves glabrous or scabrous above, generally short-hirsute below at least on the midrib and main veins, most of them petiolate, elliptic to deltoid, cordate, or hastate, coarsely dentate to occasionally shallowly lobed or subentire, only gradually reduced upward, the blade up to 25 × 20 cm. Inflorescence corymbiform-paniculiform, the heads nodding; involucre 12–16 mm high, coarsely hirsute-setose, its principal bracts (12) 13 (15), pale green or usually more or less black-tinged; reduced outer bracts sometimes as much as 6 mm long; flowers 20–35, ochroleucous. Pappus tan or sordid. Late summer. Moist woods; Ky and Tenn to Pa, Ohio, Ill, and Mo; doubtfully in w Va. *Nabalus crepidineus* (Michx.) DC—S.

2. P. aspera Michx. Stem strict, 5–17 dm tall, rough-hairy or scabrous at least above. Leaves scabrous or coarsely hirsute on the lower and often also the upper side, toothed or entire, the lower ones well developed, somewhat obovate, tapering to the petiole, but soon deciduous, the others sessile or nearly so and often clasping, oblong to elliptic or lanceolate, gradually reduced upward, the larger ones 4–10 × 1–4 cm. Inflorescence narrow and elongate, thyrsoid-racemiform, the heads crowded, loosely ascending to suberect; involucre 12–17 mm high, coarsely and usually densely long-hairy, rather pale, with (6–) 8 (–10) principal bracts; flowers mostly ca. 13 (8–19), ochroleucous. Pappus stramineous. Late summer–fall. Dry prairies; La to Okla, Neb, and Minn, e to Ohio, Ky (IP), Tenn (IP), and the black belt of Miss. *Nabalus asper* (Michx.) T. & G.—S.

3. P. roanensis (Chick.) Chick. Stem 1.5–6 (10) dm tall, glabrous below the inflorescence, or sometimes conspicuously long-hairy. Leaves mostly glabrous or nearly so, seldom evidently hairy, the lower and middle ones petiolate, with deltoid to mostly sagittate or hastate, toothed and sometimes deeply lobed or cleft blade commonly 3–10 × 2–8 cm. Inflorescence elongate and narrow, thyrsoid-racemiform, leafy-bracteate at least below, the branches all very short; heads nodding; involucre 10–12 mm high, coarsely hirsute-setose, with 5–9 principal bracts, these pale green with a darker median line and tip; flowers 5–13, ochroleucous. Pappus stramineous. Late summer. Rich woods and moist, open slopes at high elev. in the s Appalachian Mts., mainly or wholly BR; NC and Tenn, n to sw Va and reputedly Ky. *P. cylindrica* (Small) Braun—G; *Nabalus cylindricus* Small—S; *Nabalus roanensis* Chick.—S.

4. P. barbata (T. & G.) Milstead. Resembling *P. serpentaria* f. *simplicifolia*, but with a larger (12–17 mm), more obviously setose involucre, the short outer bracts as well as the principal ones beset with long, coarse hairs that tend to form a line along the midvein; middle and upper leaves tending to be broad-based, and often somewhat clasping (as in *P. aspera*), but sometimes tapering to a narrow base; leaf lobes, when present, usually small, broad-based, and tapering; flowers white or rarely ochroleucous. Late summer–fall. Prairies, barrens, and open woods, especially in calcareous regions; Ark, La, and Tex, w to Tenn (IP) and Ga.

5. P. serpentaria Pursh. LION'S FOOT. Stems mostly 5–15 dm tall, glabrous, or often rough-hairy in the inflorescence. Leaves glabrous or inconspicuously hairy, well distributed along the stem, mostly wing-petiolate and with a broad-based, usually pinnately few-lobed blade up to ca. 17 × 10 cm, often with large, more or less rounded lobes, expanded above the base, varying to occasionally merely toothed and tapering to a shortly wing-petiolar base (f. *simplicifolia* Fern.), the basal ones sometimes trifoliolate and again cleft. Inflorescence mostly rather open-paniculiform, with elongate, ascending branches; heads nodding; involucre 10–13 (15) mm high, its principal bracts ca. 8, with a few (sometimes very few) long, coarse hairs, often speckled with fine black dots, but without the larger, waxy-looking papillae of *P. trifoliolata*; reduced outer bracts averaging narrower and a little longer than in *P. trifoliolata*, often more than 2 mm long and often more than 2.5 times as long as wide; flowers ca. 13, generally ochroleucous or chloroleucous. Pappus stramineous. (n = 8) Late summer–fall. Woods, especially in sandy soil; Mass to n Fla, w to Ky (CU and encroaching onto IP), Tenn (CU and encroaching onto IP), and Miss. *Nabalus serpentarius* (Pursh) Hook.—S. *Nabalus integrifolius* Cass.—S.

6. P. trifoliolata (Cass.) Fern. Stem stout, 4.5–12 dm tall or more, glabrous. Leaves glabrous above, paler and sometimes a little hairy beneath, highly variable in size and shape, the lower ones long-petioled, pinnately or palmately few-lobed, usually rather deeply so, or trifoliolate and again cleft, varying to occasionally hastate and merely toothed, the middle and upper ones progressively smaller, less petiolate, and less cut. Inflorescence elongate-paniculiform; heads nodding; involucre 10–14 mm high, with (7) 8 (9) principal bracts, these glabrous, often purplish, generally provided with minute, white, waxy-appearing papillae (often barely visible at 10×) and sometimes black-dotted (as in *P. serpentaria*) as well; reduced outer bracts relatively short and broad, seldom more than 2 (2.5) mm long or more than 2 (2.5) times as long as wide; flowers (9) 10–11 (–13), white to ochroleucous or chloroleucous or grayish, seldom pale lavender. Pappus stramineous or pale brown. (n = 8) Late summer–fall. Woods, especially in sandy soil; coastal or subcoastal states and provinces from Nf to Md and Pa, and s along the coast to NC, and in the mts. to NC, Tenn, and n Ga. *Nabalus trifoliatus* Cass.—S.

7. P. alba L. Stem stout, 4–15 dm tall; herbage more or less glaucous. Leaves glabrous above, often hairy beneath, highly variable in size and shape, the lower long-petioled, palmately or pinnately few-lobed to sagittate or hastate-reniform and merely coarsely toothed, becoming smaller, less cut, and less petiolate upward. Inflorescence elongate-paniculiform; heads nodding; involucre 11–14 mm high, with mostly 8 principal bracts, these glabrous, generally somewhat purplish, more or less densely papillate with white, waxy-appearing cells that are usually readily distinguishable at 10×; reduced outer bracts, as in *P. serpentaria*, mostly a little longer and narrower than in *P. trifoliolata*; flowers 10–15, mostly white to pinkish to lavender. Pappus cinnamon-brown. (n = 16) Late summer. Woods; Me to Man, s to NC, WVa, and Mo. *Nabalus* *albus* (L.) Hook.—S. Plants called *P. alba* ssp. *pallida* Milstead (R—the name not validly published) may reflect hybridization with *P. trifoliolata*.

8. P. autumnalis Walter. Plants 4–14 dm tall, glabrous throughout. Basal and lowermost cauline leaves elongate, 7–35 × 1–12 cm, few-toothed to generally pinnately lobed, the lobes often rather narrow and distant; cauline leaves conspicuously reduced upward, often becoming linear and entire. Inflorescence very narrow and elongate, either subracemiform, or in more robust plants with closely ascending, subracemiform branches; heads nodding; involucre 10–13 mm high, generally purplish, without long hairs, with 8 principal bracts, the reduced outer ones often passing into those of the slender peduncle; flowers mostly 8–12, pink or pale lavender. Pappus stramineous. Fall. Sandy, usually moist places, often among pines; CP from NJ to n Fla. *Nabalus virgatus* (Michx.) DC.—S.

9. P. altissima L. Stem 4–20 dm tall, glabrous or often spreading-hirsute toward the base. Leaves thin, glabrous above, often hirsute on the midrib and main veins beneath, the lower long-petioled, highly variable in size and shape, commonly with deltoid to sagittate or cordate, merely mucronate-toothed blade 4–15 × 2.5–16 cm, or sometimes deeply few-lobed, gradually reduced upward, becoming short-petiolate or subsessile and elliptic. Inflorescence elongate-paniculiform; heads nodding; involucre glabrous, 9–14 mm high, with (4) 5 (6) principal bracts, generally dark-tipped; flowers 5–6, ochroleucous or chloroleucous. (n = 8) Late summer–fall. Woods; Nf to Ga (all), w to Mich, Mo, Ark, and La. The var. *altissima*, with whitish to pale brown pappus, occupies most of the range of the sp.; southwestward it passes into var. *cinnamomea* Fern., chiefly of Mo, Ark, and La, with bright yellow-brown or almost orange pappus. *Nabalus altissimus* (L.) Hook.—S.

115. LACTUCA L. LETTUCE

Herbs with milky juice. Leaves alternate, entire to pinnatifid. Heads mostly numerous in a more or less paniculiform inflorescence; flowers relatively few (8–56 per head in our spp.), all ligulate and perfect, the corolla tube mostly more than half as long as the yellow, blue or rarely white ligule; involucre cylindric, often broadening at the base in fruit, the bracts imbricate in most species, calyculate in some extralimital ones. Achenes more or less compressed, winged or strongly nerved marginally, with 1–several lesser nerves on each face, beaked or sometimes beakless, but in any case expanded at the summit where the pappus is attached; pappus of numerous capillary bristles, none markedly larger than the others.

1 Achenes evidently to very prominently several-nerved on each face; beak of the achene variously long and slender, or short and stout, or wanting.
  2 Achenes with a short, stout beak less than half as long as the body, or beakless; flowers mostly blue, seldom yellow or white.
    3 Pappus light brown; inflorescence elongate, narrowly paniculiform .... 1. *L. biennis*.
    3 Pappus white; inflorescence ample, open-paniculiform .... 2. *L. floridana*.
  2 Achenes with a filiform beak from nearly as long to twice as long as the body; flowers yellow, but commonly drying blue.
    4 Leaves prickly-margined; achenes spinulose or hispidulous above .... 3. *L. serriola*.
    4 Leaves not prickly-margined; achenes merely scabrous above .... 4. *L. saligna*.
1 Achenes with only a median nerve on each face, or occasionally with an additional pair of very obscure ones; beak of the achene slender, from a little more than half to fully as long as the body.
  5 Leaves fairly well distributed along the stem; flowers yellow, seldom blue.

6 Heads relatively large, the involucre 15–22 mm high in fruit; achenes 7–10 mm long (beak included).
7 Leaves more or less prickly-margined; flowers 20–56 per head .... 5. *L. ludoviciana.*
7 Leaves toothed or lobed but scarcely prickly; flowers 13–25 per head ..... 7. *L. hirsuta.*
6 Heads smaller, the involucre 10–15 mm high in fruit; achenes 4.5–6.5 mm long .... 6. *L. canadensis.*
5 Leaves basally disposed; flowers blue, seldom yellow .... 8. *L. graminifolia.*

1. L. biennis (Moench) Fern. TALL BLUE LETTUCE. Robust, leafy-stemmed annual or biennial, 6–20 dm tall, glabrous, or the leaves hairy on the main veins beneath. Leaves pinnatifid or occasionally merely toothed, 10–40 × 4–20 cm, sometimes sagittate at the base. Heads numerous in an elongate, rather narrow, paniculiform inflorescence, often crowded, with 15–34 (55) bluish to white or occasionally yellow flowers; involucre 10–14 mm high in fruit. Achenes 4–5.5 mm long, thin-edged, prominently several-nerved on each face, tapering to the beakless or shortly stout-beaked tip; pappus light brown. (n = 17) Late summer–fall. Moist places; Nf to BC, s to NC (chiefly in the mts.), Tenn, Ill, Colo, and Calif. *Mulgedium spicatum* (Lam.) Small, misapplied. A hybrid with *L. canadensis* has been called *L. morssii* Robins.—F.

2. L. floridana (L.) Gaertn. WOODLAND LETTUCE. Robust, leafy-stemmed annual or biennial, 5–20 dm tall. Leaves mostly petiolate and not sagittate at the base, often hairy along the main veins beneath, otherwise glabrous or nearly so, the blade 8–30 × 2.5–20 cm, from elliptic (or sometimes cordate) and merely toothed to evidently pinnatifid. Heads numerous in an ample, paniculiform inflorescence, with 11–17 (27) blue (white) flowers; fruiting involucre 9–14 mm high. Achenes 4–6 mm long, narrowed upward, beakless or with a stout beak sometimes a third as long as the body, several-nerved on each face, tending to be thickened on the margins; pappus white. (n = 17) Late summer–fall. Thickets, woods, and moist, open places; NY to c Fla, w to Minn, Kans, and Tex. *Mulgedium floridanum* (L.) DC.—S, *Mulgedium villosum* (Jacq.) Small—S.

3. L. serriola L. PRICKLY LETTUCE. Leafy-stemmed biennial or winter annual, 3–15 dm tall, the stem often prickly below, otherwise glabrous. Leaves prickly on the midrib beneath, and more finely prickly-toothed on the margins, otherwise generally glabrous, pinnately lobed or lobeless, commonly twisted at the base to stand in a vertical position, sagittate-clasping, oblong or oblong-lanceolate in outline, 5–30 × 1–10 cm, the upper much reduced. Heads numerous in a long, often diffuse inflorescence, with (13) 18–24 (27) light yellow flowers, drying blue; involucre 10–15 mm high in fruit. Achenes gray or yellowish-gray, the body compressed, 3–4 mm long, a third as wide, prominently several-nerved on each face, spinulose or hispidulous above at least marginally, the slender beak about equaling the body, rarely twice as long. (n = 9) Summer. A weed in fields and waste places; native of Europe, now naturalized throughout most of the US. The common form with lobeless leaves, called var. *integrata* Gren. & Godr., may be derived by introgression from the cult. lettuce, *Lactuca sativa* L., with which *L. serriola* hybridizes freely. *L. scariola* L.—F, R, S. *L. virosa* L.—S, misapplied to var. *integrata.*

4. L. saligna L. Leafy-stemmed annual or biennial, the stem glabrous, 3–10 dm tall, more slender than in *L. serriola.* Leaves conspicuously sagittate, linear and entire or with scattered, narrow, sometimes slightly toothed lobes, 6–15 cm long, the rachis (or the leaf, when entire) 3–8 mm wide, glabrous or with a few prickles on the midrib beneath and sometimes minutely stellate marginally. Heads numerous in a long inflorescence; flowers 8–16, light yellow, drying blue; involucre 12–18 mm high in fruit.

Achenes similar to those of *L. serriola*, but averaging slightly smaller and narrower, merely scabrous above, the beak commonly twice as long as the body. (n = 9) Late summer–fall. A weed in waste places; native of Europe, now found here and there in the n part of our range, and elsewhere in US.

5. L. ludoviciana (Nutt.) DC. Leafy-stemmed biennial, or possibly short-lived perennial, the stem glabrous, 3–15 dm tall. Leaves prickly or coarsely hairy on the midrib and sometimes also on the main veins beneath, more or less prickly toothed on the margins and usually also pinnately lobed, mostly 7–30 × 3–20 cm. Heads relatively large, with 20–56 yellow or occasionally blue flowers; involucre 15–22 mm high in fruit. Achenes strongly flattened, blackish, with a median nerve on each face, transversely rugulose, 7–10 mm long, the slender beak equaling or a little shorter than the body; pappus 7–10 mm long. (n = 17) Summer. Open, rather moist places; native to the prairies and plains of c US and adj. Can, reputedly extending into Ark and La, and casually intro. elsewhere.

6. L. canadensis L. WILD LETTUCE. Leafy-stemmed annual or usually biennial, 3–25 dm tall, glabrous or occasionally coarsely hirsute, often more or less glaucous. Leaves well distributed along the stem, highly variable, entire or toothed to pinnately lobed or cleft, sagittate or sometimes narrowed to the base, 10–35 × 1.5–12 cm. Heads numerous, small, with 13–22 yellow flowers; involucre 10–15 mm high in fruit. Achenes blackish, very flat, with a median nerve on each face, transversely rugulose, 4.5–6.5 mm long, the slender beak from a little more than half to about as long as the body; pappus 5–7 mm long. (n = 17) Spring–summer. Fields, waste places, and woods; se Can to n Fla, w to Sask and Tex. *L. sagittifolia* Ell.—S.

7. L. hirsuta Muhl. Similar to *L. canadensis*, often more hairy, and often with some tendency for the leaves to be basally disposed, but the stem more obviously leafy than in *L. graminifolia*; leaves pinnatisect, with broad, toothed lobes that are commonly more than 1 cm wide and often somewhat narrowed at the base; heads larger than in *L. canadensis*, the fruiting involucre 15–22 mm high, the achenes 7–9 mm long, and the pappus 8–12 mm long in fruit. (n = 17) Spring–summer. Dry, open woods and clearings; se Can to n Fla, w to Mich, Mo, and Tex. [Incl. var. *sanguinea* (Bigel.) Fern.—F, G, the less hairy phase]

8. L. graminifolia Michx. Glabrous and somewhat glaucous biennial 3–12 dm tall. Leaves narrow, not sagittate, sometimes prickly on the midrib beneath, basally disposed, the basal and lower cauline ones the largest and generally persistent, mostly 10–20 (40) cm long, linear and entire or merely toothed and up to 1 (2) cm wide, or with a few narrow, entire or subentire lobes, these up to 3.5 cm long, less than 1 cm wide, not constricted at the base; leaves otherwise entire and progressively reduced upward, the stem naked or nearly so above. Inflorescence open-paniculiform at maturity; flowers mostly 18–24, blue; fruiting involucre 13–17 mm high. Achenes blackish, very flat, with a median nerve on each face, transversely rugulose, 6–9 mm long, the beak from a little more than half to nearly as long as the body; pappus 7–9 mm long. (n = 17) Spring–summer. Open woods and disturbed sites; CP from NC to c Fla, w to Tex and to Ariz and n Mex.

116. IXERIS Cass.

1. **I. stolonifera** A. Gray. CREEPING LETTUCE. Delicate, creeping, glabrous perennial with milky juice. Leaves alternate, rather long petiolate, with broadly elliptic to orbicular, entire to obscurely toothed blade 0.5–3 cm long. Heads solitary or paired on peduncles ca. 1 dm long or less; flowers all ligulate and perfect, yellow, the tube 1/4 to 1/2 as long as the ligule; involucre 7–12 mm high, cylindric, calyculate. Achenes reddish-brown, moderately compressed, 4–6 mm long, the 10 nerves somewhat wing-elevated, the slender beak about as long as the body; pappus of slender capillary bristles. (n = 24) Summer. A weed in lawns; native of Japan, occasionally intro. in e US, as in Del, Pa, and NY.

117. LAUNAEA Cass.

1. **L. intybacea** (Jacq.) Beauv. Glabrous annual or biennial herb with milky juice, 2–12 dm tall. Leaves alternate, finely prickly-toothed and often also more or less pinnatifid, tending to be sagittate-clasping, the basal or lower cauline ones commonly the largest, mostly 3–25 × 1–14 cm, the others more or less reduced, especially in smaller plants. Inflorescence open-paniculiform, with elongate, ascending main branches and short, slender peduncles; flowers all ligulate and perfect, yellow, mostly 12–34 in each head; involucre of several unequal series of bracts, 11–15 mm high in fruit. Achenes 4–5 mm long, narrow, thick, coarsely ridged and grooved, muricate on the ridges, contracted above to a stout, smooth beak ca. 0.5–1 mm long; pappus of very numerous white capillary bristles, a few of them a little stouter and straighter than the others. All year. A tropical Am weed, reaching our range in s Fla. *Brachyrhamphus intybaceus* (Jacq.) DC.—S.

118. SONCHUS L. SOW THISTLE

Herbs (some extralimital spp. woody) with milky juice. Leaves alternate (or all basal), entire to dissected, mostly auriculate, often prickly-margined. Heads solitary to usually several or many in a terminal inflorescence; flowers all ligulate and perfect, yellow, more or less numerous (36–480, in our spp. not less than 80); involucre ovoid or campanulate, rarely narrower, its bracts generally imbricate, occasionally merely calyculate, often basally thickened in age. Achenes flattened, mostly 6–20-ribbed, merely narrowed at the tip, beakless, glabrous, often transversely rugulose; pappus of numerous white, capillary, often somewhat crisped bristles that tend to fall connected, and some stouter outer bristles that fall separately. Boulos, L. 1972–4. Révision systématique du genre *Sonchus* L. Bot. Notis. 125: 287–319; 126: 155–196; 127: 7–37, 402–451.

1 Perennial, deep-rooted and creeping below ground; heads relatively large, mostly 3–5 cm wide in flower, the fruiting involucre mostly 14–22 mm high .... 1. *S. arvensis.*
1 Annual; heads relatively small, mostly 1.5–2.5 cm wide in flower, the fruiting involucre mostly 9–13 mm high.
 2 Mature achenes transversely tuberculate-rugose as well as several-nerved; auricles of the leaves coming to a distinctly acute base .... 2. *S. oleraceus.*

2 Mature achenes merely several-nerved, not rugulose; auricles rounded throughout .... 3. *S. asper.*

1. S. arvensis L. PERENNIAL SOW THISTLE. Perennial, 4–20 dm tall, with long vertical roots and spreading by creeping roots, glabrous at least below the inflorescence and often somewhat glaucous. Leaves prickly-margined, the lower and middle ones usually pinnately lobed or pinnatifid, 6–40 × 2–15 cm, becoming less lobed and more strongly auriculate upward, the upper reduced and distant. Heads several in an open-corymbiform inflorescence, relatively large, 3–5 cm wide in flower; fruiting involucre 14–22 mm high. Achenes 2.5–3.5 mm long, with 5 or more prominent ribs on each face, strongly rugulose. Midsummer–fall. A European weed, now widely introduced in temperate N Am, including the n part of our range. Var. *arvensis* is more or less copiously provided with coarse, spreading, gland-tipped hairs on the involucre and peduncles; n = 27. Var. *glabrescens* (Guenth.) Grab. & Wimm, lacks these hairs; n = 18 *S. uliginosus* Bieb.—G.

2. S. oleraceus L. COMMON SOW THISTLE. Annual 1–20 dm tall, from a short taproot, glabrous except sometimes for a few spreading gland-tipped hairs on the involucres and peduncles. Leaves pinnatifid to occasionally merely toothed, soft, the margins rather weakly or scarcely prickly, 6–30 × 1–15 cm, all but the lowermost ones prominently auriculate, the auricles well rounded but eventually sharply acute; leaves progressively less divided upward and more or less reduced. Heads several in a corymbiform inflorescence, relatively small, 1.5–2.5 cm wide in flower; involucre 9–13 mm high in fruit; corolla tube about equaling the ligule. Achenes 2.5–3 mm long, transversely rugulose and evidently to rather obscurely 3–5-ribbed on each face. (n = 16, 18) Spring–fall, or all year southward. A cosmopolitan weed, native to Europe.

3. S. asper (L.) Hill. PRICKLY SOW THISTLE. Similar to *S. oleraceus*, commonly more prickly; leaves pinnatifid, or frequently obovate and lobeless, with rounded and conspicuously prickly-toothed, not acute auricles; corolla tube somewhat longer than the ligule; achenes with 3 (4–5) evident ribs on each face, not rugulose, although there may be minute projections from the marginal ribs. (n = 9) Spring–fall. A cosmopolitan weed, native to Europe.

119. HIERACIUM L. HAWKWEED

Fibrous-rooted perennial herbs from an elongate or very short rhizome or from a short caudex or crown; juice milky; most spp. with at least a few stellate hairs on the herbage or involucre. Leaves alternate or all basal, simple and entire or more or less toothed. Heads in a corymbiform or umbelliform or paniculiform, sometimes narrow and elongate inflorescence, or sometimes solitary; flowers all ligulate and perfect, yellow to sometimes red-orange, rarely white; involucre cylindric to hemispheric, its bracts more or less imbricate. Achenes terete or prismatic, mostly narrowed below, truncate or occasionally narrowed toward the summit, more or less strongly ribbed-sulcate; pappus of numerous whitish to more often tawny or brownish capillary bristles. (x = 9)

Our native spp. are all diploid (so far as known) and produce occasional hybrids, some of which have been named. Our intro. spp. belong to a huge polyploid-apomictic complex without clear specific boundaries.

1 Leaves mainly or in considerable part cauline, although the basal or lower cauline leaves may be larger than those above (extreme specimens approach the next group in habit); inflorescence various.
  2 Stem leafy up to the broadly open-paniculiform inflorescence; plants not very hairy, the leaves glabrous or with a few long hairs on the lower side .... 1. *H. paniculatum*.
  2 Stem leafy chiefly below the middle or toward the base, the upper part generally naked or nearly so; plants more evidently hairy, at least with regard to the leaves.
    3 Flowers numerous, mostly 40–100 per head; stem often densely stipitate-glandular upward.
      4 Stem very long-hairy below, the hairs well over 5 mm, mostly 1 cm long or more; achenes 3–4.5 mm long, evidently narrowed above .... 2. *H. longipilum*.
      4 Stem not so long-hairy, the hairs seldom 5 mm long; achenes 2–3 mm long, scarcely or not at all narrowed above .... 3. *H. scabrum*.
    3 Flowers fewer, mostly 20–40 per head; stem only sparsely or not at all stipitate-glandular upward, achenes 2.5–5 mm long, distinctly narrowed above.
      5 Inflorescence in well-developed plants mostly elongate and cylindric; involucre mostly 6–9 mm high; achenes 2.5–4 mm long .... 4. *H. gronovii*.
      5 Inflorescence broadly and openly corymbiform, not elongate; involucre mostly 8–11 mm high; achenes 3.5–5 mm long .... 5. *H. megacephalum*.
1 Leaves mainly or wholly basal, the stem otherwise naked or with only one or a few small leaves; inflorescence corymbiform, or the head solitary.
  6 Heads several or numerous, commonly 5–30, sometimes more.
    7 Flowers yellow.
      8 Native, nonweedy plants; basal leaves relatively broad, mostly 2–5 times as long as wide, often more than 3 cm wide; achenes mostly 2.2–4 mm long.
        9 Basal leaves not purple-veined; involucre with gland-tipped hairs or setae, and sometimes also with long, glandless setae.
          10 Involucral setae all gland-tipped (occasional forms of) .... 5. *H. megacephalum*.
          10 Involucre with long, glandless setae as well as shorter, gland-tipped hairs .... 6. *H. traillii*.
        9 Basal leaves evidently purple-veined above in life; involucre glabrous to sometimes evidently stipitate-glandular, but without long, glandless setae .... 7. *H. venosum*.
      8 Introduced, weedy plants; basal leaves relatively narrow, mostly 5–12 times as long as wide, seldom more than 3 cm wide; achenes 1.5–2 mm long.
        11 Herbage glaucous, sparsely hairy to subglabrous .... 8. *H. florentinum*.
        11 Herbage not glaucous, the leaves hairy on both sides .... 9. *H. caespitosum*.
    7 Flowers red-orange, becoming deeper red in drying .... 10. *H. aurantiacum*.
  6 Heads solitary, or seldom 2 (–4) .... 11. *H. pilosella*.

1. H. paniculatum L. Stems solitary (2) from a short caudex or crown, 3–15 dm tall, long-hairy below, otherwise glabrous or nearly so, leafy up to the inflorescence. Leaves thin, glabrous or with a few long hairs on the persistently glaucous lower surface, irregularly callous-toothed or subentire, the lowest ones petiolate, only slightly if at all enlarged, mostly soon deciduous, the others elliptic, often narrowly so, narrowed to a sessile or subsessile base, only gradually reduced upward, 4–12 × 1–3 cm. Inflorescence open-paniculiform, with long, flexuous, very slender peduncles, these, like the involucre, glabrous or occasionally with some gland-tipped hairs, involucre narrow, 5–9 mm high; flowers 8–30 per head. Achenes truncate, 1.8–2.5 mm long. (n = 9) Mid- and late summer. Woods; NS and Que to Minn, s to Va, Ohio, and in the mts. to NC, SC, Tenn, and n Ga. A hybrid with *H. gronovii* is *H. allegheniense* Britton—S; one with *H. venosum* is *H. scribneri* Small—S.

2. H. longipilum Torr. Stems solitary or few from a short stout caudex or crown, 6–20 dm tall, densely long-hairy below, the hairs mostly ca. 1 cm long or more, sometimes 2 cm, becoming glabrous or nearly so above. Leaves pubescent like the stem, or the hairs shorter, the basal and lower cauline leaves rather numerous, oblanceolate or narrowly elliptic, 9–30 × 1.5–4 cm, crowded, the lowest ones often deciduous, the others progressively reduced upward, the upper half of the stem commonly naked or merely bracteate. Inflorescence elongate, cylindric, the branches and peduncles stellate, long-stipitate-glandular and sometimes sparsely setose; involucre 7–10 mm high, stellate-puberulent and hispid with blackish, mostly gland-tipped hairs; flowers 40–90 per head. Achenes 3–4.5 mm long, narrowed above. Mid- and late summer. Dry prairies, open woods, and fields, especially in sandy soil; Mich and Ind to Kans and Okla; to be expected in Ark, and reported from La.

3. H. scabrum Michx. Stems mostly solitary from a short, simple caudex or crown, 2–15 dm tall, setose at least near the base with spreading hairs seldom as much as 5 mm long, becoming stellate and long-stipitate-glandular upward, densely so in the inflorescence. Leaves setose on both sides, more densely so on the petiole and on the midrib beneath; basal and often also the lowest cauline leaves ordinarily deciduous, the lower leaves broadly oblanceolate to elliptic, 5–20 cm long (including the usually short petiole) and 1–4.5 cm wide, the others progressively reduced upward, soon becoming sessile, so that the upper part of the stem does not appear to be very leafy. Inflorescence open-corymbiform (especially in smaller specimens) to more often elongate and cylindric; involucre 6–9 mm high, hispid with blackish, mostly gland-tipped hairs, especially toward the also stellate-hairy base; flowers 40–100 per head. Achenes 2–3 mm long, truncate, only very obscurely if at all narrowed upward. (n = 9) Mid- and late summer. Open ground and dry woods, especially in sandy soil; NS and Que to Minn, s to Va, Ky, and Mo, and in the mts. to NC, SC, and n Ga.

4. H. gronovii L. Stems mostly solitary on a crown or short, praemorse rhizome, 3–15 dm tall, conspicuously spreading-hairy toward the base, the hairs rarely approaching 1 cm long, becoming merely strigose-puberulent or obscurely stellate to subglabrous upward, sometimes also finely glandular. Leaves finely stellate and usually also more or less long-hairy, the basal or lowest cauline ones broadly oblanceolate to obovate or elliptic, 4–20 cm long (including the usually short petiole) and 1.2–5 cm wide, deciduous or persistent, the cauline ones progressively reduced upward, soon becoming sessile and commonly somewhat clasping, the upper part of the stem naked or nearly so. Inflorescence more or less elongate and openly cylindric, at least in well-developed plants, the peduncles puberulent and sparsely to fairly copiously long-stipitate-glandular; involucre 6–9 mm high; flowers 20–40 per head. Achenes 2.5–4 mm long, distinctly narrowed toward the top. (n = 9) Summer–fall, or spring–late fall southward. Dry, open woods, especially in sandy soil, sometimes in fields and pastures; Mass to c Fla, w to s Ont, Kans, and Tex.

5. H. megacephalum Nash. Much like *H. gronovii*, but the inflorescence short, broad, open-corymbiform; stem sometimes almost naked; heads averaging larger, the involucre mostly 8–11 mm high, the achenes 3.5–5 mm long. Fall–spring. Dry, sandy woodlands, commonly with pines, sometimes with oaks or palmettos or in hammocks; throughout peninsular Fla, n occasionally into s Ga. *H. argyraeum* Small—S.

6. H. traillii Greene. Similar to *H. venosum*, but more hairy; leaves long-setose over the surface beneath and often sparsely so above, not purple-veined; peduncles averaging stouter, densely stellate-tomentose as well as copiously provided with spreading, blackish, gland-tipped hairs; heads averaging larger, the involucre up to 12 mm high, conspicuously hairy, with long, eglandular setae, shorter, blackish, gland-tipped hairs, and also some small, stellate hairs. Late spring–summer. Dry, open woods, sometimes on shale barrens; mts. of s Pa to Va and WVa. *H. greenii* Porter & Britton—S.

7. H. venosum L. Stems 1–few from a short, praemorse rhizome, 2–8 dm tall, glabrous or very nearly so, naked or with 1–3 (6) reduced leaves. Basal leaves elliptic to ovate or broadly oblanceolate, 3–16 cm (short petiole included) × 0.8–5 cm, 1.7–5 times as long as wide, often densely long-setose along the margins and toward the base, otherwise sparsely so or subglabrous, the midrib and main veins generally reddish-purple above in life, the whole undersurface sometimes reddish purple. Inflorescence open, corymbiform, the peduncles elongate, usually slender, and often rather flexuous; involucre 7–10 mm high, glabrous or sometimes evidently stipitate-glandular, obscurely or not at all stellate, the peduncles likewise; flowers 15–40 per head. Achenes 2.2–4 mm long, truncate or more often distinctly narrowed near the summit. (n = 9) Mostly late spring–midsummer. Mostly in dry, open woods; NY to Va (all) and Ga (mt. provinces, PP, and GC), w to Mich, Ky (IP), Tenn (CU and encroaching onto IP), and Ala (PP); disjunct (?) in c Fla. *H. marianum* Willd. is thought to be a hybrid with another species, perhaps *H. scabrum* or *H. gronovii*.

8. H. florentinum All. Plants 2–10 dm tall from a usually rather short, praemorse rhizome; stems (1–) several, naked or with 1 or 2 (–5) small leaves; herbage glaucous, sparsely long-setose or subglabrous, the peduncles stipitate-glandular and somewhat stellate-hairy. Basal leaves oblanceolate, 3–18 cm (petiole included) × 0.5–2 cm, 5–12 times as long as wide. Inflorescence compact to fairly open, corymbiform; involucre 6–8 mm high, hispid with blackish, mostly gland-tipped hairs and somewhat stellate. Achenes 1.5–2 mm long, truncate. (2n = 36, 45) Late spring–summer. A weed of fields, meadows, pastures, and roadsides; native of Europe, now well established in se Can and ne US, s to n Ga.

9. H. caespitosum Dumort. KING DEVIL. Plants with a short or more often elongate rhizome and commonly with short, stout stolons; stems 1–several, 2.5–9 dm tall, long-setose, becoming stellate-tomentose and blackish-glandular-hispid above, naked or with only 1 or 2 (3) reduced leaves. Basal leaves oblanceolate or narrowly elliptic, 5–25 cm (petiole included) × 1–3 cm, mostly 5–12 times as long as wide, long-setose on both sides, sometimes sparsely so above, commonly slightly stellate beneath. Heads mostly 5–30 in a compact, corymbiform inflorescence; involucre 6–8 mm high, hispid with blackish, gland-tipped hairs, commonly also sparsely long-setose and slightly stellate; flowers bright yellow. Achenes 1.5–2 mm long, truncate. (2n = 18, 27, 36, 45) Spring (summer). A weed in fields, pastures, and along roadsides, occasionally in dry woods; native of Europe, now widespread in se Can and ne US, s to NC, n Ga, and Tenn. *H. pratense* Tausch.—F, G, R.

10. H. aurantiacum L. KING DEVIL, DEVIL'S PAINT BRUSH. Much like *H. caespitosum*, but the flowers red-orange (unique among our spp.), becoming deeper red in drying; plants 1–6 dm tall, with slender stolons and ordinarily with a slender, elongate rhizome; leaves sometimes a little wider, up to 3.5 cm; involucre 5–8 mm high, long-setose, hispid with blackish, gland-tipped hairs, and slightly tomentose. (2n = 27, 36, 45, 54, 63) Late spring–summer. A weed of fields, roadsides, and meadows; native of Europe, now widespread in se Can and ne US, s to NC and WVa.

11. H. pilosella L. MOUSE EAR. Abundantly stoloniferous, and with a long slender rhizome; stem 3–25 (40) cm, leafless or with a single generally much reduced leaf, viscid-puberulent or subtomentose, sparsely or moderately spreading-hispid with gland-tipped, usually blackish hairs, and often long-setose as well. Basal leaves oblanceolate or a little broader, 2–13 × 0.6–2 cm, tawny-tomentose with stellate hairs beneath, at least when young, and with some long glandless setae as well, green and glabrous above except for the long setae; leaves of the stolons similar but smaller. Heads solitary, or rarely 2 and long-pedunculate; involucre 7–11 mm high, stellate, short-hispid with black, sometimes gland-tipped hairs, and occasionally long-setose as well. Achenes 1.5–2 mm long, truncate. (2n = 18, 36, 45, 54, 63) Late spring–midsummer. A weed in pastures and fields; native of Europe, now widespread in se Can and ne US, s to NC and Tenn.

H. flagellare Willd., regarded as reflecting hybridization between *H. pilosella* and *H. caespitosum*, may be expected in the n part of our range. It averages coarser and more bristly than *H. flagellare*, with the leaves greener beneath, and it has 2–4 heads, at least one of them on a peduncle no more than ca. 4 cm long.

120. CREPIS L. HAWK'S BEARD

Herbs with milky juice, mostly (including all our spp.) taprooted. Leaves alternate, entire to bipinnatifid, more or less basally disposed, the cauline ones progressively reduced, the uppermost ones bractlike. Heads mostly several to numerous in an open, corymbiform or paniculiform inflorescence; flowers all ligulate and perfect, yellow; involucre cylindric or campanulate, its principal bracts in 1 or 2 equal or subequal series, the reduced outer ones few or many; receptacle naked, glabrous in our spp. Achenes terete or subterete, fusiform or nearly columnar, 10–20-ribbed, with or without a slender beak; pappus of numerous whitish capillary bristles. Babcock, E. B. 1947. Univ. Calif. Publ. Bot. vols. 21–22.

1 Achenes narrowed toward the summit, but scarcely beaked.
  2 Involucre 8–12 mm high, glabrous; achenes 4–6 mm long .... 1. *C. pulchra.*
  2 Involucre 5–8 mm high, pubescent; achenes 1.5–2.5 mm long .... 2. *C. capillaris.*
1 Achenes distinctly slender-beaked .... 3. *C. vesicaria.*

1. C. pulchra L. Annual 2–10 dm tall, the stem bearing spreading glandular hairs below, glabrous above. Leaves spreading-hairy like the lower part of the stem, the lowermost ones up to ca. 20 × 4 cm, mostly oblanceolate or spatulate, short-petiolate, and coarsely toothed or runcinate-pinnatifid. Heads mostly numerous; involucre 8–12 mm high, glabrous; inner bracts ca. 13, generally with raised and thickened, basally expanded midrib; outer bracts very short; flowers ca. 15–30. Achenes 4–6 mm long, contracted above but not beaked, stramineous or pale greenish, 10–12-ribbed, the outer ones scabrous-hirtellous. (n = 4) Spring. Native of Eurasia, locally established in waste places in e US, s to Ga, Ala, Miss, and La.

2. C. capillaris (L.) Wallr. Annual or biennial, 2–9 dm tall, often much branched, the stem hispidulous at least near the base. Leaves glabrous or hispidulous, the basal ones up to ca. 30 × 4.5 cm; petiolate, lanceolate or oblanceolate, denticulate to runcinate-pinnatifid or even bipinnatifid; cauline leaves progressively reduced, clasping and acutely auriculate. Heads several to usually numerous; involucre 5–8 mm high, tomentose and often glandular-bristly with black hairs as well; inner bracts 8–16, becoming spongy-thickened on the back; outer bracts linear, up to half as long as the inner; flowers 20–60. Achenes mostly tawny or pale brown, 1.5–2.5 mm long, narrowed at the base, ca. 10-ribbed. (n = 3) Spring–fall. Native of Europe, sparingly intro. in e US, s to NC and Tenn.

3. C. vesicaria L. Annual, biennial, or occasionally short-lived perennial; stem 1–8 dm high, purple near the base, more or less hispid or tomentulose or both. Leaves finely short-hairy on both sides, the basal ones up to ca. 20 × 4 cm, narrowly oblanceolate to spatulate or obovate, dentate to runcinate-pinnatifid or pectinately parted; cauline leaves progressively reduced, auriculate-clasping. Heads several to more or less numerous; involucre 8–12 mm high, its inner bracts 9–13, tomentose and often glandular as well, sometimes with some short black setae near the tip, becoming carinate-thickened in fruit; outer bracts up to ca. half as long as the inner; flowers numerous. Achenes pale brown, 4.5–9 mm long, gradually attenuate into a slender beak equaling or a little longer than the body, 10-ribbed. (n = 4, 8) Spring–midsummer. Native of the Mediterranean region and w Europe, sparingly intro. in e US, s to NC; Am plants are ssp. *taraxacifolia* (Thuill.) Thellung.

121. YOUNGIA Cass.

1. Y. japonica (L.) DC. Polymorphic, subscapose annual, 1–9 dm tall, glabrous or more or less hairy toward the base; juice milky. Leaves mainly or all basal, mostly lyrate-pinnatifid, or subentire in small specimens, up to ca. 20 × 6 cm. Heads small, numerous in a corymbiform or paniculiform inflorescence; flowers ca. 10–20, all ligulate and perfect, yellow, the tube about 1/4 as long as the ligule; involucre 3.5–5 mm high, glabrous, calyculate, with 5 short outer and ca. 8 longer inner bracts. Achenes brownish, 1.5–2.5 mm long, slightly compressed, fusiform, tapering above, but scarcely beaked, slightly expanded to the pappiferous disk at the summit, strongly and somewhat unequally 11–13-ribbed; pappus of numerous fine white capillary bristles, 2.5–3.5 mm long. (n = 8) All year. Native to se Asia, now a pantropical weed, and becoming common in waste places on CP in our region, especially southward. *Crepis japonica* (L.) Benth.—F, R, S.

122. LAPSANA L. NIPPLEWORT

1. L. communis L. Hirsute to subglabrous annual weed with milky juice, 1.5–15 dm tall. Leaves alternate, thin, petiolate, with ovate to subrotund, obtuse or rounded, toothed or occasionally basally lyrate blade 2.5–10 × 2–7 cm, progressively less petiolate and eventually narrowed upward. Heads several or many in a corymbiform or paniculiform inflorescence, naked-pedunculate; flowers 8–15, all ligulate and perfect, yellow; involucre 5–8 mm high, cylindric-campanulate, minutely calyculate, the inner bracts mostly 8, subequal, uniseriate, keeled at least toward the base; receptacle naked. Achenes 3–5 mm long, narrow, subterete or slightly compressed, curved, narrowed to both ends, glabrous, 18–30-nerved, 5 or 6 of the nerves stronger; pappus none. (n = 6, 7, 8) Summer. Waste places, fields, and sometimes woods; native of Eurasia, now widely naturalized in n US, s to NC and Ky.

123. CHONDRILLA L. SKELETON-WEED

1. C. juncea L. Branching, rushlike, taprooted perennial herb with milky juice, 3–15 dm tall, the stem strongly spreading-hispid near the base. Basal leaves runcinate-pinnatifid, 5–13 × 1.5–3.5 cm, often deciduous; cauline leaves alternate, reduced, linear, 2–10 cm × 1–8 mm. Heads scattered along the branches; flowers commonly 9–12, yellow, all ligulate and perfect; involucre 9–12 mm high, thinly tomentose-puberulent, cylindric, with ca. 8 principal bracts, calyculate; receptacle naked. Achenes multinerved, the body 3 mm long, muricate above and bearing a circle of small scales at the base of the long, slender beak; pappus of numerous white capillary bristles. (2n = 15, 30) Summer. Roadsides, fields, and waste places; native of Eurasia, sparingly intro. in ne US, s to Ga.

124. TARAXACUM Wiggers, nom. conserv. DANDELION

Taprooted herbs with milky juice. Leaves all basal and rosulate, entire to pinnatifid or subbipinnatifid. Heads solitary, erect, terminating the 1–several hollow scapes; flowers all ligulate and perfect, yellow, mostly numerous; involucral bracts biseriate, the outer usually shorter than the inner and often reflexed; receptacle naked. Achenes narrowly obconic, terete or 4–5-angled, longitudinally sulcate or ribbed, ordinarily muricate or tuberculate at least above, commonly topped by a smooth, conic or pyramidal cusp that tapers into a slender beak, or rarely beakless; pappus of numerous white capillary bristles. Apomixis and polyploidy are rife in the genus, and many of the species (including ours) are confluent.

1 Achenes red or purplish-red or brownish-red at maturity; leaves tending to be very deeply cut for their whole length, the lobes narrow .... 1. *T. laevigatum*.
1 Achenes brown or olivaceous or stramineous; leaves usually less deeply cut .... 2. *T. officinale*.

1. T. laevigatum (Willd.) DC. RED-SEED DANDELION. Similar to *T. officinale*, often more slender. Leaves generally very deeply cut for their whole length, the lobes narrow, the terminal lobe seldom much larger than the others. Heads a little smaller; involucre 1–2 cm high, its inner bracts 11–13, often somewhat corniculate (i.e., with a hooded appendage near the summit), the outer bracts appressed to reflexed, a third to a little over half as long as the inner. Body of the achene becoming bright red or purplish-red or brownish-red at maturity, commonly somewhat rugulose below as well as muricate above, the beak usually stramineous, 1/2–3 times as long as the body. (2n = 16–32) Early spring–late fall. Fields, pastures, lawns, and other disturbed sites, often in drier places than the next sp., but much less common; native of Eurasia, now established as a weed throughout most of s Can and n US, s to Ga and Tex. *T. erythrospermum* Andrz.—F, R. *Leontodon erythrospermum* (Andrz.) Eichw.—S.

2. T. officinale Weber. COMMON DANDELION. Leaves commonly sparsely hairy beneath and on the midrib, otherwise generally glabrous, or sometimes completely so, oblanceolate, 6–40 × 0.7–15 cm, more or less runcinate-pinnatifid or lobed, the terminal lobe tending to be larger than the others, tapering to a narrow, scarcely or obscurely winged petiolar base. Scapes 5–50 cm tall, glabrous or more or less villous, especially upward. Heads usually large, the involucre mostly 1.5–2.5 cm high, the inner bracts mostly 13–20, these at first erect, finally reflexed, the mature achenes and pappus then forming a conspicuous, easily fragmented ball; outer bracts a little shorter and scarcely wider than the inner, reflexed. Body of the achene 3–4 mm long, pale gray-brown or stramineous, muricate above or sometimes to near the base, the beak 2.5–4 times as long as the body. (2n = 16–48) Early spring–late fall. Lawns and disturbed sites; native of Eurasia, now a cosmopolitan weed of temperate climates, found throughout our range except s Fla. *Leontodon taraxacum* L.—S.

125. CICHORIUM L. CHICORY

1. C. intybus L. Perennial herb from a long taproot, 3–17 dm tall; juice milky. Leaves alternate, the lower ones oblanceolate, petiolate, toothed or more often pinnatifid, 8–25 × 1–7 cm, becoming reduced, sessile and entire or merely toothed upward. Heads sessile or short-pedunculate, borne 1–3 together in the axils of the much reduced upper leaves, the long branches thus racemiform; heads up to 4 cm wide at anthesis, the flowers blue (rarely pink or white), all ligulate and perfect; involucre 9–15 mm high, biseriate, the outer bracts loose, fewer than the inner and at least half as long, becoming callous-thickened at the base. Achenes 2–3 mm long, glabrous, striate-nerved, sub-5-angled, or the outer rounded on the back and angled within; pappus of numerous minute, narrow, commonly truncate scales. (n = 9, 10) Summer–fall. Roadsides, fields, and waste places; native of Eurasia, now a nearly cosmopolitan weed, in our range s to Ga, Miss, and La.

126. PYRRHOPAPPUS DC., nom. conserv. FALSE DANDELION

Annual or perennial herbs with milky juice. Leaves alternate or all basal, entire to pinnatifid. Heads few or solitary, long-peduncled; flowers numerous (30–150+), all ligulate and perfect, yellow to sometimes ochroleucous or white; involucre campanulate or narrower, evidently calyculate, the inner bracts corniculate-appendaged near the tip; receptacle naked. Achenes oblong or linear, subterete or fusiform, with 5 broad, cross-rugulose ridges and as many narrow grooves, contracted above into a long, slender beak, which bears a minute ring of soft, white, reflexed hairs just beneath the pappus of numerous sordid or rufescent capillary bristles.

P. grandiflorus (Nutt.) Nutt., of Tex to Kans, probably does not reach our range. It is a perennial with a slender, vertical, rhizomelike stem arising from the buried, tuberous-thickened root; the aerial stems are monocephalous and essentially naked above the cluster of leaves at groundlevel.

1 Middle and lower part of the stem generally glabrous; leaf margins mostly smooth; outer involucral bracts mostly 1/3–2/3 as long as the inner .... 1. *P. carolinianus*.
1 Middle and lower part of the stem generally with some long, loosely spreading hairs; leaf margins mostly irregularly ciliate; outer involucral bracts seldom more than 1/3 as long as the inner .... 2. *P. multicaulis*.

1. P. carolinianus (Walter) DC. Annual or winter-annual (rarely short-lived perennial) from a well-developed taproot, caulescent (seldom only shortly so), mostly 2–10 dm tall; stem commonly minutely hirtellous-puberulent under the heads, the plants otherwise generally glabrous or nearly so. Leaves entire to pinnatisect, the segments mostly slender and tending to be broadest near the base; basal leaves sometimes much the largest and persistent, up to 25 × 6 cm, sometimes deciduous and scarcely larger than the well-developed cauline ones; uppermost foliage leaves generally entire or with only one pair of slender segments near the base. Heads several or occasionally solitary; involucre 1–2 cm high at anthesis, somewhat accrescent in fruit; flowers yellow or occasionally ochroleucous or white. Body of the achene 4–6 mm long, the filiform, subapically very fragile beak longer, often twice as long. (n = 6) Spring, or continuing through summer. Fields, dry woods, bottomlands, and waste places; Del and Md to c Fla, w to Ill, Mo, and Tex. Plants from Fla and CP of Ga tend to differ from the bulk of the species in being less leafy-stemmed and in having more conspicuously dissected leaves, with long, narrow segments and narrow rachis. Occasional specimens of such plants approach *P. multicaulis* in having relatively short outer involucral bracts, or in being slightly hairy. Optimists may try to distinguish this se phase as *P. carolinianus* var. *georgianus* (Shinners) Blake.

2. P. multicaulis DC. Similar to *P. carolinianus*, especially to the phase called var. *georgianus*, but more pubescent, the middle and lower part of the stem generally with some long, spreading hairs, the leaf margins commonly irregularly ciliate, the leaf surfaces sometimes also hairy; leaves more consistently pinnatifid, the segments often broadest above the middle, the uppermost foliage leaves generally evidently pinnatifid; outer involucral bracts mostly shorter, seldom more than 1/3 as long as the inner. (n = 6) Spring. Disturbed habitats; mostly Tex and Okla, but sometimes intro. eastward, as in Miss.

127. KRIGIA Schreb., nom. conserv. DWARF DANDELION

Herbs with milky juice, commonly with some spreading, gland-tipped hairs under the heads, otherwise glabrous or sometimes with some similar hairs on the stem and leaves. Leaves alternate or less often subopposite, more or less basally disposed or even all basal, variously entire or toothed to pinnatifid. Heads solitary or several, mostly long-pedunculate; flowers all ligulate and perfect, yellow or orange; involucre campanulate, with 5–18 essentially equal bracts, not calyculate; receptacle naked. Achenes oblong or turbinate, 10–20-nerved or -ribbed, tending to be transversely rugulose; pappus of 5 or more prominent to sometimes very inconspicuous scales and usually 5–40 longer bristles, or obsolete. Shinners, L. H. 1947. Revision of the genus *Krigia* Schreber. Wrightia 1: 187–206.

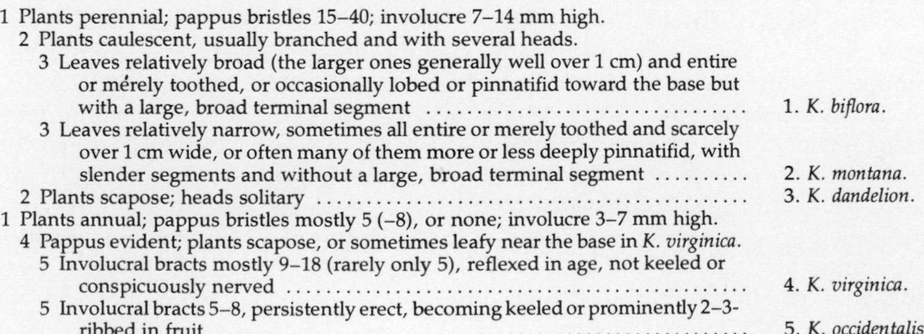

1 Plants perennial; pappus bristles 15–40; involucre 7–14 mm high.
  2 Plants caulescent, usually branched and with several heads.
    3 Leaves relatively broad (the larger ones generally well over 1 cm) and entire or merely toothed, or occasionally lobed or pinnatifid toward the base but with a large, broad terminal segment .... 1. *K. biflora*.
    3 Leaves relatively narrow, sometimes all entire or merely toothed and scarcely over 1 cm wide, or often many of them more or less deeply pinnatifid, with slender segments and without a large, broad terminal segment .... 2. *K. montana*.
  2 Plants scapose; heads solitary .... 3. *K. dandelion*.
1 Plants annual; pappus bristles mostly 5 (–8), or none; involucre 3–7 mm high.
  4 Pappus evident; plants scapose, or sometimes leafy near the base in *K. virginica*.
    5 Involucral bracts mostly 9–18 (rarely only 5), reflexed in age, not keeled or conspicuously nerved .... 4. *K. virginica*.
    5 Involucral bracts 5–8, persistently erect, becoming keeled or prominently 2–3-ribbed in fruit .... 5. *K. occidentalis*.

4 Pappus wanting, or a mere minute scaly vestige; leaves cauline as well as basal .... 6. *K. oppositifolia*.

1. K. biflora (Walter) Blake. Fibrous-rooted perennial from a short caudex, 2–8 dm tall, glabrous except generally under the heads, somewhat glaucous. Basal leaves oblanceolate to broadly elliptic, mostly 4–25 cm (petiole included) × 1–5 cm, entire or toothed, or sometimes somewhat pinnatifid toward the base but with a large, broad terminal segment; cauline leaves few, sessile and clasping, often much reduced, the uppermost often subopposite and with several long peduncles in their common axil. Heads several; involucre 7–14 mm high, much surpassed by the orange flowers; bracts 9–18, narrow, reflexed in age. Pappus of 20–35 very fragile, unequal bristles, the longer ones ca. 5 mm long, and ca. 10 inconspicuous hyaline scales less than 0.5 mm long. (n = 5) Spring–fall. Woods, roadsides, and fields; Mass to Va and (mainly in the mts.) to n Ga, w to Man, Colo, Ark (OU), Miss, and Ariz. *Cynthia virginica* (L.) D. Don—S.

2. K. montana (Michx.) Nutt. Similar to *K. biflora*, but averaging more slender, often branched mainly toward the base and with elongate, scapiform peduncles, or sometimes unbranched and single-headed; leaves relatively narrow, the principal ones either more or less deeply pinnatifid, with slender segments, or undivided and scarcely over 1 cm wide; pappus of ca. 15–25 bristles and ca. 5 scales, the latter 0.6–1 mm long. Spring–summer. Moist cliff crevices and streambanks; at upper altitudes in BR of NC, Tenn, and Ga. *Cynthia montana* (Michx.) Standley—S.

3. K. dandelion (L.) Nutt. Colonial perennial with slender, tuberiferous rhizomes, generally glabrous except under the heads; scape 1–5 dm tall, usually solitary. Leaves all basal (sometimes a single pair present just above the base), linear to oblanceolate, entire or pinnately few-lobed, 3–20 cm × 2–25 mm. Head solitary, 3–4.5 cm wide in flower; involucre 9–14 mm high, much surpassed by the golden-yellow flowers; bracts 9–18, narrow, reflexed in age. Pappus of 25–40 unequal capillary bristles, the longer ones ca. 5 mm long or more, and ca. 10 inconspicuous scales 0.6–1 mm long. (polyploids based on x = 5) Spring. Low prairies, fields, and other moist, chiefly open places, often in sandy soil; NJ to c Fla, w to Kans and Tex, chiefly on CP and PP. *Cynthia dandelion* (L.) DC.—S.

4. K. virginica (L.) Willd. Slender annual, 3–40 cm tall, scapose or leafy only near the base, the several scapes or scapiform peduncles generally spreading-glandular-hairy above, sometimes throughout. Leaves linear to oblanceolate or obovate, 1.5–12 cm × 1–12 mm, entire to pinnatifid, loosely villous-hirsute to glabrous, the hairs sometimes glandular. Involucre 4–7 mm high, its (5) 9–18 bracts lanceolate or narrower, reflexed in age, not keeled or ribbed. Pappus of 5 short, thin scales alternating with as many scabrous bristles several times as long. (n = 5, 10) Spring–midsummer. Sandy places; Me and Vt. to n Fla, w to Mich, Mo, Ark, La, and Tex.

5. K. occidentalis Nutt. Delicate scapose annual, 2–20 cm tall, sometimes with spreading, gland-tipped hairs along the scapes and on the leaves, as well as under the heads. Leaves oblanceolate or broader, entire to pinnatifid, 1.5–8 cm × 2–9 mm. Involucre 3.5–6 mm high; bracts 5–8, lanceolate or broader, persistently erect, becoming keeled or occasionally 2–3-ribbed in fruit. Pappus of 5 (–8) broad, thin scales less than 1 mm long, alternating with as many scabrous bristles, or the bristles rarely wanting. (n = ca. 4) Spring. Prairies and other dry, open places, often in sandy soil; sw Mo and adj. Ark to s La, Kans, and Tex.

6. K. oppositifolia Raf. Slender, branching, usually several-stemmed annual, 5–45 cm tall. Leaves basal and cauline, linear-oblong to oblanceolate or a little broader, entire to pinnatifid, 1.5–15 cm × 2–20 mm; some of the upper internodes commonly so shortened that the leaves appear opposite, with several long, sometimes similarly again branched and bracteate peduncles in their common axil. Heads 8–15 mm wide in flower, yellow with a golden center; involucre 3–5 mm high, its bracts 6–10, lanceolate or broader, persistently erect, becoming keeled in fruit. Pappus wanting, or a mere scaly vestige. Spring. Generally in moist, low places, either open or shaded; s Va to Fla, w to s Ill, Kans, and Tex. *Serinia oppositifolia* (Raf.) Kuntze—F, S.

128. HYPOCHOERIS L. CAT'S EAR

Similar to *Leontodon*, from which it is distinguished primarily by having long, chaffy bracts on the receptacle; stem sometimes with some more or less well-developed leaves, especially toward the base; flowers yellow or sometimes white.

1 Pappus of an outer series of short, merely barbellate bristles and an inner series of much longer, plumose bristles; stem naked or only sparsely and minutely bracteate.
  2 Plants mostly annual, essentially glabrous; heads not very showy, the ligules about equaling the involucre and only about twice as long as broad; outermost achenes usually beakless .... 1. *H. glabra*.
  2 Plants perennial, with evidently hispid leaves; heads showy, the ligules evidently surpassing the involucre and about 4 times as long as wide; outermost achenes usually evidently beaked, like the inner .... 2. *H. radicata*.
1 Pappus entirely of long, plumose bristles; stem with a few more or less well-developed leaves, at least toward the base.
  3 Flowers yellow; middle and outer involucral bracts in ours hispid; heads relatively broad, the involucre campanulate, mostly 5–8 mm wide at the middle at anthesis .... 3. *H. brasiliensis*.
  3 Flowers in ours white; involucral bracts glabrous or inconspicuously tomentose-puberulent; heads narrow, the involucre cylindric, mostly 2–4 mm wide at the middle at anthesis .... 4. *H. microcephala*.

1. H. glabra L. Taprooted annual or winter annual, essentially glabrous; stem 1–4 dm tall, simple or sparingly branched, naked or only sparsely and minutely bracteate. Basal leaves oblanceolate, toothed or pinnatifid, 2.5–15 × 0.7–3.5 cm. Heads several or solitary, terminating the branches, not very showy, opening only in full sun, the ligules about equaling the involucre and only about twice as long as broad; involucre mostly 8–10 mm high at anthesis, up to 17 mm in fruit, its bracts imbricate. Body of the achenes mostly 4–5 mm long, multinerved, the nerves muricate upward; outermost achenes usually beakless, the others with a well-developed, slender beak; shorter outer pappus bristles commonly merely barbellate. (n = 4, 5, 6) Mostly spring, or continuing until fall. Disturbed and waste places, especially in sandy soil; native of Europe, intro. in our area from NC to n Fla and w to La and Ark. Although *H. glabra* differs from *H. radicata* in several well-correlated characters, all of the differences are subject to failure, perhaps as a result of past hybridization and segregation.

2. H. radicata L. Perennial from a caudex, fibrous-rooted, or more often several of the roots enlarged; stem 1.5–6 dm tall, branched above or in small plants simple, naked or only sparsely and minutely bracteate, often spreading-hispid below. Basal leaves hispid, oblanceolate, toothed or pinnatifid, 3–35 × 0.5–7 cm. Heads usually several, terminating the branches, rather showy, open in bright or dull weather, the ligules surpassing the involucre and about 4 times as long as wide; involucre 10–15 mm high at anthesis, up to about 25 mm in fruit, its bracts imbricate, glabrous or hispid. Body of the achenes mostly 4–7 mm long, multinerved, the nerves muricate; achenes all with a slender beak from a little shorter to more often much longer than the body, or the outer achenes sometimes only short-beaked; shorter outer pappus bristles commonly merely barbellate. (n = 4) Spring–midsummer, or continuing until fall. A weed in lawns, pastures, and other disturbed sites; native of Eurasia, now widely established in the US and s Can, in our range s to n Fla, Ga, Miss, and reputedly La.

3. H. brasiliensis (Less.) Hook. & Arn. Rather coarse perennial from a well-developed, simple or branched, vertical caudex or a stout root, mostly 4–10 dm tall. Leaves irregularly ciliate and often with scattered long hairs on the surface, especially along the midrib beneath; basal leaves oblanceolate, mostly 7–30 × 1.5–8 cm, from merely toothed to often lobed halfway to the midrib, or deeper; cauline leaves few, mostly sessile, progressively reduced upward. Heads several or rather numerous, terminating the branches; involucre campanulate, at anthesis mostly 12–15 mm high and 5–8 mm wide near the middle (as pressed), elongating to 15–20 mm in fruit, its bracts imbricate; flowers numerous, yellow. Body of the achenes 5–6 mm long, transversely roughened, longitudinally 5-grooved, tapering into a beak about as long or a little longer; pappus a single series of plumose bristles, the basal part merely barbellate and slightly flattened. (n = 4) Spring. A weed along roadsides and in waste places, often in sandy soil; native of warm-temperate S Am, now well established in our range from NC to n Fla and w to La. *H. elata* (Weddell) Griseb.—R; *Crepis foetida* L.—S, misapplied. Our plants differ from typical *H. brasiliensis* in having the middle and outer bracts sparsely to moderately hispid (as opposed to glabrous or merely inconspicuously tomentose-puberulent), and may be distinguished, if desired, as var. *tweediei* (H. & A.) Baker.

4. H. microcephala (Schultz-Bip.) Cabrera. Similar in most respects to *H. brasiliensis*, but smaller, and with notably narrower, fewer-flowered heads, the involucre at anthesis 9–12 mm high and only 2–4 mm wide at the middle, elongating to 13–16 mm in fruit; plants 1–6 dm tall; stem often with a few spreading bristles; cauline leaves more reduced; involucre glabrous or inconspicuously tomentose-puberulent; body of the achene mostly 4–5 mm long. Spring. A weed of roadsides and waste places; native of warm-temperate S Am, now intro. in s Tex and spreading into La. Our plants have white flowers and represent the var. *albiflora* (Kuntze) Cabrera.

129. LEONTODON L., nom. conserv.

Scapose herbs with milky juice. Leaves basal, entire to pinnatifid, scape naked or merely scaly-bracted. Heads solitary or several; flowers all ligulate and perfect, yellow; involucre ovoid or oblong, imbricate or calyculate; receptacle alveolate or fimbriately villous, not chaffy-bracted. Achenes narrow, subterete, several- or many-nerved, long-beaked or merely narrowed upward; pappus of plumose bristles, sometimes with some shorter nonplumose outer bristles or scales, or that of the marginal achenes sometimes wholly of the latter type.

1 Pappus of the outer flowers, like that of the inner, wholly of plumose bristles; heads usually several .... 1. *L. autumnalis*.
1 Pappus of the outer flowers reduced to a crown; head solitary .... 2. *L. taraxacoides*.

1. L. autumnalis L. FALL DANDELION. Fibrous-rooted perennial from a short caudex or crown. Scapes 1–8 dm tall, commonly decumbent at the base, scaly-bracted at least above, ordinarily branched, commonly tomentose-puberulent at the summit, otherwise glabrous. Basal leaves oblanceolate, 4–35 × 0.5–4 cm, glabrous or moderately hirsute, deeply, narrowly, and rather distantly lobed to occasionally entire. Heads terminating the branches; involucre 7–13 mm high, with narrow, imbricate bracts, scarcely elongating in fruit. Achenes fusiform-columnar, not beaked, weakly nerved, transversely rugulose, 4–7.5 mm long; pappus wholly of plumose bristles, these chaffy-flattened at the base. (n = 6, 12) Midsummer–fall. Roadsides, pastures, fields, meadows, and waste places; native of Eurasia, now widely established in e Can and ne US, s to Del.

2. L. taraxacoides (Vill.) Mérat. Fibrous-rooted, chiefly perennial, 1–3.5 dm tall, the scapes simple, curved-ascending, ordinarily naked. Basal leaves oblanceolate, hispid-hirsute, 4–15 × 0.6–2.5 cm, usually shallowly lobed. Head solitary; involucre shortly calyculate, 6–11 mm high, scarcely elongating in fruit, glabrous or hairy. Achenes fusiform, scarcely or shortly beaked, 3–6 mm long, scabrous; pappus of the inner flowers partly of plumose bristles, partly of shorter outer scales that may be tipped with a scabrous bristle; outer flowers with the pappus reduced to a short, laciniate crown. (n = 4) Summer. A weed in lawns and waste places; native of Europe, intro. at scattered localities in e US, s to NC, Tenn, and Ala. *L. leysseri* (Wallr.) G. Beck—F, G; *L. nudicaulis* (L.) Banks—R, probably misapplied.

130. PICRIS L. BITTERWEED, OX-TONGUE

1. P. hieracioides L. Biennial or short-lived perennial herb with milky juice, resembling *Hieracium* in aspect, spreading-hirsute or -hispid, many of the hairs, especially of the stem and involucre, shortly bifurcate, anchor-shaped (visibly so at 10×). Leaves alternate, entire to coarsely toothed or with slender lobe teeth, the basal and lowest cauline ones oblanceolate, 7–30 × 0.5–5 cm, often deciduous, the others lanceolate or oblong, sessile and often clasping, gradually or abruptly reduced. Heads several in a corymbiform inflorescence; flowers all ligulate and perfect, yellow; involucre campanulate or ovoid-urceolate, 8–15 mm high, its bracts all narrow, less than 3 mm wide, the inner ones uniseriate, subequal, the outer narrow and successively shorter; receptacle naked. Achenes evidently cross-rugulose, longitudinally grooved, 3.5–6 mm long, narrowed above but scarcely or not at all beaked; pappus of plumose capillary bristles, readily deciduous as a unit. (n = 5) Mid- and late summer. Fields, meadows, roadsides, and waste places; native of Eurasia, established here and there in ne US, s sometimes to NC. *P. echoides*, L., another European species, is adventive in ne US, s to DC. It is a more coarsely spiny-hispid annual, the hairs 3–several-parted at the tip, like a grappling hook; involucre biseriate, the outer bracts leafy, ovate or lance-ovate, 3.5–8 mm wide; achenes with a slender, basally very fragile beak, the outer ones woolly.

131. TRAGOPOGON L. GOAT'S BEARD

Taprooted herbs with milky juice. Leaves alternate, linear, entire, commonly somewhat grasslike. Heads solitary, or terminating the branches; flowers all ligulate and perfect, yellow or purple; involucre cylindric or campanulate, its bracts uniseriate and equal; receptacle naked. Achenes linear, terete, or angled, 5- to 10-nerved, narrowed at the base, slender-beaked, or the outer occasionally beakless; pappus of a single series of plumose bristles, united at the base, the plume branches interwebbed, several of the bristles commonly longer than the others and naked at the tip.

1 Flowers purple .... 1. *T. porrifolius*.
1 Flowers yellow.
  2 Peduncles enlarged and fistulous above in flower and fruit; bracts surpassing the rays .... 2. *T. dubius*.
  2 Peduncles not enlarged in flower, scarcely so in fruit; bracts not surpassing the rays .... 3. *T. pratensis*.

1. T. porrifolius L. SALSIFY, VEGETABLE OYSTER. Glabrous biennial, 4–10 dm tall. Leaves up to 30 cm long and nearly 2 cm wide, tapering rather gradually from the base, not recurved at the tip; peduncles evidently enlarged and fistulous under the heads in flower and fruit. Involucral bracts mostly ca. 8, 2.5–4 cm long in flower, slightly to strongly surpassing the purple rays, elongating to 4–7 cm in fruit. Achenes 25–40 mm long, the body thicker than in *T. dubius* and usually only 10–16 mm long, abruptly contracted to the long, slender beak; pappus brownish. (n = 6) Spring–summer. Roadsides and waste places, mostly in rather moist soil; European cultigen, established as a weed here and there over much of the US, in our range s to Ga.

2. T. dubius Scop. Mostly biennial, 3–10 dm tall. Leaves elongate, generally tapering fairly uniformly from base to apex, not recurved, evidently floccose when young, later more or less glabrate except commonly in the axils; peduncles evidently enlarged and fistulous under the heads in flower and fruit. Involucral bracts typically 13, sometimes only 8 on later heads or in small plants, 2.5–4 cm long in flower, distinctly surpassing the rather pale lemon-yellow rays, elongating to 4–7 cm in fruit. Achenes slender, 25–36 mm long, gradually narrowed to the stout beak; pappus whitish. (n = 6) Spring–midsummer. Roadsides and waste places; native of Europe, now widely established in the US, in our range s to NC, Tenn, and Ark. *T. major* Jacq.—F.

3. T. pratensis L. Mostly biennial, 1.5–8 dm tall. Leaves elongate, up to 30 × 2 cm, rather abruptly narrowed a little above the base, and tending to have the margins somewhat crisped, cirrhose-recurved at the tip, slightly floccose when young, soon glabrous; peduncles not enlarging in flower and scarcely so in fruit. Involucral bracts typically 8, 12–24 mm long in flower, equaling or shorter than the chrome-yellow rays, elongating to 18–38 mm in fruit. Achenes 15–25 mm long, rather abruptly contracted to the slender, relatively short beak, the body not much shorter than in our other spp.; pappus whitish. (n = 6) Spring–summer. Roadsides, fields, and waste places, commonly in slightly moister habitats than *T. dubius*; native to Europe, now widely established in n US, s in our range to Tenn.

INDEX